Photoelasticity

Proceedings of the International Symposium
on Photoelasticity, Tokyo, 1986

Editor: M. Nisida
Coeditor: K. Kawata

With 287 Figures

Springer-Verlag
Tokyo Berlin Heidelberg New York

Masataka Nisida

Faculty of Science and Technology
Science University of Tokyo
Noda, Chiba, 278 Japan

Kozo Kawata

Faculty of Science and Technology
Science University of Tokyo
Noda, Chiba, 278 Japan

ISBN-13: 978-4-431-68041-3   e-ISBN-13: 978-4-431-68039-0
DOI: 10.1007/978-4-431-68039-0

Library of Congress Card Number: 86-6721

© Springer-Verlag Tokyo 1986

Softcover reprint of the hardcover 1st edition 1986

Preface

Thirty-five papers were presented at the International Symposium on Photoelasticity, Tokyo, 1986, representing fifty-five authors. Eighteen of these papers were presented by Japanese photoelasticians and seventeen by leading foreign authorities from eleven countries (Austria, Canada, Czechoslovakia, F.R. of Germany, France, Greece, India, Switzerland, UK, USA and USSR).

This is the first symposium on photoelasticity of international scope held in Japan. The primary objectives of this symposium are to help bridge the gap between photoelastic researchers around the world, to promote mutual understanding and communications and to facilitate exchange of newly acquired knowledge in theories and techniques. In addition, it is important that these valuable results are communicated effectively to engineers who can apply them in practice in industry.

The papers presented at this symposium cover all branches of photoelasticity in a broad sense, including, in addition to long established photoelasticity, newly developed moire, interferometric, and holographic photoelasticity, caustics and speckle. Therefore, from an optical stress analysis perspective, this volume is the latest comprehensive collection of photoelastic expertises.

Masataka Nisida

Chairman of the Organizing
Committee of
the International Symposium on
Photoelasticity, Tokyo, 1986
and the Editor of the Proceedings

Organizing Committee of the International Symposium on Photoelasticity, Tokyo, 1986

Chairman

    Masataka Nisida   Professor, Dr. Eng., Science University of Tokyo

Vice Chairman

    Kozo Kawata   Professor, Dr. Eng., Science University of Tokyo

    Takeshi Kunio   Professor, Dr. Eng., Keio University

    Hiroyuki Okamura   Professor, Dr. Eng., University of Tokyo

    Atsushi Saito   Professor, Dr. Eng., Kinki University

    Masao Shibahara   Professor, Dr. Eng., Kanazawa University

    Heihachi Shimada   Professor, Dr. Eng., Tohoku University

    Susumu Takahashi   Professor, Dr. Eng., Kanto Gakuin University

    Sakutaro Nakamura   President, The North Japan Soil Research Company

    Takuo Hayashi   Professor, Dr. Eng., Osaka University

    Gengo Matsui   Professor, Dr. Eng., Waseda University

    Satoshi Miki   Professor, Dr. Eng., Kurume Institute of Technology

    Minoru Yamamoto   Professor, Dr. Eng., Tokyo Metropolitan University

    Keiji Yoshikawa   Professor, Dr. Eng., Saitama University

(Secretariat)   Yoshiharu Masuda, Yoshiaki Sawa

List of Contributors

Aben, H.K.
   Institute of Cybernetics, Academy of Sciences of the Estonian SSR,
   Akadeemiatee 21, Tallinn 200108 Estonia, USSR

Abo-El-Ezz, A.E.
   Kyushu University, 6-1 Kasuga, Fukuoka 816, Japan

Aono, H.
   Aero-Engine & Space Operations, Ishikawajima-Harima Heavy Industries, Co., Ltd., 3-1-15 Toyosu, Koto-ku, Tokyo 135-91, Japan

Broadbent, T.P.
   Department of Mechanical Engineering, The University of Nottingham,
   University Park, Nottingham NG7 2RD, UK

Dally, J.W.
   University of Maryland, College Park, Maryland 20742, USA

Danyluk, H.T.
   University of Sasukatchewan, Sakatoon, S7N OWO, Canada

Date, K.
   Miyagi National College of Technology, Natori, Miyagi 981-12, Japan

Dietz, P.R.
   Institut für Maschinenwesen, Technische Universität Clausthal,
   D-3392 Clausthal-Zellerfeld, F.R. of Germany

Fessler, H.
   Department of Mechanical Engineering, The University of Nottingham,
   University Park, Nottingham NG7 2RD, UK

Funayoshi, T.
   Kwansei Gakuin University, 1-1-115 Uegahara, Nishinomiya 662, Japan

Hashimoto, S.
   Institute of Interdisciplinary Research, The University of Tokyo,
   4-6-1 Komaba, Meguro-ku, Tokyo 153, Japan

Hayashi, T.
   Osaka University, 1-1 Machikaneyama, Toyonaka 560, Japan

Hyer, M.W.
   Department of Mechanical Engineering, University of Maryland,
   College Park, MD 20742, USA

Inamura, T.
   Research Institute, Ishikawajima-Harima Heavy Industries, Co., Ltd.,
   3-1-15 Toyosu, Koto-ku, Tokyo 135-91, Japan

Jávor, T.
   Research Institute of Civil Engineering / VUIS/, Lamacska 8,
   Bratislava, Czechoslovakia

Kalthoff, J.F.
   Fraunhofer-Institüt für Werkstoffmechanik, Wohlerstr. 11, D-7800
   Freiburg, F.R. of Germany

X

Kamiyama, T.
Research Institute, Ishikawajima-Harima Heavy Industries, Co., Ltd.,
3-1-15 Toyosu, Koto-ku, Tokyo 135-91, Japan

Kawashima, T.
Research Institute, Ishikawajima-Harima Heavy Industries, Co., Ltd.,
3-1-15 Toyosu, Koto-ku, Tokyo 135-91, Japan

Kawata, K.
Science University of Tokyo, 2641 Yamazaki, Noda 278, Japan

Kunio, T.
Keio University, 3-14-1 Hiyoshi, Kohoku-ku, Yokohama 223, Japan

Kuramoto, M.
The Institute of Vocational Training, 1960 Aihara, Sagamihara,
Kanagawa 229, Japan

Lagarde, A.
Universite de Poitiers, Laboratoire de Mecanique des Solides 40,
Avenue du Recteur Pineau, 86022 Poitiers Cedex, France

Matsui, G.
Waseda University, 3-4-1 Okubo, Shinjuku, Tokyo 160, Japan

Miki, S.
Kurume Institute of Technology, 2228 Kamitu-machi, Kurume 830,
Japan

Mitsui, Y.
Shinshu University, 500 Wakasato, Nagano, Japan

Miyano, Y.
Kanazawa Institute of Technology, 7-1 Ogigaoka, Nonoichi,
Ishikawa 921, Japan

Mönch, E.
Technische Universität München, Arcisstrasse 21, D-8000, München 2
F.R. of Germany

Morimoto, Y.
Osaka University, 1-1 Machikaneyama, Toyonaka 560, Japan

Nisida, M.
Science University of Tokyo, 2641 Yamazaki, Noda 278, Japan

Okamura, H.
The University of Tokyo, Faculty of Engineering, 7-3-1 Hongo,
Bunkyo-ku, Tokyo 113, Japan

Parks, V.J.
The Catholic University of America, Washington D.C. 20064, USA

Pindera, J.T.
Department of Civil Engineering, University of Waterloo, Waterloo,
Ontario, Canada N2L 3GI

Post, D.
Department of Engineering Science & Mechanics, Virginia Polytechnic
Institute and State University, Blacksburg, Virginia 24061, USA

Rossmanith, H.P.
Institute of Mechanics, Technical University of Vienna, Karlsplatz
13, A-1040 Vienna, Austria

Sakai, S.
The University of Tokyo, Faculty of Engineering, 7-3-1 Hongo,
Bunkyo-ku, Tokyo 113, Japan

Sawa, Y.
Science University of Tokyo, 2641 Yamazaki, Noda 278, Japan

Schumann, W.
    Swiss Federal Institute of Technology, Zürich, Laboratorium für
    Photoelastizität CH-8092 Zürich, Switzerland

Shimada, H.
    Tohoku University, Aoba, Aramaki, Sendai 980, Japan

Shimizu, K.
    Kanto Gakuin University, 4834 Kanazawa-ku, Yokohama 236, Japan

Shintani, R.
    Kwansei Gakuin University, 1-1-15 Uegahara, Nishinomiya 662, Japan

Simokohge, K.
    Research Institute, Ishikawajima-Harima Heavy Industries, Co., Ltd.,
    3-1-15 Toyosu, Koto-ku, Tokyo 135-91, Japan

Srinath, L.S.
    Indian Institute of Technology, Madras 600036, India

Sugimori, S.
    Kanazawa Institute of Technology, 7-1 Ogigaoka, Nonoichi,
    Ishikawa 921, Japan

Takahashi, K.
    Kyushu University, 6-1 Kasuga, Fukuoka 816, Japan

Takahashi, S.
    Kanto Gakuin University, 4834 Kanazawa-ku, Yokohama 236, Japan

Takeda, N.
    Kyushu University, 6-1 Kasuga, Fukuoka 816, Japan

Tanaka, T.
    Kokushikan University, 4-10-28 Higashi Shibuya, Tokyo 150, Japan

Taylor, C.E.
    Department of Engineering Science, University of Florida,
    Gainesville, Florida 32611, USA

Theocaris, P.S.
    Department of Engineering Science, Athens National Technical
    University, 5 Heroes of Polytechnion Avenue, GR 157 73 Athens,
    Greece

Tsuboi, Y.
    Hosei University, Koganei, Tokyo 184, Japan

Uchino, K.
    Research Institute, Ishikawajima-Harima Heavy Industries, Co., Ltd.,
    3-1-15 Toyosu, Koto-ku, Tokyo 135-91, Japan

Yamaguchi, I.
    The Institute of Physical and Chemical Research, Wako, Saitama
    351-01, Japan

Yokomise, H.
    Kokushikan University, 4-10-28 Higashi Shibuya, Tokyo 150, Japan

Yoshida, S.
    Shinshu University, 500 Wakasato, Nagano, Japan

Yoshikawa, T.
    Kwansei Gakuin University, 1-1-15 Uegahara, Nishinomiya 662, Japan

Contents

Part I   Survey
(Chairmen: K. Kawata, W. Schumann)

E. Mönch (Technische Universität München, F.R. of Germany)
A Historical Survey of the Development of Photoelasticity
in Germany, Especially in Munich 1

M. Nisida (Science University of Tokyo, Japan)
Photoelasticity in Japan: A Survey 9

Part II   Visco-Elasto-Plastic Behaviour
(Chairmen: P.R. Dietz, S. Miki)

S. Miki (Kurume Institute of Technology, Japan)
Determination of Sign and Components of Artificial
Birefringence of Visco-Elasto-Plastic Materials 15

S. Sugimori, Y. Miyano (Kanazawa Institute of Technology,
Japan), T. Kunio (Keio University, Japan)
A Simplified Optical Method for Measuring Residual Stress
by Rapid Cooling in Thermosetting Resin Plate 23

P.R. Dietz (Technische Universität Clausthal, F.R. of Germany)
Silver-Chloride, a Model Material for Photoelastic
Investigations into Plastic-Elastic Behaviour of
Machine-Elements 31

Part III   Moire and Speckle
(Chairmen: D. Post, S. Takahashi)

D. Post (Virginia Polytechnic Institute and State University,
USA)
Advances in Moire Interferometry 39

Y. Morimoto, T. Hayashi (Osaka University, Japan)
Scanning-moire Method 47

W. Schumann (Swiss Federal Institute of Technology, Zurich,
Switzerland)
Modification at the Reconstruction in Holographic
Interferometry 53

I. Yamaguchi (The Institute of Physical and Chemical Research,
Japan)
Strain Measurements by Laser-Speckle 65

Part IVa   High Speed Photoelasticity
(Chairmen: J.W. Dally, H. Shimada)

K. Kawata (Science University of Tokyo, Japan), S. Hashimoto
(University of Tokyo, Japan)
Dynamic Stress Concentration Analysis in High Velocity
Tension of Strips with Notches or Hole by Means of High
Speed Photoelasticity 73

S. Hashimoto (University of Tokyo, Japan), K. Kawata (Science
University of Tokyo, Japan)
Analysis of Impact Bending of Cantilevers with Various
Depth/Span Ratios by Means of High Speed Photoelasticity 81

Part IVb   High Speed Photoelasticity (continued)
(Chairmen: H.K. Aben, Y. Miyano)

J.W. Dally (University of Maryland, USA)
Photoelastic Analysis of Dynamic Fracture Behaviour 89

H. Shimada (Tohoku University, Japan), K. Date (Miyagi National
College of Technology, Japan)
Visualization of Pulsed Ultrasound Using a Combined
Photoelastic-Schlieren System 103

Part Va   Caustics and Stress Intensity Factor
(Chairmen: P.S. Theocaris, K. Yoshikawa)

J.F. Kalthoff (Fraunhofer-Institüt für Werkstoff-mechanik,
F.R. of Germany)
The Shadow Optical Method of Caustics — An Overview on
Its Applications in Stress-Concentration Problems 109

K. Shimizu, S. Takahashi (Kanto Gakuin University, Japan),
H.T. Danyluk (University of Sasukatchewan, Canada)
Some New Trials on the Technique of the Method of Caustics ... 121

K. Takahashi, N. Takeda, A.E. Abo-El-Ezz (Kyushu University,
Japan)
Application of the Caustic Method to an Environmental
Crack-Craze Growth Problem 129

Part Vb   Caustics and Stress Intensity Factor (continued)
(Chairmen: J.F. Kalthoff, K. Takahashi)

H.P. Rossmanith (Technical University of Vienna, Austria)
Topics in Photomechanics - Crack, Wave and Contacts 135

P.S. Theocaris (Athens National Technical University, Greece)
The Pseudocaustics for the Evaluation of the Order of
Singularity in Stress Fields 145

Part VIa   New Techniques in Photoelasticity
(Chairmen: H. Okamura, L.S. Srinath)

L.S. Srinath (Indian Institute of Technology, India)
Some New Developments in Photoelasticity 159

R. Shintani, T. Yoshikawa, T. Funayoshi (Kwansei Gakuin
University, Japan)
Micro-Photoelasticity and Its Picture Processing 167

M. Nisida, Y. Sawa (Science University of Tokyo, Japan)
A Stress Freezing Procedure for Less Photoelastic Model
Distortion 173

Part VIb   New Techniques in Photoelasticity (continued)
(Chairmen: H. Fessler, J.T. Pindera)

G. Matsui (Waseda University, Japan), T. Tanaka, H. Yokomise
(Kokushikan University, Japan)
On Wooden Column under Torsion and Deflection of Membrane 179

S. Sakai, H. Okamura (University of Tokyo, Japan)
A Method for Detecting Isoclinics by Modified
Isochromatics 187

J.T. Pindera (University of Waterloo, Canada)
New Research Perspectives Opened by Isodyne and Strain
Gradient Photoelasticity 193

Part VIIa   Stresses in Structures
(Chairmen: G. Matsui, V.J. Parks)

G. Matsui (Waseda University, Japan), Y. Tsuboi (Hosei
University, Japan)
Stresses and Deformations at the Beam-to-Wall Joints of
Shear Wall Structures 203

K. Uchino, T. Kamiyama, T. Inamura, K. Simokohge, H. Aono,
T. Kawashima (Ishikawajima-Harima Heavy Industries Co.,
Ltd., Japan)
Three-Dimensional Photoelastic Analysis of Aeroengine
Rotary Parts 209

Part VIIb   Stresses in Structures (continued)
(Chairmen: E. Mönch, K. Uchino)

M.W. Hyer (University of Maryland, USA)
Stresses in Pin-Loaded Glass-Epoxy Plates Using Transmission
Photoelasticity 215

V.J. Parks (The Catholic University of America, USA)
Photoelastic Analysis of Stresses in Composites 225

T. Jávor (Research Institute of Civil Engineering,
Czechoslovakia)
Photostress and New Photo-elastic Coating Technique by
in Situ Testing of Bridges 233

Part VIII   Propagation of Polarized Light
(Chairmen: T. Kunio, C.E. Taylor)

H.K. Aben (Institute of Cybernetics, Academy of Sciences of the
Estonian SSR, USSR)
Integrated Photoelasticity as Tensor Field Tomography 243

M. Kuramoto (The Institute of Vocational Training, Japan)
Stress Analysis for Axi-Symmetrical Problems by Scattered
Light Photoelasticity 251

C.E. Taylor (University of Florida, USA)
Applications of Coherent Optics to Experimental Mechanics 259

Part IX   Numerical Method of Stress Analysis
(Chairmen: T. Hayashi, A. Lagarde)

Y. Mitsui, S. Yoshida (Shinshu University, Japan)
   Separation of Principal Stresses Using Boundary Element
   Method 263

A. Lagarde (Universite de Poitiers, France)
   Non Destructive Three Dimensional Photoelasticity, Finite
   Strain Applications 269

T.P. Broadbent and H. Fessler (The University of Nottingham, UK)
   An Automatic Micropolariscope Used to Study A Cracked
   Thread 281

A Historical Survey of the Development of Photoelasticity in Germany, Especially in Munich

E. Mönch

Technische Universität München, Arcisstraße 21, D-8000 München 2, Federal Republic of Germany

When we survey the historical development of photoelasticity this must be done of course on international range because photoelasticity of to-day is a result of the research efforts of various nations.

In Fig. 1 some remarkable publications of the world literature are re-gistered, in chronological order. Books on photoelasticity are marked by framing. Each item is indicated by the name of the respective author and the year of publication.

In the middle column of this time-table, the German constribution is treated a little more in detail. If the author dares to deal prefer-ateley with German research work, the reader may forgive this because, on the one hand, the author thinks that the share constributed by Ger-man researchers is not quite unimportant. On the other hand, the author believes that he is able to report one period of the development - the time of L. Föppl and his co-workers in Munich - which is especially significant and may perhaps find greater interest because the author was an immediate witness of it.

As our time-table shows, the investigations of the fundamentals of photoelasticity begin with the well known famous experiments of Sir David Brewster (1816). When we inspect the time which then follows we can state that during about 80 years the scientific work concen-trated itself merely to the investigation of the physical facts of the stress-optic effect. What we nowadays mean by the conception of photo-elasticity, namely the application to engineering problems, begins only later. One must be aware that the revolution of technology which had begun with James Watt (1736-1819) was yet in the state of development. On the other hand mathematical physics were already highly consolidated, especially in France. Very significant publications about stress bire-fringence such as those of Fresnel (1822) and Wertheim (1851) appeared there.

In Germany, this period of research of the physical fundamentals is marked by the preeminent work of Franz Neumann (1798-1895). Some de-tails of the life of this outstanding man (cf. Timoshenko 1953) should be mentioned.

In the war against the French dictator Napoleon he joined, 17 years old, the Prussian army as a volunteer and was severely wounded 1815 and left for dead on the battlefield. The next day, he was found to be alive. After having recovered he remained in the army up to the end of the campaign. In 1817 Neumann entered Berlin University but had to master personal economical difficulties and gave lessons in order to earn his living. At the beginning he attended lectures on theology and laws, but soon turned to natural sciences. He learned mathematics by self-teach-ing. In 1820 he made a trip to the mountains in order to collect mi-

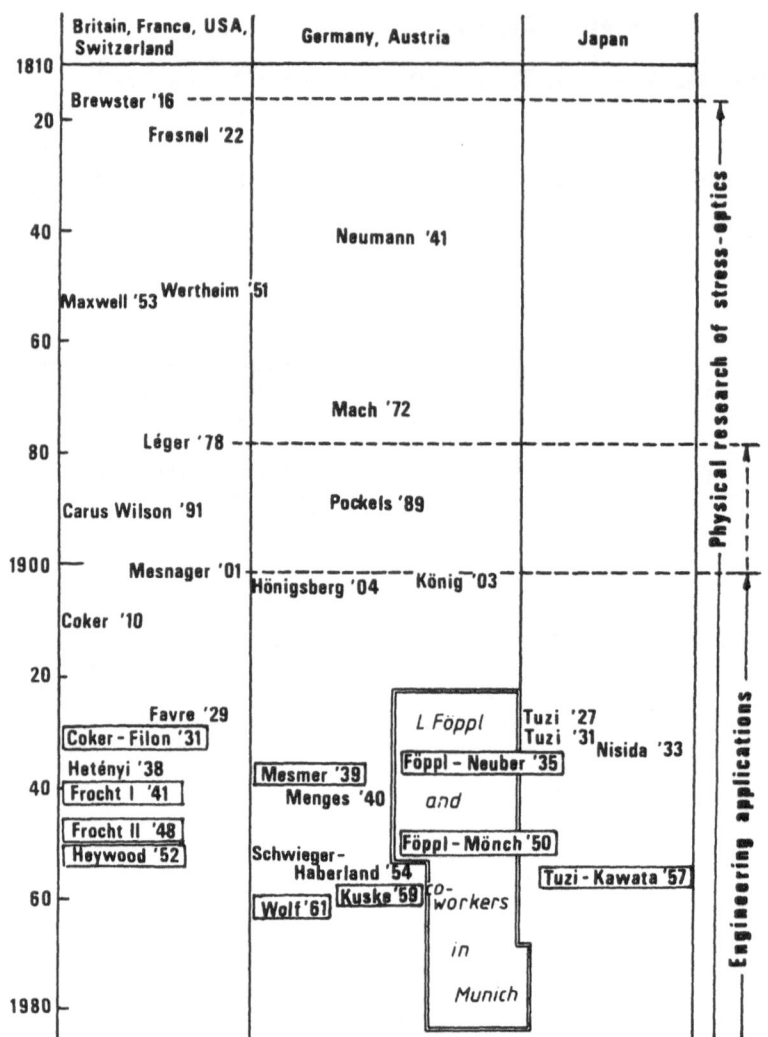

Fig. 1. Some remarkable publications on photoelasticity

▭ Books

nerals and fossils for Berlin's museum of natural sciences. In connection with this activity resulted his interest for mineralogy and crystal physics. 1826 he got a lectureship for mineralogy at Königsberg University and became professor there in 1829. At Königsberg he expanded his activities to other branches of physics, such as geophysics, theory of heat, theory of sound, optics and electricity. Thus at Königsberg University the first seminar of theoretical physics in Germany was initiated by Neumann. Neumann died at Königsberg, 97 years old.

In connection with crystal physics Neumann investigated the double refraction in stressed transparent bodies. As a result, he published, in 1841, the famous Neumann equations, which describe, in differential form, the changes which polarized light travelling through a stressed body of arbitrary shape suffers in every point. The changes are stated to be proportional to the difference of principal strains. The Neumann equations, enounced 145 years ago, are until now the generally acknowledged fundament of all photoelasticity.

Later on, intensive research work on the physics of stress birefringence was carried out by the great Maxwell, who 1853 first enounced its dependence on the principal stress difference. In the German language zone, we may mention the physicist and philosopher Ernst Mach (1872), otherwise known for his work on supersonic motion, F. Pockels (1889), an expert in crystal physics, and W. König (1903) and several pupils of them.

Only towards the beginning of the 20th century people tried to apply photoelasticity to engineering problems. Perhaps a first step in this direction is a paper of the French engineer A. Léger (1878). In England, a remarkable paper is due to Carus Wilson (1891), who first determined the stress trajectories in a bent beam photoelastically. However, commonly the work of the French engineer Mesnager is considered to be the begin of the application of photoelasticity in practical technology, first in his detailed paper of 1901.

In Austria Hönigsberg, a railway engineer, made experiments and wrote a paper (1904) on photoelasticity in which he pointed out its practical importance.

In England, in 1910 and the following years Coker started his well known work investigating many practical engineering problems. Already before, about 1902, Filon had begun a series of investigations under more physical view. The work of both culminated 1931 in the extraordinary book "A treatise on photoelasticity".

About 1930 a world-wide boom of interest of the engineers in photoelasticity can be noticed. Out of the very great number of publications which now appeared only some remarkable are indicated in Fig.1. Maybe the exemplary work of Coker and Filon has shown what a help photoelasticity can really be for the engineer. Moreover, perhaps the use of highly active model materials, first celluloid, introduced by Coker, then other plastic materials, had simplified the experiments and made them more effective. On the other hand, perhaps in those times the engineers began to calculate more economically and for this reason were forced to look for the help of the photoelastic experiment.

In England, many researchers continued the work of Coker and Filon and a great many papers and several books on photoelasticity were written. Only that of Heywood: "Designing by photoelasticity" may be mentioned. In the United States, out of the multitude of papers we remember only that of Hetényi (1938) on the three-dimendional technique, and, of course the famous two volumes "Photoelasticity" by Frocht (1941 and 1946).

In Switzerland, the excellent work of Favre (1929) and his pupils, based on the interferometric method, must be mentioned.

The admirable photoelastic work done at that time in Japan is due to the outstanding personage of Ziro Tuzi and his co-workers. Out of the publications written in English language we only mention Tuzi's report which appeared 1927 in Japan and that of 1931, published in England, and subsequently, Nisida's first paper of 1933, dealing with the properties of phenolite. In 1954, the book of Tuzi and Kawata "Photoelastic experiment" appeared in Tokyo.

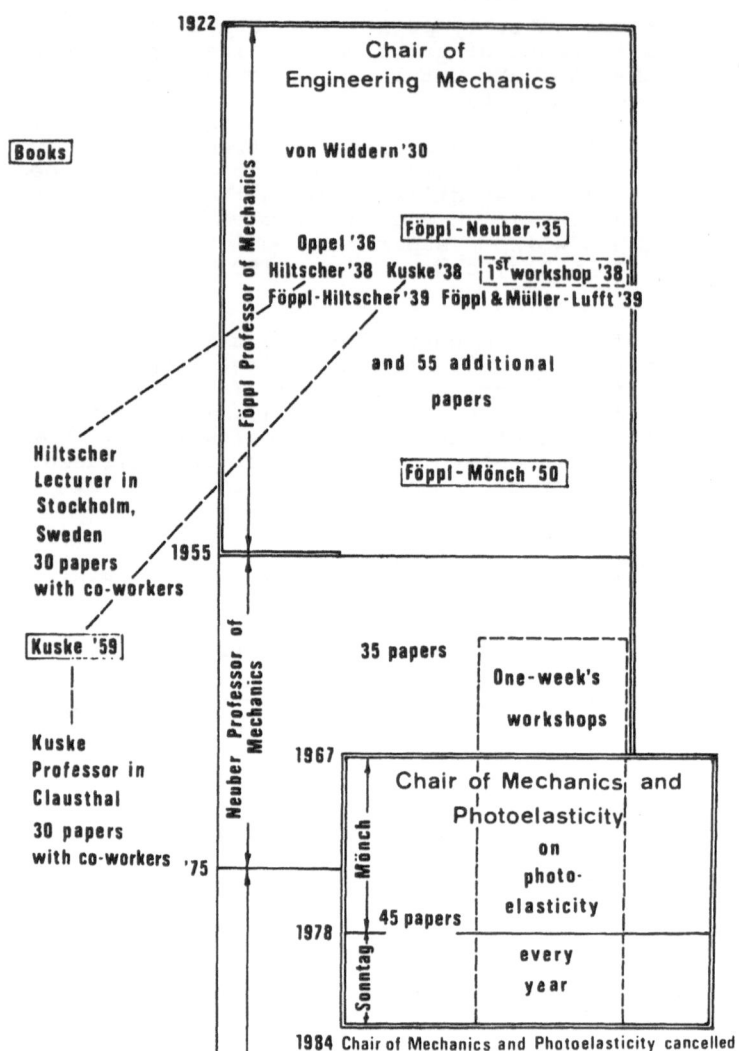

Fig. 2. Publications and activities in photoelasticity of L. Föppl and co-workers in Munich

In Germany, besides Föppl and co-workers, whose activities will be dealt with separately, also scientists in other places did much work on photoelasticity, independently of the Munich school: Mesmer at Aachen presented his book "Photoelasticity" 1939. A series of papers, beginning in 1954, chiefly on plate problems and dynamical experiments, were written by Schwieger, Haberland and co-workers in Berlin. Also one early but remarkable paper on the scattered light method by Menges, Darmstadt 1940 must be mentioned. 1961 the voluminous book of Wolf, Mülheim, was published.

The author may be allowed now to depict, by Fig. 2, more in detail the work of L. Föppl and his co-workers in Munich, because he was one of them since 1937.

Ludwig Föppl became professor of engineering mechanics in 1922. The lecture of Maxwell's work motived him to begin with photoelastic experiments. A scientific assistant of him, von Widdern, constructed a polariscope of the type which was then conventional, with Nicol polarizers and wrote the first doctorate thesis on photoelasticity in Munich (1930). Four others followed in the next years, using the same polariscope. These activities led to the first book on photoelasticity in Germany, written 1935 by Föppl and Neuber, who was also a scientific assistant of Föppl at that time.

Between 1935 and 1940 some most important progresses in photoelasticity were achieved in Munich. 1936 Oppel invented the stress freezing method. The world wide importance of this novelty is well known and must not be explained. Hiltscher had the idea to use those polarizing sheets of large diameter, which at that time were becoming commercially available, in the polariscope instead of Nicol prisms. In his new type of polariscope, originated in this way, lenses were no more necessary. It was described in a short paper of 1939. Müller-Lufft simultaneously found a method for manufacturing large diameter quarter wave sheets for this polariscope. In using the new simple polariscope the visual inspection of the photoelastic models became very much easier and thus the experimental processing much more effective. In all experiments carried out in Munich in the following time we used only the new apparatus. Another remarkable innovation of that period was presented by Hiltscher in his doctorate (1938) dealing with the conoscopic method for analysing frozen stress patterns.

Professor Föppl now was convinced that, through the spectacular progresses attained in his institute, photoelasticity was ripe to be introduced in the engineering practice. For this purpose he held a one-week-workshop in 1938 in which people from industry and also from Universities participated. The author remembers very well how since this time inquiries from the industry to the Munich institute increased asking for help in unsolved calculation problems of engineering. We must consider that the possibilities of calculating engineering stress problems, especially stress concentrations were still limited. Neuber's famous book on the theory of notch stresses was published only 1937 and its significance was recognized only later.

During Föppl's professorship about 65 papers on photoelasticity were published by himself and his pupils. The experiences of the Munich institute were collected in the book "Practical photoelasticity" written by Föppl and the author in 1950.

Föppl retired in 1955 from his chair of mechanics. Neuber became his successor. Photoelasticity was continued by the author.

In 1967, an additional chair named "for Mechanics and Photoelasticity" was installed at the Technical University of Munich after a suggestion of Professor Föppl and the author was entrusted with its guidance. This was the first time in Germany that a University chair bore the title of photoelasticity. We could now continue research and teaching activities in photoelasticity with more personal and financial resources. We did research in various special fields, e.g. handling of model materials. shells, photoplasticity. The one-week-workshops, initiated by Föppl, were repeated almost every year, for propagating our experiences to industry laboratories. The author believes that these workshops contributed much to the fact that now in many industrial companies of Germany independently working photoelasticity laboratories exist. In most of them the photoelastic methods of the Munich school are applied.

Also the polariscope of Föppl-Hiltscher, nowadays manufactured commercially by Mr. Tiedemann, a former assistant of Föppl, is used almost everywhere in Germany. A list of branches of industry in which photoelasticity is applied at present in Germany is shown in the table.

Table. Present activities in photoelasticity in Germany

in the University of: Berlin, Clausthal, Dresden, Dortmund, Karl-Marx-Stadt, Karlsruhe, Kassel, Munich, Paderborn, Stuttgart, Wuppertal; Academy of Sciences Berlin	in ~ 25 industrial companies of motor-cars, motor-engines, turbines, power stations, aeronautics, machine tool, building construction, etc.

After the author had retired, Professor Sonntag became his successor until he retired in 1984.

Photoelasticity in the manner of Föppl's school was also practised and taught in other engineering universities by former co-workers: Hiltscher who was mentioned before, went to Stockholm (Sweden) in 1949 where he did much work based on the Munich school. Chiefly his successes with the lateral extensometer may be mentioned. Kuske who received his doctor degree in Munich 1938 became professor for machines in Clausthal where he worked on photoelasticity together with many pupils. The work of Kuske is continued by his successor, Professor Dietz.

When we try to analyse the astonishing success of Föppl in the area of photoelasticity in Germany it is not enough to appreciate his 20 scientific papers on the subject. The question is what enabled him to gather such excellent people as Neuber, Oppel, Hiltscher around him. In the author's opinion it was the fascination as a teacher emanating from him. All his co-workers were former students who had listened to his unforgettable, high-spirited and humourous lectures on engineering mechanics and were glad when they succeeded in joining the team of such a man. Moreover, the generous intercourse between chief and co--workers encouraged successful work.

It might be added that Professor Föppl never had the ambition to manage a big institute of superficial glamour with many people and rooms etc. In this regard, the situation was rather modest. His co-workers were mostly the assistants for the mechanics lectures for whom photoelasticity was rather an additional occupation. It might happen e.g. that a visitor came to the institute and one of us had to show him some photoelastic experiment. He did this using a polariscope installed in his office. The visitor wondered why he was not guided for this purpose to a well equipped studio for experiments, and could not imagine that such a studio did not exist. However, mere representation was not considered to be important. Only efficiency was essential. A successful experiment of a co-worker was the greatest delight for the chief.

The author has dealt with the development of photoelasticity at the Engineering University of Munich in detail. As to other places in Germany, the interest in research on photoelasticity has also increased considerably. Very active photoelastic laboratories in many engineering Universities now exist, as is to be seen in the table. The subjects beeing treated are chiefly the farther development of the methods by modern means such as digital image processing, holography etc. and special applications, e.g. photoviscoelasticity.

In Munich, the Faculty of Mechanical Engineering had the chair of Mechanics and Photoelasticity cancelled in 1984, inspite of the protest of the author, taking for reason the actual economical difficulties of the University. Perhaps, the actual authorities of our University believe that photoelasticity has not more its importance for engineering sciences as before. To be sure, this problem can also be seen from another point of view. In any case, it is deplorable that this chair which was in certain respect the endpoint of Föppl's successful work in photoelasticity, has ceased to exist.

Of course photoelasticity is not stopped totally in Munich. The laboratory is carried on by Dr. Ficker who was taking care of it already since 1955 and has written several significant papers on photoelasticity himself. However, it is no more the same thing as before when the laboratory was incorporated in a chair which bore expressly the title of "photoelasticity". One must know that in Germany a University chair is connected with a guarantee of a certain number of posts of scientific staff and certain financial resources, so that formerly photoelasticity was expressly fixed in the research programme of the University. Now, in the future the activities in photoelasticity will probably diminish in Munich and the center of photoelasticity perhaps move to another place of Germany. The author is glad that he had at least the opportunity in this Symposium to point out the important work of Ludwig Föppl and his school in Munich.

REFERENCES

Brewster D (1816) On the communication of the structure of doubly refracting crystals to glass, muriate of soda, fluor spar and other substances, by mechanical compression and dilatation. Phil Trans Roy Soc London (1816): 156-178
Coker EG (1910) The optical determination of stress. Phil Mag VI 20: 740-751
Coker EG, Filon LNG (1931) A treatise on photo-elasticity. Cambridge Univ Press
Favre H (1929) Sur une nouvelle méthode optique de détermination des tensions intérieures. Revue d'Optique 8: 193-213,241-261,289-307
Föppl L (1928) Untersuchung ebener Spannungszustände mit Hilfe der Doppelbrechung. Abh Bayr. Akad Wiss 1928: 247-265
Föppl L, Hiltscher R (1939) Die neue spannungsoptische Apparatur des mechanisch-technischen Laboratoriums der Technischen Hochschule München. Der Bauingenieur 20: 231-232
Föppl L, Mönch E (1950) Praktische Spannungsoptik. Springer, Berlin Göttingen Heidelberg
Föppl L, Müller-Lufft E (1939) Spannungsoptische Einrichtung mit Polarisationsfiltern. Arch Techn Messen V: 137-1
Föppl L, Neuber H (1935) Festigkeitslehre mittels Spannungsoptik. Oldenbourg, München Berlin
Fresnel A (1822) Note sur la double réfraction du verre comprimé. Ann Chim Phys II 20: 376-383
Frocht MM (1941, 1948) Photoelasticity vol 1,2 Wiley, New York
Hetényi (1938) The fundamentals of three-dimensional photoelasticity. J Appl Mech 5: 149-155
Heywood RB (1952) Designing by photoelasticity. Chapman & Hall, New York
Hiltscher R (1938) Polarisationsoptische Untersuchung des räumlichen Spannungszustandes im konvergenten Licht. Forschung Geb Ing Wesens 9: 91-103
Hönigsberg (1904) Über unmittelbare Beobachtung der Spannungsverteilung und Sichtbarmachung der neutralen Schichte an beanspruchten Körpern. Ztschr Österr. Ing Arch Vereines 56: 165-173

König W (1903) Doppelbrechung in Glasplatten bei statischer Biegung.
    Ann Phys IV 11: 842-866
Kuske A (1938) Das Kunstharz Phenolformaldehyd in der Spannungsoptik.
    Forschung Geb Ing Wesens 9: 139-149
Kuske A (1959) Einführung in die Spannungsoptik, Wissenschaftliche
    Verlagsgesellschaft, Stuttgart
Léger MA (1878) Transmission des forces extérieures aux travers des
    corps solides. Mém Soc Ing Civils (1878): 252-262
Mach E (1872) Über die temporäre Doppelbrechung der Körper durch ein-
    seitigen Druck. Ann. Phys II 146: 313-316
Maxwell JC (1853) On the equilibrium of elastic solids. Roy Soc Edin
    Trans 20: 87-120
Menges HJ (1940) Die experimentelle Ermittlung räumlicher Spannungszu-
    stände an durchsichtigen Modellen mit Hilfe des Tyndalleffektes. Z
    angew Math Mech 20: 210-217
Mesmer G (1939) Spannungsoptik. Springer, Berlin
Mesnager A (1901) Contribution à l'étude de la déformation élastique
    des solides. Ann Ponts Chauss 4: 128-190
Neuber H (1937) Kerbspannungslehre. Springer, Berlin
Neumann FE (1841) Über die Gesetze der Doppelbrechung des Lichtes in
    komprimierten oder ungleichförmig erwärmten unkrystallinischen Kör-
    pern. Abh königl Akad Wiss Berlin. In: Neumann C, Voigt W, Wan-
    gerin A (1912) (eds) Franz Neumanns gesammelte Werke vol 3. Teubner,
    Leipzig, p 5-254
Nisida M (1933) On the physical properties of the photoelastic mate-
    rial "phenolite". Sci Pap Inst Phys Chem Research Tokyo 22: 269-283
Oppel G (1936) Polarisationsoptische Untersuchung räumlicher Spannungs-
    - und Dehnungszustände. Forschung Geb Ing Wesen 7: 240-248
Pockels F (1889) Über den Einfluss elastischer Deformationen, speciell
    einseitigen Druckes, auf das optische Verhalten Krystallinischer
    Körper. Ann Phys III 37: 144-172
Schwieger H, Haberland G (1954) Bestimmung des Biegespannungszustandes
    elastischer, quadratischer Platten. Bauplanung Bautechn 8: 358-366
Timoshenko S (1953) History of strength of materials. McGraw-Hill,
    New York Toronto London p 246
Tuzi Z (1927) A new material for the study of photo-elasticity. Photo-
    -elastic study of stress in a specimen of three-dimensional form.
    Photoelastic study of stress in heat-treated column. Sci Pap Inst
    Phys Chem Res Tokyo 7: 79-96
Tuzi Z (1931) Optical stress analysis. Engineering 131: 116-117,
    193-196
Tuzi Z, Kawata K (1957) Photo-elasticity experiment (In Japanese)
    Riken Keiki Fine Instrument, Tokyo
Wertheim G (1851) Sur les effets optiques de la compression du verre.
    Compres Rendus 32: 144-145
Widdern H von (1930) Polarisationsoptische Spannungsmessungen an Stab-
    ecken. Mitt Mech Techn Lab Techn Hochschule München 34: 4-16
Wilson C (1891) On the influence of surface loading on the flexure of
    beams. Phil Mag V 32: 481-503
Wolf H (1961) Spannungsoptik. Springer, Berlin Göttingen Heidelberg

Photoelasticity in Japan
A Survey

M. Nisida

Faculty of Science and Technology, Science University of Tokyo, Noda, Chiba, 278 Japan

INTRODUCTION

The studies of photoelasticity in Japan date back to 1920 when Z. Tuzi started his pioneer work and a small descriptive book on photoelasticity was published. Since then, steady efforts have been continued by many successors to study photoelasticity in its various aspects.

Photoelastic studies in Japan may roughly be grouped into two categories: Fundamental research from the purely scientific or educational point of view at universities and official institutes, and the applied investigations necessary for designing structures which are light yet of sufficient strength at various industrial companies.

Although the utilization of the finite element method for determining stress distribution in structures has spread dominantly over the country with the rapid progress of computers, photoelasticity is still playing an important part for its unique characteristics of visual proof and versatility, and is often used to check the results obtained by the finite element method to meet the increasing requirement of confirming structural strength.

It is presumed that a good many photoelastic investigations are carried out at various industrial companies but, to our regret, only a few results are made public.

PHOTOELASTIC RESEARCHERS AND TOPICS IN JAPAN

Photoelastic researchers who have published papers or made interim reports during the last several years and the topics are listed as follows:

(1) Photoelastic fringe processing and computer-aided stress analysis

   a. Y. Seguchi, Y. Tomita and M. Watanabe; Exp. Mech., 19-10(1979) 362.

   b. Y. Mitsui and S. Yoshida; PJSP., 3-1(1981)15, P. Civil E., 66-5(1981)62.

   c. H. Okamura and S. Sakai;

   d. M. Nisida, M. Kotani, Nakajima and S. Honda et al.; PJSP., 4-1(1982)20.

    e. E. Umezaki and S. Takahashi; J. NDI., 33-8(1984)603.

    f. T. Hayashi et al.; RJSP, 7(1985)43.

    g. S. Arai; PJSP., 5-2(1984)17.

(2) Scattered light photoelasticity and propagation of polarized light

    a. Kuramoto; PJSP, 1-1(1979)13.

    b. T. Kihara, H. Kubo and R. Nagata; PJSP., 4-1(1982)1, 5-1 (1984)25.

    c. **M. Chiba and Shimada; Exp. Mech., 15-4(1975)142, T. JSME., 44-384(1978)2528.

(3) High-speed photoelasticity

    a. K. Kawata and S. Hashimoto; Exp. Mech., 24-4(1984)316, PJSP., 1-1(1979)33.

    b. M. Shibahara and T. Ueda; PJSP., 5-1(1984)15.

    c. ***M. Nisida, Y. Sawa and M. Kurabe; PJSP., 1-1(1979)1.

(4) Stress freezing technique

    a. M. Nisida and Y. Sawa; T. JSME., 47-418(1981)611, PJSP., 2-1 (1980)31.

    b. Kishi and J. Kawagoe;

    c. M. Nisida, S. Miyasono and Y. Sawa PJSP., 1(1979)53.

(5) Viscoelastic properties of photoelastic materials

    a. Miki and T. Hirano; PJSP., 1-1(1979)7, 2-1(1980)26, 3-1 (1981)36.

    b. Sugimori, Y. Miyano, and T. Kunio; PJSP., 3-1(1981)36.

    c. M. Yasui; PJSP., 1(1979)83.

(6) Thermal stresses

    a. K. Miyao and A. Iwaki; Exp. Mech., 8-8(1984)20.

    b. N. Hirai and A. Saito; JTS., 6(1983)153.

    c. M. Nisida and Y. Sawa; T. JSME., 44-383(1978)2264.

    d. Furukawa et al.; Kikai K. 25-11(1973)1404.

    e. M. Tsuji et al.; T. JSME., 43-369(1977)1592.

f. Y. Tsukada and A. Amada; T. JSME., 40-332(1977)926.

(7) Isoclinic lines, singular points and separation of principal stresses

a. S. Sakai and H. Okamura; T. JSME., 50-452(1984)700.

b. Y. Masuda; PJSP., 5-2(1948)9.

c. M. Nisida and H. Takeishi; PJSP., 3-1(1981)1.

(8) Analysis of structural stresses and stress concentration

a. Y. Sawa; B. JSME., 25-209 (1982)1662, 27-229(1984)133.

b. S. Nakamura et al.; PJSP., 2-1(1980)14.

c. M. Nisida and H. Kobayashi; RJSP., 3(1981)9, 5(1983)37.

d. G. Matsui and H. Seya; RJSP., 3(1981)29.

e. Tsuboi; RJSP., 3(1981)25.

f. M. Totsuka and T. Shiino; RJSP., 3(1981)21.

g. C. Miyata et al.; PSSP., 4-1(1982)15.

h. S. Takahashi, A. Shimamoto and F. Nogata; PJSP., 2-1(1980)7.

i. J. Tokuhiro and M. Ono; PJSP., 3-1(1984)30.

j. K. Jinbo;

k. T. Morioka;

l. M. Yamamoto and R. Yamazaki;

(9) Stress intensity factor

a. T. Nagai and S. Matsumoto; PJSP., 1-1(1979)21.

b. S. Takahashi, A. Shimamoto, F. Nogata and T. Izumi; PJSP., 3-1(1981)24.

c. F. Nogata, K. Seo and J. Masaki; PJSP., 5-2(1984)1.

(10) Miscellaneous

a. R. Shintani, T. Kubota and T. Hoshikawa; PJSP., 5-1(1984)1 (microphotoelasticity)

b. *K. Mori; PJSP., 2-1(1980)1 (thickness effect of photoelastic coatings).

c. K. Kawata et al.; RJSP., 1(1979)57, 2(1980)66. (stress in anisotropic plates).

d. M. Nisida and Y. Tsuruta; P. 8th AUCP., (1979)126, PJSP., 5-1
    (1984)9. (contact problem - interferometric photoelasticity).

e. M. Nisida and H. Takabayashi; BJSME., 19-138(1976)711.
    (rectangular and skew plates under bending - laminate plate
    method).

Here, a few papers other than those to be presented at the symposium
will be briefly described.

*K. Mori studied analytically the "thickness effect or end effect" of
photoelastic coating. When a photoelastic coating with thickness h
and length $2l$ is cemented to the surface of a metal body and the body
is subjected to a uniform strain ε_0, the distribution of strain in the
coating ε_x which is related to the birefringence can be exactly calculated by the equation derived by Mori. Fig. 1.   Fig. 2 shows the
relation between the mean stress along the optical path in the coating
σ_m/σ_0 and the abscissa x/l with the thickness of the coating h/l as a
parameter. It is interesting to note that, since the increase of the
birefringence of fringe order measured upon reflection at the center
diminishes in rate with the thickness and the fringe order becomes
saturated for large thickness, Fig. 2, the optimum thickness exists
from the measuring sensitivity view point, Fig. 3. (h/l=0.5)

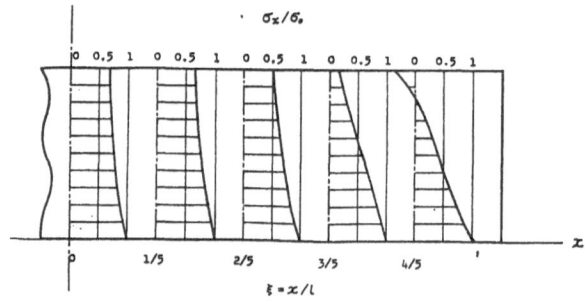

Fig. 1

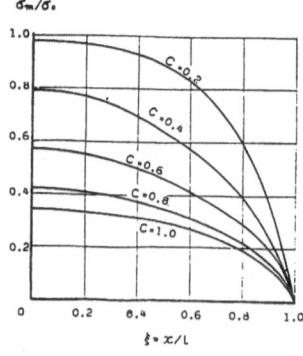

Fig. 2

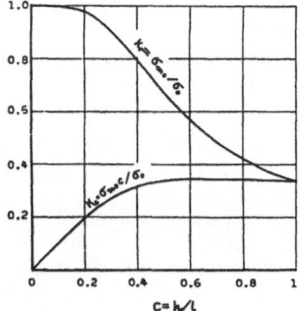

Fig. 3

**M. Chiba and H. Shimada have contributed to clarify experimentally
the effects of rotation of secondary principal stress axes in three-
dimensional stress state. Using specimens of bending and membrane
stress (σ_b and σ_m) combined so as to rotate the directions of the
resultant principal stress along the optical path, it has been found
that the directions of secondary principal stresses and the phase retardation obtained at the back of the specimen agree quite well with
the values obtained by the numerical calculation based on H. Aben's
equation, Fig. 4.

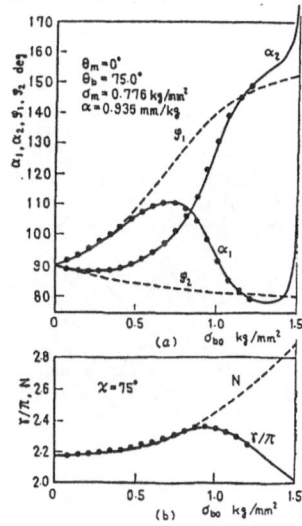

Fig. 4

***M. Nisida, Y. Sawa and M. Kurabe studied the distributions of plas-
tic strains in Charpy test pieces during impact using photoelastic
coatings and a high speed recording technique. Colored photoelastic
patterns of the coating film (epoxy rubber, thickness 0.1 mm, strain
sensitivity for green light β = 38.3-64.1 mm^{-1}, E = 1.0-2.0 kgf/mm^2)
which relate to the difference of principal strains $\varepsilon_1 - \varepsilon_2$ were taken
at various stages of impact by a delay-Xenon-micro-flash system with
a resolving power of about 2 μsec., though the number of patterns
taken in an impact of about 1/1,000 sec. duration is limited to three
due to the roughness of the metal surface in plastic state.
    For example, Fig. 5 shows the photoelastic pattern taken at
T = 300 μsec. (test piece; material 25% carbon steel, dimensions

JIS standard No. 3 with a U-notch). It has been found that the strain
energy absorbed in the test piece is notably localized on two regions:
Near the contact point of hammer "A" and the bottom of notch "B" at
which cracks usually originate, Fig. 6.

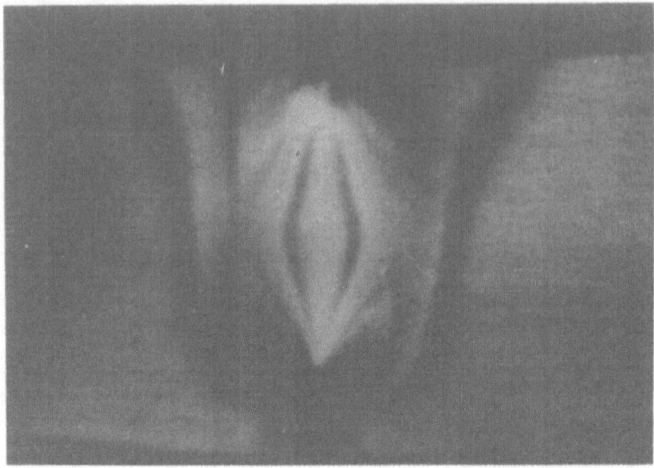

Fig. 5   (original in color)

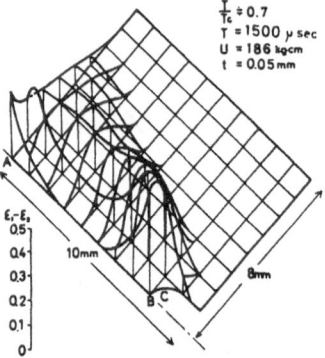

Fig. 6

Many stress freezing photoelastic experiments have been conducted to
analyze stresses in various machine parts and engineering structures.
For example, S. Takahashi et al. investigated the stress distributions
in a rotating impeller of an axial-flow type compressor, especially
the stress concentration in the dove-tail type connection.
    T. Morioka and N. Kikuno et al. studied respectively detailed
stress distributions in various parts of diesel engines in connection
with design improvement.

Determination of Sign and Components of Artificial Birefringence of Visco-Elastic-Plastic Materials

Satoshi Miki

Kurume Institute of Technology, Kamitsu-machi, Kurume, 830 Japan

INTRODUCTION

The artificial birefringence behavior of visco-elasto-plastic polymers under stress generally exhibit creep effect. The author has often reported studies on creep and sign of birefringence. In those papers[1-14], it was reported that positive and negative signs of birefringence depend on the temperature and the loading time. In this paper, the variational quantity of principal refractive index and the birefringence of polymers during the creep under a constant stress and a constant temperature were investigated in the temperature range from -100°C to 130°C using the Mach-Zehnder interferometer and photoelectric apparatus. The sign and components of artificial birefringence which depend on the temperature and the loading time were determined, and the birefringence behavior of polymers are classified into four types by combinations of the component signs. Moreover, the visco-elasto-plastic stress and strain in polymers were analyzed with relations among loading time, birefringence, stress and strain which were obtained by tensile creep testing for birefringence and strain of materials.

EXPERIMENTAL METHOD

Isopachics was measured using the Mach-Zehnder interferometric method[15]. In this interferometer apparatus, two identical immersion tanks which were filled with the immersion liquid are placed in optical path I and II . In this experiment, the refractive index of these liquids is equal to the specimen's refractive index. A polarizer is placed in front of the observing point at the interferometer, the variation of principal refractive index of the specimen in a form of strip during the creep under a constant tensile stress in the immersion tank in optical path I is then measured continuously by the photoelectric method at a constant temperature. At the same time, the artificial birefringence of the specimen during the creep is measured in this apparatus. This present experiment was conducted at a constant temperature in a range from -100 °C to 130°C. Positive and negative of isopachics and isochromatics are determined using the Senarmont Compensator method by the direction of rotation of the polarized plane of the polarizer. The positive sign of the fringe is determined with the direction of rotation of the polarized plane when the difference of principal refractive index is positive.

SIGN AND COMPONENTS OF CREEP BIREFRINGENCE

Notations

$[N]_t$ Isochromatic fringe order at loading time t

$[N_E]_0$	Instantaneous isochromatic fringe order due to the perfect elasticity in the moment of loading
$[N_R]_t$	Delayed isochromatic fringe order at loading time t due to retarded elasticity
n_0	Refractive index of the specimen under no loading
n	Refractive index of air
n_1, n_2	Principal refractive index under the principal stress at loading time t
$\Delta n_{1E}, \Delta n_{2E}$	Variational quantity of principal refractive index in the moment of loading
$\Delta n_{1Rt}, \Delta n_{2Rt}$	Variational quantity of delayed principal refractive index at loading time t
$\Delta n_{1t}, \Delta n_{2t}$	Variational quantity of principal refractive index at loading time t

$$
\begin{aligned}
[N]_t \cdot \lambda &= [n_1 - n_2]_t \\
&= [n_0 \pm \Delta n_{1t}] - [n_0 \pm \Delta n_{2t}] \\
&= [\pm \Delta n_{1t}] - [\pm \Delta n_{2t}] \\
&= [(\pm \Delta n_{1E}) + (\pm \Delta n_{1Rt})] - [(\pm \Delta n_{2E}) + (\pm \Delta n_{2Rt})] \\
&= [(\pm \Delta n_{1E}) - (\pm \Delta n_{2E})]_0 + [(\pm \Delta n_{1R}) - (\pm \Delta n_{2R})]_t \quad (1)
\end{aligned}
$$

$$
\therefore \quad [\pm N]_t = [\pm N_E]_0 + [\pm N_R]_t \quad (2)
$$

where λ is the wavelength of Hg·5460.5 Å light.

It can be seen from equation (2) that the birefringence fringe order $[\pm N]_t$ of polymers at loading time t under creep is the algebraic summation of the instantaneous isochromatic fringe order $[\pm N_E]_0$ which is due to the perfect elasticity in the moment of loading and delayed isochromatic fringe order $[\pm N_R]_t$ at the loading time t which occurs with the retarded elasticity as loading time elapses.

It is clear from eq. (2) that the birefringence $[\pm N]_t$ behavior of polymers is classified into four types by combination of positive and negative sign of $[\pm N_E]_0$ and $[\pm N_R]_t$. Similarly, behavior of the instantaneous birefringence $[N_E]_0$ is classified into six types by the signs of $(\Delta n_{1E})_0$ and $(\Delta n_{2E})_0$, and delayed birefringence $[N_R]_t$ is classified into six types by the signs of $(\Delta n_{1R})_t$ and $(\Delta n_{2R})_t$. These are shown in Tables 1 through 3.

Table 1. Four types of creep birefringence $[N]_t$ at loading time t

Type	$[N_E]_0$	$[N_R]_t$	$[N]_t = [N_E]_0 + [N_R]_t$	
1	+	+	+	
2	+	−	$+ \rightarrow 0 \rightarrow -$	as loading time elapses
3	−	+	$- \rightarrow 0 \rightarrow +$	as loading time elapses
4	−	−	−	

The positive sign is indicated by +, and the negative sign by −. Signs of $[N_E]_0$ and $[N_R]_t$ depend on temperature.

Table 2. Six types of instantaneous birefringence $[N_E]_0$ of component of creep birefringence $[N]_t$

Type	$(\Delta n_{1E})_0$	$(\Delta n_{2E})_0$	$[N_E]_0 = (\Delta n_{1E})_0 - (\Delta n_{2E})_0$	
1	+	+	+	$\lvert(\Delta n_{1E})_0\rvert > \lvert(\Delta n_{2E})_0\rvert$
2	+	+	−	$\lvert(\Delta n_{1E})_0\rvert < \lvert(\Delta n_{2E})_0\rvert$
3	+	−	+	
4	−	+	−	
5	−	−	−	$\lvert(\Delta n_{1E})_0\rvert > \lvert(\Delta n_{2E})_0\rvert$
6	−	−	+	$\lvert(\Delta n_{1E})_0\rvert < \lvert(\Delta n_{2E})_0\rvert$

Table 3. Six types of delayed birefringence $[N_R]_t$ of component of creep birefringence $[N]_t$

Type	$(\Delta n_{1R})_t$	$(\Delta n_{2R})_t$	$[N_R]_t = (\Delta n_{1R})_t - (\Delta n_{2R})_t$	
1	+	+	+	$\|(\Delta n_{1R})_t\| > \|(\Delta n_{2R})_t\|$
2	+	+	−	$\|(\Delta n_{1R})_t\| < \|(\Delta n_{2R})_t\|$
3	+	−	+	
4	−	+	−	
5	−	−	−	$\|(\Delta n_{1R})_t\| > \|(\Delta n_{2R})_t\|$
6	−	−	+	$\|(\Delta n_{1R})_t\| < \|(\Delta n_{2R})_t\|$

DETERMINING MATERIAL CONSTANT A AND B

The variation of specimen thickness in a plane stress state $(d'- d)$ is
$- \frac{\nu}{E}(\sigma_1 + \sigma_2)d$.
where ν and E are the Poisson's ratio and Young's modulus of specimen, σ_1 and σ_2 are principal stresses.
Therefore, in pure tensile stress (σ_1 state),

$$\Delta n_{1t} = n_{1t} - n_0 = A\sigma_1$$
$$\Delta n_{2t} = n_{2t} - n_0 = B\sigma_1 \tag{3}$$
$$[N]_t \lambda = \Delta n_{1t} - \Delta n_{2t}$$

where A and B are the material constants of the specimen.

$$n_{1t} = n_0 + \frac{d'-d}{d} n_0 - \frac{d'-d}{d} n + A\sigma_1 + B\sigma_2$$

$$= n_0 - \frac{\nu}{E}(\sigma_1 + \sigma_2)n_0 + \frac{\nu}{E}(\sigma_1 + \sigma_2)n + A\sigma_1 + B\sigma_2$$

$$= n_0 + \left\{ A - \frac{\nu}{E}(n_0 - n)\right\} \sigma_1 + \left\{ B - \frac{\nu}{E}(n_0 - n)\right\} \sigma_2$$

$$= n_0 + A'\sigma_1 + B'\sigma_2 \tag{4}$$

Similarly
$$n_{2t} = n_0 + B'\sigma_1 + A'\sigma_2 \tag{5}$$
where material constants A' and B' are

$$A' = A - \frac{\nu}{E}(n_0 - n) \tag{6}$$

$$B' = B - \frac{\nu}{E}(n_0 - n) \tag{7}$$

In this experiment, two identical immersion tanks which were filled with the immersion liquid are placed in optical paths I and II of the interferometer, and the specimen is loaded in the immersion tank in optical path I. Therefore, from eqs. (6) and (7)
$$A' = A, \quad B' = B$$
The polarized plane of polarizer which is placed in front of the observing point coincides with or makes a right angle to the direction of the pure tensile stress σ_1 in the specimen, the intensities of the two polarized lights after the interference $[I_{/\!/}]$ and $[I_\perp]$ are

$$[I_{/\!/}] = \frac{1}{4} + \frac{1}{4} \cos\left[\frac{2\pi}{\lambda}\left\{ A'\sigma_1 + (n_0 - n)\right\} d\right] \tag{8}$$

$$[I_\perp] = \frac{1}{4} + \frac{1}{4} \cos\left[\frac{2\pi}{\lambda}\left\{ B'\sigma_1 + (n_0 - n)\right\} d\right] \tag{9}$$

and the isopachics fringe order $N_{PA'}$ and $N_{PB'}$ are
$$N_{PA'} = A'\sigma_1 d/\lambda \tag{10}$$

$$N_{PB'} = B'\sigma_1 d/\lambda \tag{11}$$

From eqs. (10) and (11), the values of A' and B' can be obtained as
$$A' = \frac{\lambda}{\sigma_1 d} N_{PA'} \tag{12}$$

$$B' = \frac{\lambda}{\sigma_1 d} N_{PB'} \tag{13}$$

EXPERIMENTAL RESULTS

Creep Behavior of Birefringence Patterns

Creep behavior of birefringence patterns in a 5-5 mixed polyester plane-parallel plate beam specimen during creep under a constant uniform bending and after unloading at temperature 25°C is shown in Fig. 1. The birefringence fringe order distributions obtained in this experiment during creep are shown in Fig. 2. From Photo (6) in Fig. 1, it can be seen that the instantaneous birefringence due to perfect elasticity disappears in the moment of unloading, and only the delayed birefringence which originated with the retarded elasticity appears with the residual state. This birefringence, however, decreases as unloading time elapses and nearly disappears about 17 hr. after unloading as shown in Photo (8) of Fig. 1. In the 5-5 mixed polyester, the instantaneous birefringence sign is positive and delayed birefringence is negative. Therefore, it can be seen that the sign of birefringence $[N]_t$ changes from positive to negative in all range about 30 min. after loading, as shown in Figs. 1-(5) and 2-(30').

Measuring $[N]_t$, Four Types of Creep Birefringence, and Positive and Negative Signs of $[\Delta n_1]_t$, $[\Delta n_2]_t$, $[N]_t$, $[N_E]_0$ and $[N_R]_t$

The variation of birefringence light intensity which was measured by an oscillograph, during the creep under a constant axial tension of a specimen, is shown in Figs. 3(a) and (b).

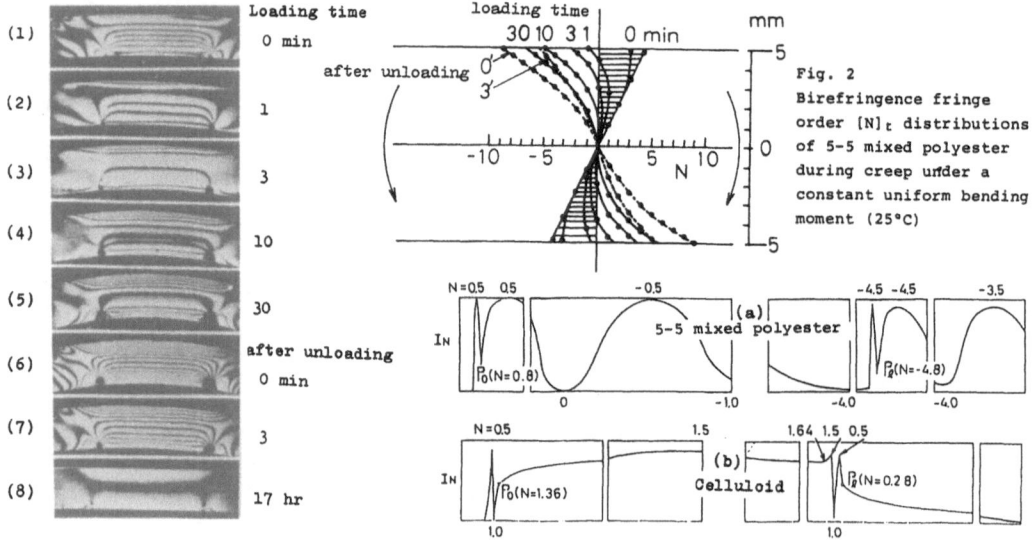

Fig. 2
Birefringence fringe order $[N]_t$ distributions of 5-5 mixed polyester during creep under a constant uniform bending moment (25°C)

Fig. 1
Birefringence patterns of 5-5 mixed polymers during the creep under a constant uniform bending moment (25°C)

Fig. 3 Variation of intensity of birefringence light during the tensile creep
P_0 in the moment after loading
P_1 in the moment after unloading
(by an oscillograph)

As shown in Table 1, the creep birefringence behavior of visco-elasto-plastic polymers is classified into four types by combinations of the component's signs. Variations of the birefringence coefficient $(\alpha)_t$

$(= \frac{[N]t}{d\sigma})$ of four typical polymers for the sign during the creep are shown in Fig. 4.

As shown in eq. (2), if $[N_E]_0$ and $[N_R]_t$ signs are different, the sign of birefringence $[N]_t$ changes during the creep. In celluloid, as shown in Fig. 5,

$$|(\Delta n_{1E})_0| < |(\Delta n_{2E})_0|$$
$$\therefore \quad [N_E]_0 = (-\Delta n_{1E})_0 - (-\Delta n_{2E})_0 > 0 \quad \text{positive sign} \quad (14)$$

and

$$|(\Delta n_{1R})_t| < |(\Delta n_{2R})_t|$$
$$\therefore \quad [N_R]_t = (-\Delta n_{1R})_t - (-\Delta n_{2R})_t > 0 \quad \text{positive sign} \quad (15)$$

therefore

$$[N]_t = [N_E]_0 + [N_R]_t > 0. \quad \text{positive sign} \quad (16)$$

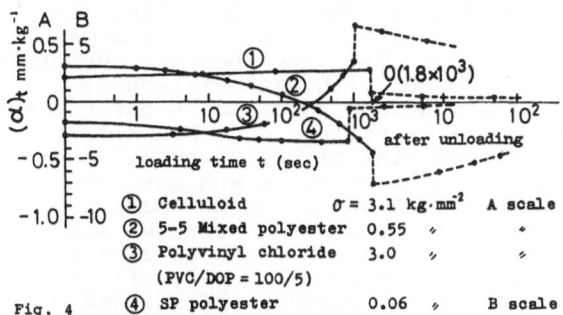

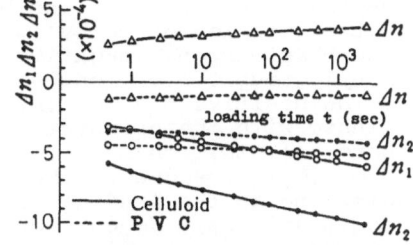

Fig. 4 Variations of birefringence coefficient $(\alpha)_t = \frac{[N]_t}{d\sigma}$ of four typical polymers for the sign during creep under a constant tensile stress (25°C)

Fig. 5 Variational quantity of principal refractive index Δn_1, Δn_2 and its difference Δn of celluloid and polyvinyl chloride during creep under a constant tensile stress (25°C, $\sigma = 2$ kg·mm^{-2})

As shown in Fig. 6, the behavior of epoxy is similar to celluloid, but $[N_R]_t$ is very small. The Author has already reported the photo-viscous studies of birefringence of a viscous flow[13]. In the paper,

$$|(\Delta n_{1E})_0| < |(\Delta n_{2E})_0|$$
$$\therefore \quad [N_E]_0 = (-\Delta n_{1E})_0 - (-\Delta n_{2E})_0 > 0 \quad (17)$$
$$[N_R]_t \fallingdotseq 0 \quad (18)$$

therefore

$$[N]_t \fallingdotseq [N_E]_0 > 0. \quad \text{positive sign} \quad (19)$$

Next, with regard to 5-5 mixed polyester, as shown in Fig. 7,

$$[N_E]_0 = (\Delta n_{1E})_0 - (-\Delta n_{2E})_0 > 0 \quad (20)$$
$$[N_R]_t = (-\Delta n_{1R})_t - (\Delta n_{2R})_t < 0 \quad (21)$$
$$[N]_t = [+N_E]_0 + [-N_R]_t = + \rightarrow 0 \rightarrow - \quad (22)$$

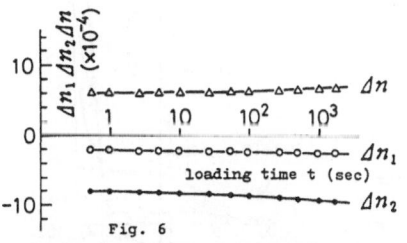

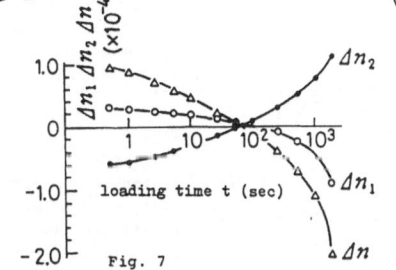

Fig. 6 Variational quantity of principal refractive index Δn_1, Δn_2 and its difference Δn of epoxy resin during creep under a constant tensile stress (25°C, $\sigma = 2$ kg·mm^{-2})

Fig. 7 Variational quantity of principal refractive index Δn_1, Δn_2 and its difference Δn of 5-5 mixed polyester during creep under a constant tensile stress (25°C, $\sigma = 0.5$ kg·mm^{-2})

as the loading time elapses, the sign of $[N]_t$ changes from positive to negative. In polyvinyl chloride, as shown in Fig. 5,

$$|(\Delta n_{1E})_0| > |(\Delta n_{2E})_0|$$
$$\therefore \quad [N_E]_0 = (-\Delta n_{1E})_0 - (-\Delta n_{2E})_0 < 0 \tag{23}$$
$$|(\Delta n_{1R})_t| < |(\Delta n_{2R})_t|$$
$$\therefore \quad [N_R]_t = (-\Delta n_{1R})_t - (-\Delta n_{2R})_t > 0 \tag{24}$$
therefore
$$[N]_t = [-N_E]_0 + [N_R]_t = - \longrightarrow 0 \longrightarrow +. \tag{25}$$

As loading time elapses, the sign of $[N]_t$ changes from negative to positive sign.

Effect of Molecular Behavior Caused by the Birefringence

It is understood that $[N_E]_0$ is dependent on the intramolecular rotation, and $[N_R]_t$ is due to the unfolding of molecular chain. Generally, it seems that the effect of the birefringence positive sign originates with the rotation of methyl group CH_3 around C-O coupling, and the effect of the birefringence negative sign originates with the rotation of carbonyl group C=O or benzene ring ◯ around C-C coupling or ester bond -O-. The positive sign of $[N_E]_0$ and $[N_R]_t$ in celluloid is known to originate with the orientation of pyranose ring of the principal chain. In hard polyvinyl chloride, the negative sign of $[N_E]_0$ and positive sign of $[N_R]_t$ are due to the orientation of H and Cl at the moment of loading and as the loading time elapses. In epoxy, the positive sign of $[N_E]_0$ and $[N_R]_t$ depends on the orientation of the benzene ring and methyl group; and in mixed polyester, the negative sign of $[N_R]_t$ is due to the orientation of the benzene ring and carbonyl group; and in soft mixed polyester, the negative sign of $[N_E]_0$ and $[N_R]_t$ is due to the orientation of the benzene ring and carbonyl group.

The Sign of the Birefringence Coefficient $(\alpha)_t = \dfrac{[N]_t}{d\sigma}$ and its Components' Coefficient $(\alpha_E)_0 = \dfrac{[N_E]_0}{d\sigma}$ and $(\alpha_R)_t = \dfrac{[N_R]_t}{d\sigma}$ during Creep under a Constant Tensile Stress σ in the Temperature Range from -100°C to 130°C, and Birefringence Coefficient α_T and α_C

From eq. (2), $(\pm\alpha)_t = (\pm\alpha_E)_0 + (\pm\alpha_R)_t$
(α_E), $(\alpha_R)_{30min}$ and $(\alpha)_{30min}$ of 5-5 mixed polyester, celluloid and polyvinyl chloride are shown in Figs. 8 and 9 (-100°C to 130°C).

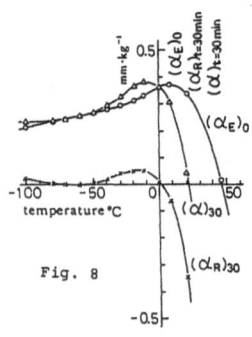

Fig. 8

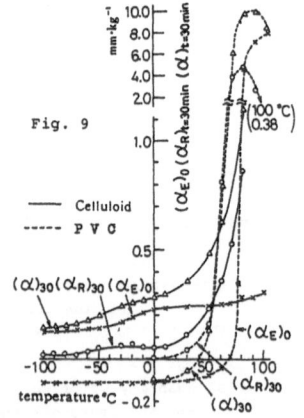

Fig. 9

Fig. 8
Birefringence coefficient $(\alpha)_t = \dfrac{[N]_t}{d\sigma}$, instantaneous birefringence coefficient $(\alpha_E)_0 = \dfrac{[N_E]_0}{d\sigma}$ and delayed birefringence coefficient $(\alpha_R)_t = \dfrac{[N_R]_t}{d\sigma}$ of 5-5 mixed polyester (-100°C to 100°C, loading time t = 30 min)

Fig. 9
Birefringence coefficient $(\alpha)_t = \dfrac{[N]_t}{d\sigma}$, instantaneous birefringence coefficient $(\alpha_E)_0 = \dfrac{[N_E]_0}{d\sigma}$ and delayed birefringence coefficient $(\alpha_R)_t = \dfrac{[N_R]_t}{d\sigma}$ of celluloid and polyvinyl chloride (-100°C to 100°C, loading time t = 30 min)

— Celluloid
---- P V C

(α_E), $(\alpha_R)_t$ and $(\alpha)_t$ under tensile stress σ_T or compressive stress σ_C in 5-5 mixed polyester in a temperature range from -100°C to 30°C are shown in Fig. 10 where it is clear that the absolute value of $(\alpha_{TE})_0$ and $(\alpha_{TR})_t$ under the tensile stress σ_T and the absolute value of $(\alpha_{CE})_0$ and $(\alpha_{CR})_t$ under compressive stress σ_C change in opposition at a temperature of about -20°C.

Relations among Loading Time, Birefringence [N]$_t$, Stress σ, and
Strain ε — Analysis of Stress and Strain Distributions during Creep

The variations of birefringence [N]$_t$ and strain ε of 5-5 mixed polyester
during creep under a constant tensile stress σ of the strip specimen
are shown in Figs. 11 and 12 respectively. Using these diagrams of
birefringence [N]$_t$ and strain ε under creep, relations between stress
σ and birefringence [N]$_t$, stress σ and strain ε, and strain ε and
birefringence [N]$_t$ with loading time t can be obtained as shown in
Figs. 13, 14 and 15. Next, with these relations, birefringence [N]$_t$,
stress σ, and strain ε distributions in 5-5 mixed polyester strip
having a circular hole, for loading time t = 0, 1, and 3 min. during
creep under a constant tensile loading could be obtained. These results
and their fringe patterns are shown in Figs. 16(a), (b) and (c).

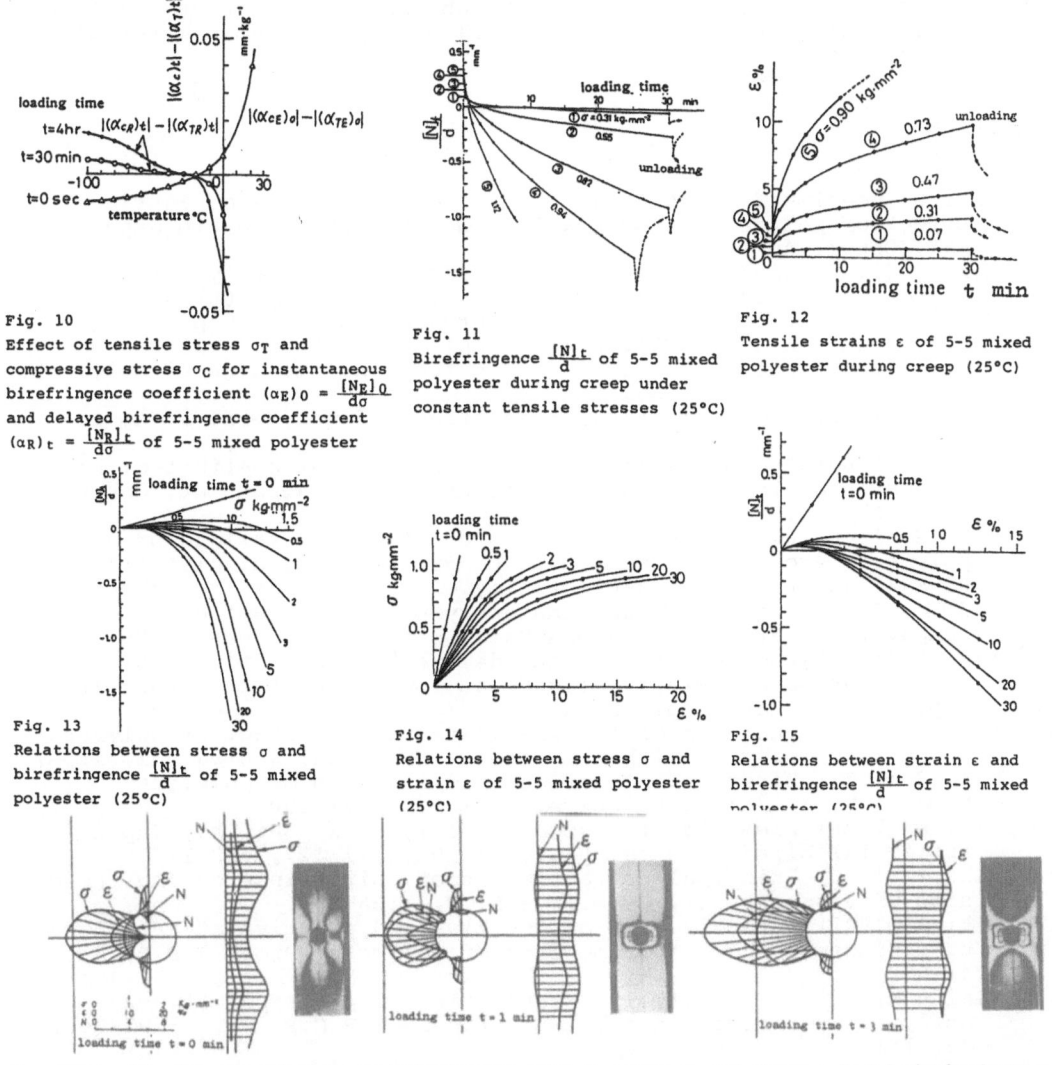

Fig. 10
Effect of tensile stress σ$_T$ and
compressive stress σ$_C$ for instantaneous
birefringence coefficient $(\alpha_E)_0 = \frac{[N_E]_0}{d\sigma}$
and delayed birefringence coefficient
$(\alpha_R)_t = \frac{[N_R]_t}{d\sigma}$ of 5-5 mixed polyester

Fig. 11
Birefringence $\frac{[N]_t}{d}$ of 5-5 mixed
polyester during creep under
constant tensile stresses (25°C)

Fig. 12
Tensile strains ε of 5-5 mixed
polyester during creep (25°C)

Fig. 13
Relations between stress σ and
birefringence $\frac{[N]_t}{d}$ of 5-5 mixed
polyester (25°C)

Fig. 14
Relations between stress σ and
strain ε of 5-5 mixed polyester
(25°C)

Fig. 15
Relations between strain ε and
birefringence $\frac{[N]_t}{d}$ of 5-5 mixed
polyester (25°C)

Fig. 16(a), (b), (c) Distributions of birefringence order N, stress σ, and strain ε in 5-5 mixed polyester
plate having a circular hole during creep under a constant tensile loading, and its fringe patterns (25°C)

CONCLUSION

The results obtained from the present investigation in the temperature
range from -100°C to 130°C are summarized as follows:
Artificial birefringence of visco-elasto-plastic polymers consists of
two components with positive and negative signs. The two components
are the instantaneous birefringence which is due to the perfect
elasticity in the moment of loading and the delayed birefringence
which originates with the retarded elasticity occurring as loading
time elapses.
Thus, the birefringence behavior of polymers is classified into four
types by combinations of its components' signs. When the sign of the
two birefringence components differ, the birefringence sign changes
during creep.
Relations among loading time t, birefringence $[N]_t$, stress σ and
strain ε under the creep state could be obtained. With these relations,
distributions of stress and strain in visco-elasto-plastic materials
could be analyzed.

REFERENCES

1. Miki S, Soejima Y (1956-1961) Studies on photo-elasto-plastic
     effect of polymers, from No. 1 to No. 8. Proc. 3rd-8th Japan
     Nat. Congr. Appl. Phys.
2. Miki S (1957) Sign of photo-elasto-plastic sensitivity. Proc.
     Symposium on photo-elastic-plasticity of J.S.A.P.:66
3. Miki S (1959) Photo-elasto-plastic studies on the creep of high
     polymers. 8th Japan Nat. Congr. Appl. Mech. "1958":231-234
4. Miki S (1962) Photo-elasto-plastic effect and its sign of polymers.
     J. Appl. Phys. 31, 4:262-268
5. Miki S (1962) Photo-elasto-plastic effect of polymers. J. Appl.
     Phys. 31, 10:844-845
6. Miki S (1967) Photoelastic effect and photoplastic effect of mixed
     polyester at low temperature (I). Proc. 14th Japan Nat. Congr.
     Appl. Phys. :229
7. Miki S (1970) Photo-visco-elastic studies on the fatigue of high
     polymers. Proc. 18th Japan Nat. Congr. Appl. Mech. "1968":183-188
8. Miki S (1968) Photo-visco-elasticity of polymers at low temperature
     (II) , Birefringence of gum and epoxy-polysulfideamine at low
     temperature. Proc. 15th Japan Nat. Congr. Appl. Phys. :189
9. Miki S (1969) Photoelasticity and delayed photoelasticity of
     celluloid and polyvinyl chloride at low temperature. Proc. 16th
     Japan Nat. Congr. Appl. Phys. :55
10. Miki S (1971) The sign of stress-birefringence effect of polymers
     and stress analysis with the effect. J. Stress & strain 1, 1:9-17
11. Miki S (1971) Birefringence effect and its sign of polymers.
     Memoirs F. E. Yamaguchi University 22, 11:91-100
12. Miki S (1974) Elements of photoelasticity. Riko Shinsha, Japan,
     artificial birefringence, 5-28, 91-105, 111-127, 162-175
13. Miki S (1980) Principal refractive index, birefringence and these
     sign of viscous flow. Proc. J.S.P. 2, 1:26-30
14. Miki S (1981) Determination of positive and negative of artificial
     birefringence and its components of polymers. Proc. J.S.P. 3, 1:
     43-53
15. Nisida M, Saito H (1961, 1962) Experimental principal stress
     analysis by interferometry. Sci. Papers I.P.C.R. 37, 5:297-305,
     38, 1:112

A Simplified Optical Method for Measuring Residual Stress by Rapid Cooling in Thermosetting Resin Plate

S. Sugimori[1], Y. Miyano[1] and T. Kunio[2]

[1] Materials System Research Laboratory, Kanazawa Institute of Technology, Ohgigaoka, Nonoichi, Ishikawa, 921 Japan
[2] Department of Mechanical Engineering, Keio University, Hiyoshi, Kohoku-ku, Yokohama, 223 Japan

INTRODUCTION

Recently, thermosetting plastics have been more widely used in parts requiring high reliability, such as precision mechanical elements of electric apparatus, structural members on the aircraft and so on. The residual stress and shrinkage of thermosetting plastics in molding are becoming a serious problem these cases. Therefore, it has become necessary to investigate the growth mechanism, method of prevention and evaluation of residual stress and strain.

Thermosetting plastics are produced by combining liquid or powder resin with a hardener and an accelerator then heating and pressing in a metal die. The residual stress and strain of thermosetting plastics develop mainly during the cooling process after curing when the plastics are rapidly cooled and the material shows viscoelastic behavior over a range of temperatures.

It is well known that the photoviscoelastic technique can be used as an experimental technique for viscoelastic stress and strain analysis. The practicality of this technique has demonstrated by Arenz et al. (1963, 1967) and by the authors (Kunio et al. 1968; Miyano et al. 1969). In a previous paper (Sugimori et al. 1984), the transient thermal stresses and strains in an epoxy resin strip subjected to rapid cooling from both sides were analyzed by photoviscoelastic technique. The experimental values agreed closely with the theoretical values calculated from the linear viscoelastic theory. The photoviscoelastic technique is thus a viable method for solving the thermoviscoelastic problem.

However, this technique is far too complicated if only the residual stresses and strains by rapid cooling in thermosetting resin are needed, since the transient birefringence and temperature should be continuously measured during cooling to obtain the transient thermal stresses and strains as well. It is the purpose of this paper to propose a simplified optical method based on photoviscoelasticity for measuring residual stress by rapid cooling in thermosetting resin plate. Actually, the residual stress in an epoxy resin plate by rapid cooling from both surfaces is measured by two methods here, the simplified optical method and the well-known layer removal method. The effectiveness of the simplified optical method can be assessed by discussing and comparing the results obtained by the two methods.

FUNDAMENTAL EQUATION

The fundamental equation for calculating the residual stress from the residual birefringence by rapid cooling in a thermosetting resin plate is derived using the linear photoviscoelastic theory.

The thermosetting resin plate which shows linear photoviscoelasticity and thermorheologically simple behavior is shown in Fig. 1. It is assumed that lengths l and l' of the plate are sufficiently larger than the thickness d, and that the plate is rapidly cooled on both surfaces at the moment t = 0 without constraint from the holding temperature T_h in rubbery state down to the cooling temperature T_c in glassy state. The transient temperature distribution T in the plate is considered to be a function of the direction z and the elapsed time t. The thermal stresses σ and strains ε_σ are generated during cooling and are only functions of z and t.

The constitutive equation is expressed by the following hereditary integral using the linear viscoelastic theory and the time-temperature superposition principle.

$$\sigma(z,t) = \int_0^t 2G_r(t'-\tau',T_0) \frac{d}{d\tau} [\varepsilon_\sigma(z,\tau)-\varepsilon_{\sigma z}(z,\tau)] d\tau \tag{1}$$

$$t' = \int_0^t \frac{du}{a_{T_0}[T(z,u)]} \tag{2}$$

where σ is σ_x (= σ_y) and ε_σ is $\varepsilon_{\sigma x}$ (= $\varepsilon_{\sigma y}$), and t' is the reduced time and T_0 is the reference temperature, and $G_r(t',T_0)$ is the shear relaxation modulus and $a_{T_0}(T)$ is the time-temperature shift factor.

The birefringence generated by double refraction of resin during cooling are also a function of z and t, the relation between fringe order per unit passage of light n (= n_x = n_y) and strain ε_σ is expressed by the following hereditary integral using the linear photoviscoelastic theory.

$$n(z,t) = \int_0^t C_{\varepsilon r}(t'-\tau',T_0) \frac{d}{d\tau} [\varepsilon_\sigma(z,\tau) - \varepsilon_{\sigma z}(z,\tau)] d\tau \tag{3}$$

where $C_{\varepsilon r}(t',T_0)$ is the relaxation birefringence-strain coefficient at the reference temperature T_0.

$G_r(t',T_0)$ and $C_{\varepsilon r}(t',T_0)$ are constant at the region of short and long reduced time where the resin is glassy and rubbery state respectively. Then $G_r(t',T_0)$ and $C_{\varepsilon r}(t',T_0)$ are expressed by the following equations using the normalized shear relaxation modulus $F_G(t',T_0)$ and the normalized relaxation birefringence-strain coefficient $F_C(t',T_0)$.

$$G_r(t',T_0)= G_G+(G_G-G_R) F_G(t',T_0) , \quad C_{\varepsilon r}(t',T_0)= C_G+(C_G-C_R) F_C(t',T_0) \tag{4}$$

where G_G and G_R are $G_r(t',T_0)$ at the glassy and rubbery states and C_G and C_R are $C_{\varepsilon r}(t',T_0)$ at the glassy and rubbery states.

It is assumed that $F_G(t',T_0)$ is equal to $F_C(t',T_0)$, and that $\varepsilon_\sigma(z,0)$ and $\varepsilon_{\sigma z}(z,0)$ are equal to zero because the whole plate is kept at the holding temperature of rubbery region during time t < 0. Here, $\sigma(z,t)$ should be the residual stress $\sigma_r(z)$, n(z,t) the residual birefringence $n_r(z)$ and $\varepsilon_\sigma(z,t)$ the residual strain $\varepsilon_{\sigma r}(z)$ when the temperature of the entire plate reaches the temperature of the cooling water T_C after an extended period of time. The following relation is obtained.

$$\sigma_r(z) = \frac{2(G_G-G_R)}{C_G-C_R} n_r(z) - \frac{2(G_G C_R - G_R C_G)}{C_G-C_R} [\varepsilon_{\sigma r}(z) - \varepsilon_{\sigma z r}(z)] \tag{5}$$

The strain $\varepsilon_\sigma(z,t)$ parallel to the plane of the plate is expressed as

$$\varepsilon_\sigma(z,t) = \varepsilon(t) + \kappa(t)z - \int_{T_h}^{T(z,t)} \alpha(T)dT \tag{6}$$

where $\varepsilon(t)$ is the expansion and $\kappa(t)$ the curvature of the central plane $z=0$ in the plate, and $\alpha(T)$ is the coefficient of thermal expansion.

The strain $\varepsilon_{\sigma x}(z,t)$ is expressed using $\varepsilon_\sigma(z,t)$, assuming that the relaxation Poisson's ratio does not change with temperature and time very much.

$$\varepsilon_{\sigma z}(z,t) = -2\bar{\nu}\varepsilon_\sigma(z,t) \tag{7}$$

where $\bar{\nu}$ is the mean Poisson's ratio.

The residual stresses $\sigma_r(z)$ in the x and y directions can be obtained substituting equations (6) and (7) into equation (5).

$$\sigma_r(z) = \frac{2(G_G-G_R)}{C_G-C_R} n_r(z) - \frac{2(G_G C_R - G_R C_G)}{C_G-C_R} (1+2\nu)[\varepsilon_r + \kappa_r z - \int_{T_h}^{T_C} \alpha(T)dT] \tag{8}$$

On the other hand, the following equilibrium equations with respect to the force parallel to the plane of the plate in the y direction and the moment must be satisfied for the free constraint condition after cooling:

$$\int_{-d/2}^{d/2} \sigma_r(z)dz = 0, \qquad \int_{-d/2}^{d/2} \sigma_r(z)zdz = 0 \tag{9}$$

The following equation can be obtained by substituting equation (8) into equations (9):

$$\sigma_r(z) = \frac{2(G_G-G_R)}{C_G-C_R} [n_r(z) - \frac{1}{d}\int_{-d/2}^{d/2} n_r(z)dz - \frac{12}{d^3}\int_{-d/2}^{d/2} n_r(z)zdz \; z] \tag{10}$$

The residual fringe order distribution $n_r{}^*(z)$ in the strip which is cut from the plate after cooling as shown in Fig. 1 can be measured expediently by the ordinary photoelastic technique. And the relation between $\sigma_r(z)$ and $n_r{}^*(z)$ can be easily derived from equation (10).

$$\sigma_r(z) = \frac{2G_G(G_G-G_R)}{G_G(C_G-C_R)-\nu_G C_G(G_G-G_R)} [n_r{}^*(z) - \frac{1}{d}\int_{-d/2}^{d/2} n_r{}^*(z)dz - \frac{12}{d^3}\int_{-d/2}^{d/2} n_r{}^*(z)zdz \; z] \tag{11}$$

where ν_G is Poisson's ratio in the glassy state of resin.

Equation (11) is the fundamental equation to obtain the residual stress distribution $\sigma_r(z)$ in the plate using the residual fringe order distribution $n_r{}^*(z)$ measured in the strip being cut from the plate after rapid cooling. The mechanical and optical coefficients G_G, G_R, C_G, and C_R in equation (11) have been previously characterized for thermosetting resin which indicates linear photoviscoelastic behavior. This simplified optical method for measuring residual stress in photoviscoelastic plate proceeds in the same straight-forwards manner as the ordinary photoelastic method.

EXPERIMENTS

Mechanical and Optical Characterization

Before determining the photoviscoelastic coefficients, this resin was found to be both mechanically and optically linear in the strain range up to 1.5 percent and in the temperature range from room temperature to 180°C.

The master curves of the shear relaxation modulus $Gr(t',T_0)$, and the relaxation birefringence-strain coefficient $C_{\varepsilon r}(t',T_0)$ at reference temperature T_0 of epoxy resin are shown in Fig. 2.

Two time-temperature shift factors, $a_{T_0}(T)$ for mechanical and optical behaviors are shown in Fig. 3. These $a_{T_0}(T)$ are exactly the same. They can also be approximated with two Arrhenius' equations with different activation energies ΔH (Sugimori et al. 1984).

The Poisson's ratio of epoxy resin measured at room temperature is ν_G = 0.35. The bulk modulus K regarded to be constant over the range of time and temperature can be easily obtained from ν_G and tensile relaxation modulus $E_r(t',T_0)$ at the reduced time $t' = 0$.

The normalized shear relaxation modulus $F_G(t',T_0)$ and the normalized relaxation birefringence-strain coefficient $F_C(t',T_0)$ of epoxy resin are shown in Fig. 4.

G_G and G_R of epoxy resin in Table 1 are the shear relaxation moduli $G_r(t',T_0)$ which are constant in the short reduced time region (glassy state) and in the long reduced time region (rubbery state) shown in Fig. 2. C_G and C_R in this table are the relaxation birefringence-strain coefficients $C_{\varepsilon r}(t',T_0)$ in glassy state and rubbery state shown in Fig. 2.

Method of Rapid Cooling

The lengths 1 and 1' of specimen shown in Fig. 1 are equal to 100 mm and thickness d is 10 mm.

The temperatures of 180°C, 150°C, and 120°C were selected as the holding temperature T_h because the glass transition temperature T_g for the resin is 132°C as shown in Fig. 2. The temperature of 10°C was selected as the cooling temperature T_C because the resin used is in glassy state at this temperature.

This specimen was held at the holding temperature for a sufficiently long time. Both surfaces of the specimen were then exposed to the cooling water ($T = T_C$) under no constraint until it cooled down to the cooling water temperature. The transient temperatures on both surfaces of the specimen were measured by thermocouples. It was confirmed that the temperature of the surfaces instantaneously reached to the cooling water temperature.

Optical and Mechanical Methods for Measuring Residual Stress

First, the residual stress distribution $\sigma_r(z)$ in the plate after rapid cooling was measured by the simplified optical method.

Strips were cut from the plate as shown in Fig. 1. The residual fringe order distributions $N_r^*(z)$ in the width direction (z direction) of the strip were determined by the ordinary photoelastic technique using monochromatic light (light wavelength $\lambda = 546$ nm) and white light. The residual stress distribution $\sigma_r(z)$ in the plate was obtained by substituting $N_r^*(z)$ and G_G, G_R, C_G, CR and ν_G in Table 1 into equation (11).

The residual stress distributions $\sigma_r^*(z)$ in the strip cut from the plate were measured by the layer removal method. The residual stress distributions $\sigma_r(z)$ in the plate were then calculated from $\sigma_r^*(z)$. The details of measurement by this method have already been described in a previous paper (Miyano et al. 1982).

Experimental Results

Figure 5 shows the photographs of the isochromatic fringe patterns in dark field of the epoxy resin strips cut from the plates which were cooled down from $T_h = 120°C$, $150°C$ and $180°C$ to $10°C$. These isochromatic fringes are parallel to the longitudinal direction in the entire specimen except near both ends. Therefore, it was clear that the residual stress is distributed in only the z direction.

Figure 6 shows the residual fringe order distributions $N_r^*(z)$ for three temperature conditions. The values of fringe order increase relatively as z moves from the surface to the center of strip for all temperature conditions. All these distributions are almost symmetrical to the central plane. The fringe order becomes largely distributed as the holding temperature T_h increases, and the fringe patterns show remarkable unbalance of positive and negative values which are the unique phenomena of photoviscoelasticity.

Figure 7 shows residual stress distributions $\sigma_r(z)$ in plates under three temperature conditions measured by two methods. The residual stress distributions $\sigma_r(z)$ obtained by the optical method and that obtained by the mechanical method show a similar tendency.

The measurement of $\sigma_r(z)$ is as easy as the ordinary two dimensional photoelastic method and is more expedient than the layer removal method which is essentially a destructive method. Therefore, we are confident that the simplified optical method in which the proposed fundamental equation is used is very effective for the experimental evaluation of residual stress by rapid cooling in thermosetting resin plates.

There are slight disagreements between $\sigma_r(z)$ obtained by optical method and that obtained by the mechanical method. The fundamental equation (11) was derived from photoviscoelastic theory assuming that the normalized shear relaxation modulus $F_G(t',T_0)$ equals the normalized relaxation birefringence-strain coefficient $F_C(t',T_0)$. There is, however, a considerable difference between $F_G(t',T_0)$ and $F_C(t',T_0)$ as shown in Fig. 6. Therefore, it is supposed that the disagreements between $\sigma_r(z)$ obtained by optical method and $\sigma_r(z)$ obtained by mechanical method are caused by the difference between $F_G(t',T_0)$ and $F_C(t',T_0)$. One will be able to measure more exactly the residual stress by rapid cooling in plate by using the simplified optical method if the $F_G(t',T_0)$ and $F_C(t',T_0)$ of material used coincide.

CONCLUSIONS

The simplified optical method based on linear photoviscoelastic theory was proposed for measuring the residual stress by rapid cooling in thermosetting resin plates.

First, the fundamental equation for calculating the residual stress from the residual birefringence was obtained. Then, thermosetting resin plates were actually subjected to rapid cooling from both surfaces. The residual stress of the plate was measured by two methods, the simplified optical method and the well-known mechanical method, and the results were similar. The results indeeds strongly confirm that the simplified optical method is a very effective means of measuring the residual stress by raid cooling in photoviscoelastic plate and relatively easy to use.

The authors wish to express their thanks to T. C. Woo, M. Shimbo, M. Takashi, A. Misawa, and H. Yokono for their helpful advice and useful discussions.

REFERENCES

Arenz RJ, Ferguson CW, Williams ML (1967) The mechanical and optical characterization of a Solithane 113 composition. Exp Mech 7:183-188

Arenz RJ, Ferguson CW, Kunio T, Williams ML (1963) The mechanical and optical characterization of Hysol 8705 with application to photo-vicoelastic analysis. GALCIT SM 63-31, California Institute of Technology

Kunio T, Miyano Y (1968) Photoviscoelastic analysis by use of polyurethane rubber. Appl Mech(Proceedings of the Twelfth International Congress of Applied Mechanics):269-276

Miyano Y, Tamura T, Kunio T (1969) The mechanical and optical characterization of polyurethane with application to photoviscoelastic analysis. Bull JSME 12:26-31

Miyano Y, Shimbo M, Kunio T (1982) Viscoelastic analysis of residual stress in quenched thermosetting resin beam. Exp Mech 22:310-316

Sugimori S, Miyano Y, Kunio T (1984) Photoviscoelastic analysis of thermal stress in a quenched epoxy beam. Exp Mech 24:150-156

Table 1. Epoxy resin mechanical and optical properties

Item	Glassy region	Rubbery region
Young's modulus E	3.53 GPa	0.0372 GPa
Poisson's ratio ν	0.35	0.50
Modulus of rigidity G	1.31 GPa	0.0124 GPa
Strain sensitivity[a] C	129×10^3 fr/m	40.6×10^3 fr/m
Coefficient of thermal expansion α	61×10^{-6} K^{-1}	166×10^{-6} K^{-1}
Glass transition temperature Tg	$132\,^\circ$C	

[a] Light wavelength $\lambda = 546$ nm

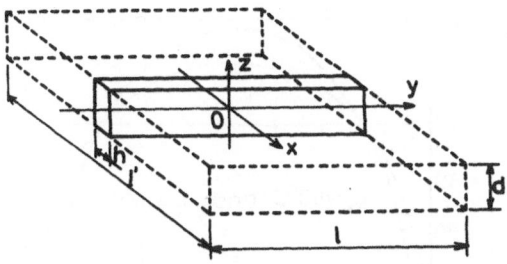

Fig. 1 Photoviscoelastic plate and coordinates

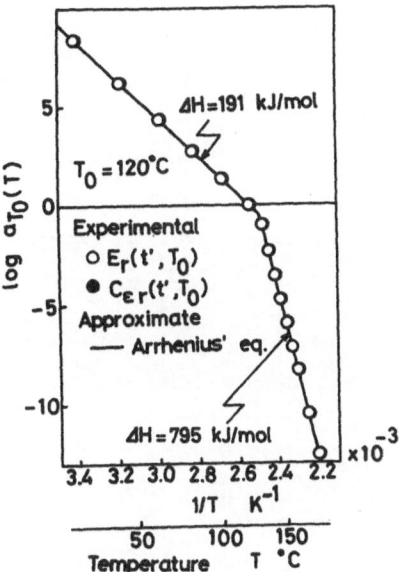

Fig. 3 Time-temperature shift factor $a_{T_0}(T)$ of epoxy resin

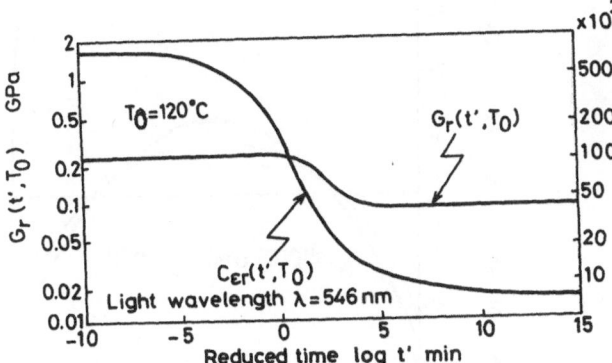

Fig. 2 Master curves of shear relaxation modulus $G_r(t', T_0)$ and relaxation birefringence-strain coefficient $C_{\varepsilon r}(t', T_0)$ of epoxy resin

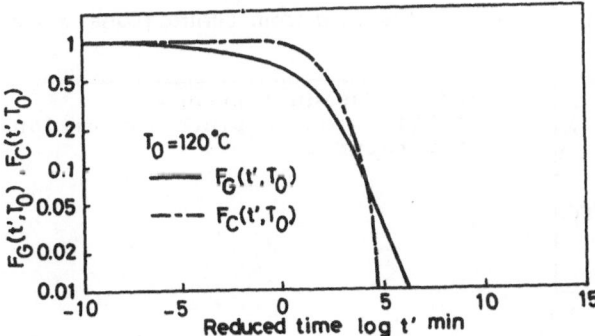

Fig. 4 Normalized master curves of shear relaxation modulus $F_G(t', T_0)$ and relaxation birefringence-strain coefficient $F_C(t', T_0)$ of epoxy resin

$T_h = 180°C$ and $T_c = 10°C$

$T_h = 150°C$ and $T_c = 10°C$

$T_h = 120°C$ and $T_c = 10°C$

Fig. 5 Isochromatic fringe pattern of epoxy strip after rapid cooling (Light wavelength λ = 546 nm, Thickness h = 3 mm)

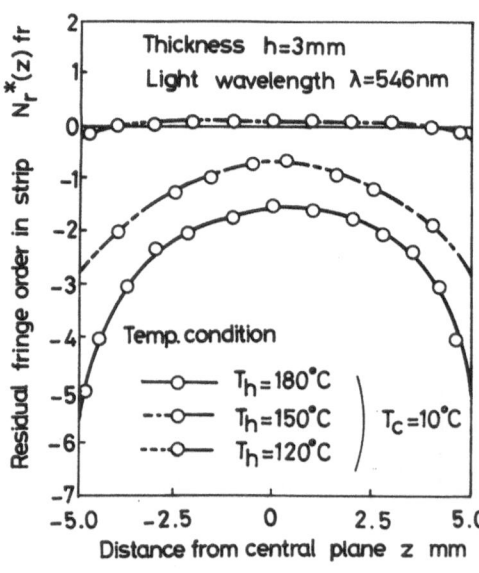

Fig. 6 Residual fringe order distribution in strip for various temperatures

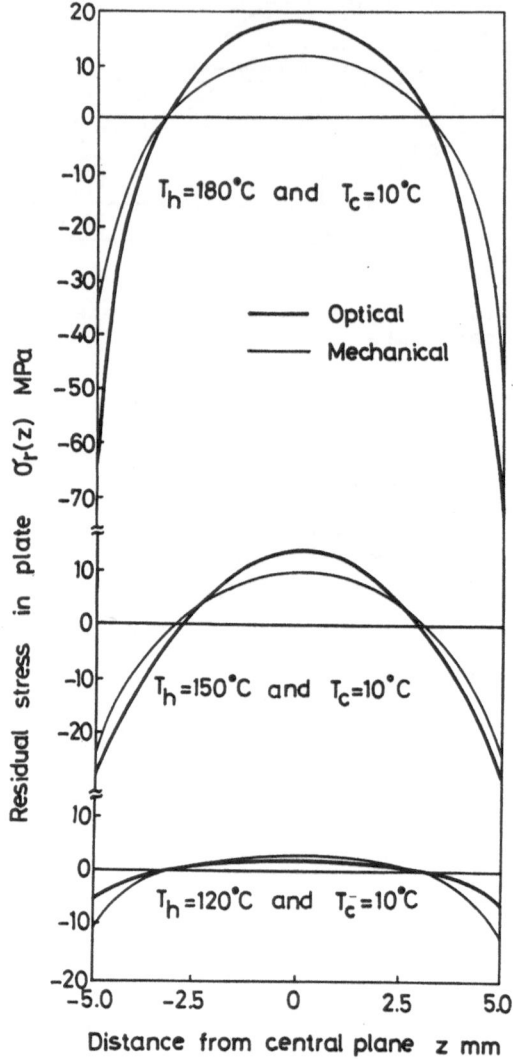

Fig. 7 Residual stress distribution in plate for various temperatures

Silver-Chloride, a Model Material for Photoelastic Investigations into Plastic-Elastic Behaviour of Machine-Elements

Peter R. Dietz

Institut für Maschinenwesen der Technischen, Universität Clausthal, Federal Republic of Germany

1. Indroduction, photoplasticity and materials used

When developing products and constructions in mechanical and civil engineering areas today, one is forced to fully exploit the entire strenght capacity of the material, thus also by knowingly applying partially plastics phenomena. Another important field of industrial application is the increased application of manufacturing processes using deformation techniques. This leads to fundamental investigations into the mechanical behaviour and properties of materials past the elastic region, where both the internal stresses remaining in structural members and the stresses created in deformation tools during deformation processes are of interest. Since the form of the structural member and the material properties, which are influenced by the local stresses acting, play a role in these cases, such investigations mostly lead to a determination of the design strenght of existing structural members, except in the case of a few fundamental research projects.

The fact that physical processes during plastic deformation are complex and depend upon the material being used, usually prevents one from making generalized statements on a mathematical basis. In the simplest theoretical investigations the stress above the elastic limit is considered to be a constant. Other investigations use parametric methods to represent the stress-strain curve outside the linearly elastic area. Fig. 1 shows that the choice of such laws simulated for the stress-strain function must be individually made for each design material and each type of treatment. Added to this, microscopic structural changes occur during the plastic phenomenon, which make the validity of these lumped-together laws seem limited.

The desire to expand the experimental methods to cover processes which deviate from the linear laws of elasticity leads to investigations within the socalled "photoplasticity" field. There are three fundamental methods of investigating "photoplasticity" according to a suggestion by Javornicky with respect to the choice of material under investigation. The first is concerned with the spontaneous deformation of amorphous behaviour under stress, we are dealing here more specifically with viscoelastic bodies and the phenomena associated with them such as creep, relaxation, etc. The second method concentrates on the plastic behaviour of polycrystalline materials, especially of metals, and their representation by using an experimental model. The simulation of plastic-elastic and plastic phenomena in metals is possible by using a model material which has similar structural properties. The halides of silver and thallium are the major model materials available which, due to their crystal structure and mechanical behaviour, are referred to be "transparent metals".

The third method is based on introducing a coating of elastic photo-active material on to an original specimen; the specimen is then to be plastically deformed, while the coating remains in elastic deformation. It is worthwhile mentioning the application of photoplasticity whilst investigating orology problems, where gelatine has proved its worth as a model because of its low elasticity-modulus value.

However, in all these investigations, one must notice that photoplastic investigations are effective specific to a certain structural part and defined load. Thus photoplasticity investigations are substantially individual solutions which only allow the results to be generalized under certain conditions. Furthermore, since microscopic deformation phenomena of materials can only be simulated by the model materials under strictly limited conditions, the predictions obtained from experimental models are only interpreted macroscopically and considered within fringe limits such as inhomogenity, deformation speed, dynamic effects, stress concentrations, etc.

2. The material silver chloride, its mechanical und optical properties

As early as 1934, the Russian researcher Stepanov investigated the relationships between the mechanical properties of crystals, their atomic structure and their optical properties. He used the halides of silver and thallium for this on the whole, because, due to their crystal structure, these material have a mechanical behaviour type like that of most metals, are transparent crystals and, as a result of their cubic system, behave optically neutral when not being loaded.

The Russian researchists Zhitnikov and Stepanov made intensive investigations with this material over a number of years, which ranged from looking at the behaviour of monocrystalline structures to observations using quasi-homogeneous specimens with fine polycrystalline structures. The obvious difficulties in making specimens with a sufficiently accurate predictability led to the fact that a further intensive development to apply it for technical problems of mechanical engineering was not carried out. The smelting and processing techniques described in this report are the first to make a sensible resuming of these experiments possible with regard to their potential for solving practical problems.

2.1 Mechanical properties

Due to the aim of practical application the following account will only cover investigations into finely grained polycrystalline bodies. For this reason the behaviour of monocrystals will not be investigated further.

Javornicky points out that the stress-strain relationship of polycrystalline silver chloride is heavily dependent on its past and that experimental conditions are important influencing factors on the stress-strain behaviour. As a result of our experiments, Fig. 2 shows the tensile testing of a specimen which had been especially developed to simulate the behaviour of pure aluminium. The grain diameter is 0.02 - 0.04 mm; the finely grained homogeneous structure was obtained by deforming by 90 - 95 % and final recrystallisation. If one compares the stress-strain curve of silver chloride in this picture with that of 99.5 % pure aluminium, then the similarity of material behaviour becomes obvious, showing discrepancies from a linear similarity law of under 1.5 %. The photoelastic constants were gauged simultaneously during tensile testing, the result is a value of $S = 32$ mm^{-2}, which was conformed in all later experiments within an accuracy of $\pm$ 2 %.

For a knowledge of the deformation behaviour of material it is necessary to observe thermal and time effects, the distortion work as well as the external frictions during the deformation process. We used here the cylindrical compression test for which one can find plentiful experimental results for 99.57 % aluminium in literature.

The pictorial representation of the yield stress K_f of a material as dependent logarithmically on the formchanging relationship at constant speed of formability and constant temperature is called the ideal flow curve of material. <u>Fig. 3</u> shows at different temperatures typical results of these measurements together with a comparison using the corresponding values for 99.5 % Al. The shape of the curves and a comparison of them show that it is fundamentally possible to set up laws of similarity. The temperature similarity is subject to the same considerations (total conversion of the deformation work, friction in bolts and extrusion canal). The dependence on the deformation speed of metallic materials is expressed in the form of an empirical exponent law. <u>Fig. 4</u> shows, that this law also holds for the material AgCl. The linear type of illustration visible in this diagram makes clear that model laws are simplifiable into exponent formuli. However, when investigating technical processes, one must note that there is mostly a local speed distribution. For intermittent processes, furthermore, the speed distribution is also time-dependent, so that the values arrived at here are limited to the ideal case of the cylindrical compression test.

The third important physical phenomenon in particular in deformation processes is friction, essentially the description of frictional relationship between workpiece and deforming tool. The model laws forcibly require the friction coefficients in the experimental model to correlate to those of the original. From the large number of measuring procedures to measure the frictional relationships during deformation processes the ring compression test was chosen. The experiments were done at temperatures between 20 and 220° C, the temperatures in the lubrication gap were examined. <u>Fig. 5</u> shows a synopsis of the coefficient of friction measurements done on silver chloride at room temperature. With reference to photoelastic investigations, such materials which allow light to pass through them and possess the same refractive index as silver chloride are, of course, increasingly selected as friction partners and lubricants. Fundamentally, from these experiments, the area of $\mu = 0.04 - 0.3$ during aluminium deformations can be covered by the model material silver chloride.

Finally in this section, a summary of the most important physical and technical properties is shown (<u>Fig. 6</u>).

2.2 Optical properties

Silver chloride has face centred cubic crystals which are originally neutral and only act doubly refractive when externally loaded. Then the main axes of the optical spheroid do not follow proportionally with those of the stress spheroid because of the crystal structure. And this means that the optical isoclinal line measured does not decrease in proportion with the main stress isoclinal line. Due to the different refractive indexes according to the position of the crystallographic main axis, one obtains different photoelastic constants.

In the previous instance of applying model materials for the phenomena in macroscopic bodies, it is necessary to optically obtain the macroscopic stress distribution as a function of the external forces. For this, the orientation effects in the trajectory as described for the crystal must disappear. In order to accomplish this, it is necessary for the beam of polarised light to pass through a large number of grains with irregular orientation in order to ascertain the individual effects of separate crystals.

Two special optical effects by working with AgCl are demonstrated in the following feature. In Fig. 7 one can note the appearance of slip bands under elastic-plastic pressure conditions often intensified by the piling up of isochromates; Fig. 8 shows the socalled "tigerskin effect" under tensile stress leading to experimental results which cannot be evalued.

3. The processing and production of silver chloride specimens

The largest difficulty associated with applying this material is the production of the specimens, especially on obtaining a finely grained and homogenous structure. In the instable molten state, silver chloride reacts readily with oxygen, which has a strongly corrosive effect on base metals with which it comes into contact. It can only be used in the solid state when there are few enough impurities present to impair the transparency for photoelastic applications. Another disturbance lies in the high sensitivity to light, because there is a strain in the lattice resulting from the increased disorder when light acts upon the system.

3.1 Preparation of the blank

The starting material for silver chloride specimen production for photoelastic investigations consists of white granules which are initially melted in a tube furnace and then converted to big zones of crystalites by the Bridgman process (Fig. 9).

The silver chloride cylinder produced in this way consists of big, single crystallites and must first of all be subjected to a further treatment aimed at obtaining a finely grained structure without texture or internal stress. The crystal size is decreased by a deformation method with a high degree of deformation, followed by a recrystallisation process. Texture and grain size can be seen on etched specimens after each step of the procedure.

Fig. 10 describes the dependence of the grain size achieved in the manner described on the degree of deformation and annealing temperature, the results of numerous experiments. The mechanical and optical homogeneity of the experiments relies upon as uniform a grain size of under 0.04 mm as possible.

The mechanical manufacture and splinter shaping operations are compareable with the treatment of aluminium. One should beware of the strongly corrosive attack of the material on tools and machine tools. Finally, glaring light should not be allowed and the specimen should be kept carefully.

4. Investigations with silver chloride

Some silver chloride investigations will now be reported upon, serving as preliminaries to further research in the field of machine elements which are either loaded in the plastic-elastic area, or which have suffered distortion as a result of their production and show possible internal stresses.

4.1 Notched rod test

One of the classic examples of explaining photoelasticity and transacting problems of notches is the depicting of the stress state using a holed tensile rod. We wish to seize on the phenomena in this example in order to comparitively study the usefulness of the material AgCl for plastic-elastic stress states.

The tensile procedure will be clarified in the following diagrams. Whilst with small stresses, the load picture familiar to araldite models does appear in the region of elastic stressing, one finds deviations, i.e.: plastifications in the region of the notch treating the rod with increasing loads. Fig. 11 shows the loaded tensile rod with partially plastic regions; the diagram shows quite clearly a counterbalancing of the reduction of peak stresses during plasticizing. Incidentally, it is possible in this diagramm to recognize the strong deviation with the naked eye because of the elliptic distortion of the hole. The residual stresses after unloading are to be seen in Fig. 12, it is a case of the stresses observed in the elastic region turning around - as also predictable from theoretical studies.

Numerical evalutions of this test have already been performed by the Russian researchists in their investigations of the time. Fig. 13 shows a diagram of a comparison between theoretically and experimentally achieved stress distributions around the edges of the hole.

4.2 Examination of partial plastifications of the thread of a screw

The next example is taken from applied mechanical engineering. Photoelasticity is particularly suitable for a model process to investigate threads, because the stress state is simultaneously perceived over the entire cross-section. The use of a plastic photoelastic material enables us in this example furthermore to research the redistributional processes of the load for plastic compliance of individual courses of thread.

In the experiment two flat thread parts are brought into operation and loaded with a tensile force in a loading apparatus whose method of operation is shown in Fig. 14. One thread half is made of silver chloride, the other of araldite. With this apparatus, it is possible to monitor exactly the loads induced in the sides of the thread; from the model law one obtains a coupling of a decarburized screw bolt with a standard nut here.

In the fully elastic region (small loads) the first thread course essentially bears all the load (Fig. 15). With increasing loads, due to the heavy overloading, the formerly load-free neighbouring notch is affected. As the load is raised further over the elasticity limit, the overload in the first course decreases, because the other courses begin to bear part of the load by plastically confirming. Fig. 16 shows this for a load of approximately thrice that in the precious diagram. A further load increase results in a larger adjustment and considerable geometrical deviations. Fig. 17 shows that the courves tense up by plastic compliance of the silver chloride part. The results are considerable with practical experiences with screw bolts made of ductile materials.

One should mention that in the investigations to improve thread geometry, the thread notch area and the possibility of lowering stresses by shaping the thread filler radius and angle of thread are particularly interesting, in order to thus be able to exploit the effect of load redistributions in permissible plastic deformations.

4.3 The use of this material for research on full plastic deformation
    processes in "cold-forming" production

The last used example is representative of procedures which demonstrate the process involved in plastic forming of metals to come to an end
component form. The aim of researching these processes lead to two
ways. First, it is used to optimize the production process and the design of wear resistant dies and tools. Second, it is of great interest
to know how the components strenght is influenced by the production
process, where the residual stress plays a dominant roll. Studies with
silver cloride to optimize extrusion tools and to evaluate the stress
contours of rolled threads on bolts deal with this problem.

The example presented deals with the cold-forming process for producing the profile of the boss of a shaft-connection. Fig. 18 shows the
principle of the fabrication process. To achieve the profile, material
is kneades around the form tool. To simulate this by a planer strain-
system in the experiment an apparatus with two glass plates for con-
tainment is used, like in the previous example (Fig. 19). The experiment simulates single steps in the fabrication process. These give information over the major stresses in component and tool in this step.
Fig. 20 shows an isochromate picture of a "kneaded" polygon profile,
in which one recognizes the high deformation grade of one step causing
cracking and double-layers. Further, the area of the profile having
the highest strain concentration in its later use is pre-stressed by
compression during the fabrication process.

5. Prospects of different applications

The model process described leads to a knowledge of stresses during
and after plastic-elastic and plastic deformation. The stress of deformation tools can be investigated for practical application and one
also obtains data on the stresses ruling in a structural member, which
are enforced upon it during the deformation process.

In mechanical engineering, an increasing number of deformation processes are being employed to produce readymade parts in which the deformation can, in most cases, cause an increase in fatigue strenght. Besides the production of screws by thread rolling already observed, this
is true of the production of splinded shaft profiles by rolling or accurate forging profiles or gearwheels. The process described can help
here to optimize structural members regarding to their strenght: opti-
mizing material flow in non-splintering production means a considerable reduction in manufacturing costs and an increase in tool life.

The second major field of application of this process is the extension
of the notch effect theory into the plastic-elastic area, whereby the
specific behaviour of various materials past the yield limit can be
subject of calculations. The mechanical properties of silver chloride
can be influenced by the presence of different constituents (eg: so-
dium chloride), so that we can presumably witness the material behaviour of different materials by model processes.

There is certainly still a long way to go before statements on the behaviour and model laws are satisfactorily clarified. However, the large possibilities associated with this particular material and with the
solving of urgenst technical application problems are so manifold,
that it is worth the considerable effort of producing models and making evaluations.

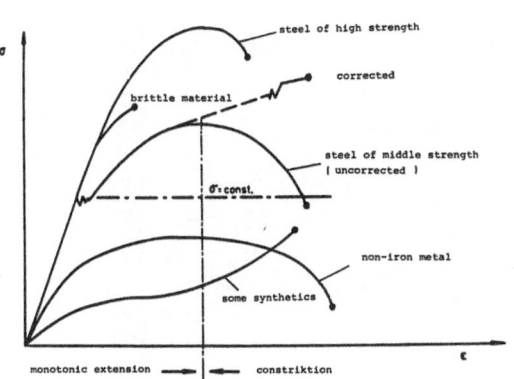

Fig.1:Stress-strain diagrams from tensile tests on various materials

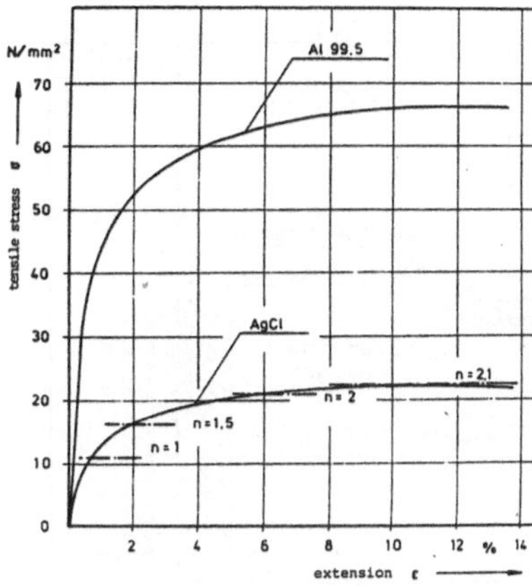

Fig.2:Comparison of stress-
strain curves for AgCl and Al99

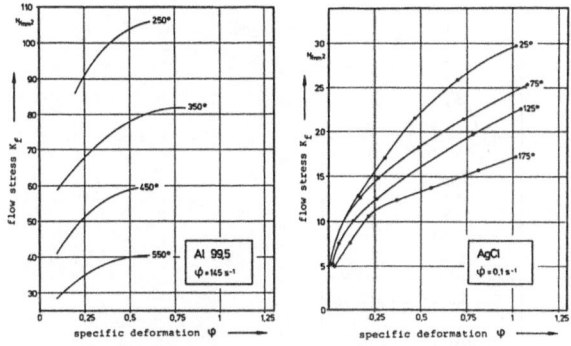

Fig.3:Flow-stress K_f in dependance of
specific deformation with AgCl and Al99

	friction partner	steel polished	glass	plexiglass
lubricant				
unlubricated		0,3	0,25	0,3
hydrocollag		0,16	-	-
vaseline		0,12	0,08	0,13
molybdic sulfide		0,07	-	-
silicone paste		0,04	0,02	0,06

Fig.5:Friction coefficient from
ring compression test with AgCl

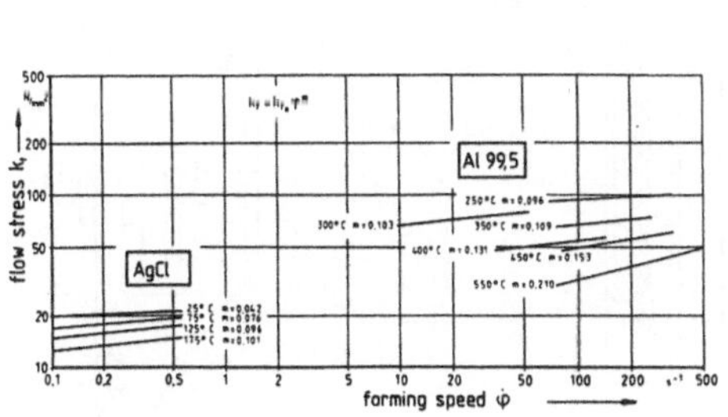

Fig.4:Flow-stress in dependence of
forming speed of AgCl and Al99

physical properties	unit	AgCl	Al 99,5
tensile strength Rm	N/mm^2	22,5*	70
app. yielding point $Rp0,2$	N/mm^2	9*	28
breaking extension ε	%	13*	22
modulus of elasticity E	N/mm^2	700...4500	72700
modulus of rigidity G	N/mm^2	980	27000
factor of transverse extension ν	-	0,41	0,34
specific density ρ	kg/m^3	$5,56 \cdot 10^3$	$2,7 \cdot 10^3$
specific heat c	kcal/kg grd	0,085	0,22
coefficient of expansion α	m/m grd	$2,98 \cdot 10^{-5}$	$2,4 \cdot 10^{-6}$
melting point	$^\circ C$	457,5	660

* sample material prepared for experimental tests

Fig.6:Summary of the most important
physical prperties of AgCl and Al99

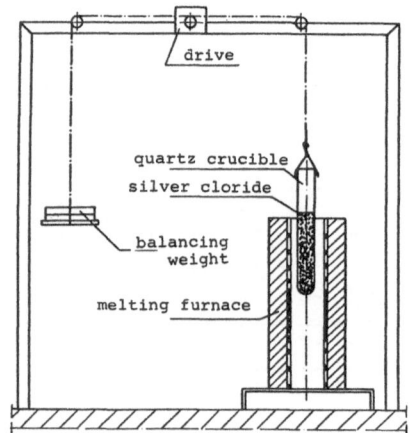

Fig.9:Principle of zone melting

Fig.11:Notched rod test.Isochromates within elastic deformation field

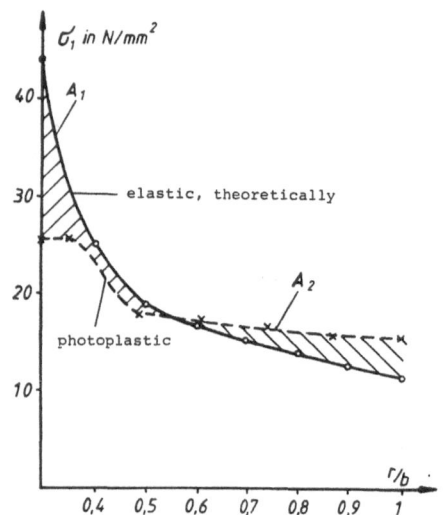

Fig.13:Notched rod test. Comparison experimental/theoretical

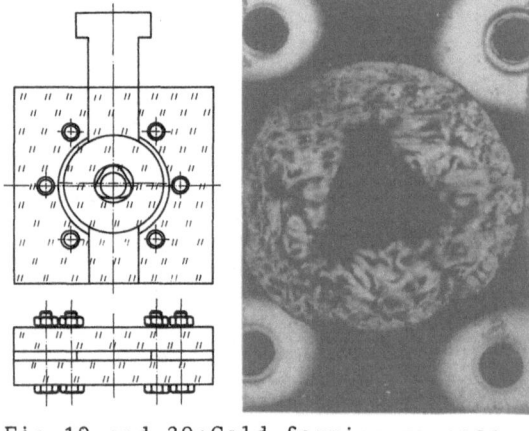

Fig.19 and 20:Cold-forming process Test-apparatus and isochrometes of the formed specimen

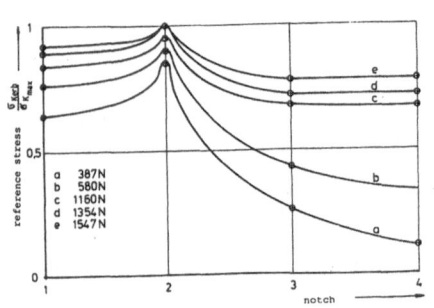

Fig.17:Load-distribution in a thread connction with elastic-plastic deformations

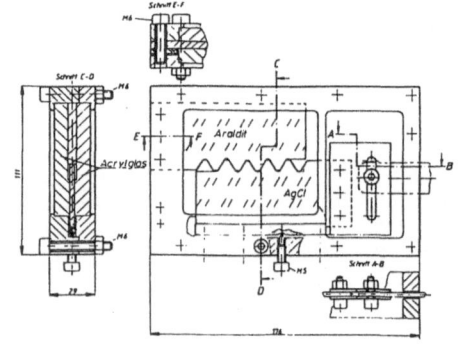

Fig.14:Loading attachment for experiments with thread connections

Advances in Moire Interferometry

Daniel Post

Engineering Science and Mechanics Department, Virginia Polytechnic Institute and State University, Blacksburg, VA 24061, USA

INTRODUCTION

Moire interferometry is a relatively new technique[1], but it has already been applied to the solution of several practical problems. An extensive review and bibliography is given in Ref. 2.

Moire interferometry responds only to geometric changes of the specimen, and thus it is effective for diverse engineering materials, including the currently important anisotropic and nonlinear materials. It provides whole-field contour maps of in-plane deformation fields -- precisely the experimental counterpart to the primary output of theoretical studies by finite element methods and related computer analysis methods. The sensitivity of traditional geometric moire has been inadequate for most engineering applications, but recent developments in moire interferometry provide increased sensitivity by nearly two orders of magnitude. Now, moire interferometry provides the needed sensitivity and promises to be an extraordinarily useful method of experimental solid mechanics.

A brief description of the method is given here. Then, applications that illustrate unique capabilities are presented.

MOIRE INTERFEROMETRY, BRIEF DESCRIPTION

In moire interferometry, a grating is applied to the surface of the specimen and it deforms with the underlying specimen. This specimen grating is a reflection-type phase grating of frequency f/2. It is formed on the specimen by the replication method illustrated in Fig. 1. The mold represented in the figure is a crossed-line holographic grating, overcoated with a metallic film of evaporated aluminum. The mold is pried off after the adhesive -- usually epoxy -- has been polymerized, but the metallic film remains bonded to the adhesive. The result is a thin (about 0.025 mm) crossed-line diffraction grating on the specimen.

The specimen grating is viewed together with a superimposed reference grating, just as in geometrical moire. Usually, a virtual reference grating is used as illustrated in Fig. 2a. It is formed by two beams of coherent light, wherein one reaches the specimen directly at angle $-\alpha$, while its companion beam is reflected by a plane mirror to reach the specimen at angle $+\alpha$. The two intersecting beams form alternating "lines" of constructive and destructive interference on the specimen surface. These lines comprise a <u>virtual reference grating</u>, where the grating pitch, g, and frequency, f, is given by

$$f = \frac{1}{g} = \frac{2}{\lambda} \sin\alpha \qquad (1)$$

where λ is the wavelength of coherent light employed here.

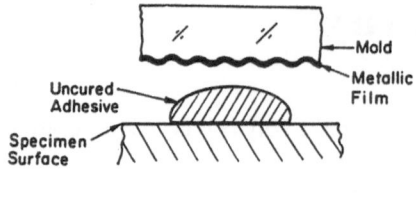

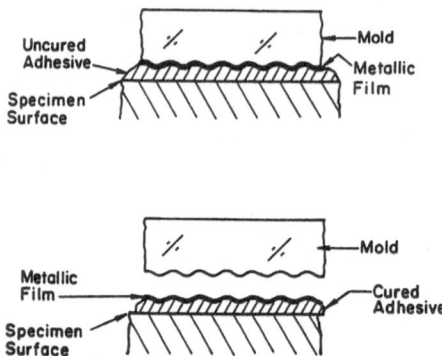

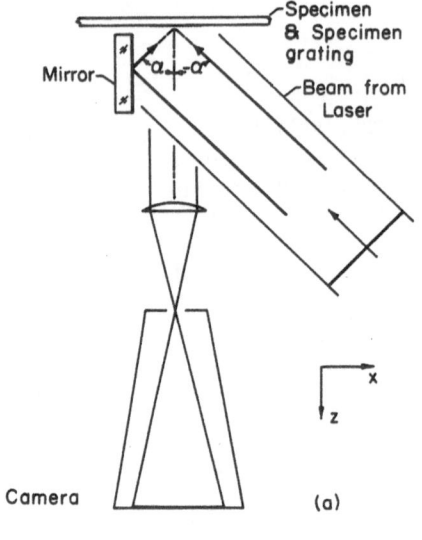

Fig. 1. Replication technique to
form a high frequency, high
reflectance grating on the specimen.

(a)

Fig. 2. (a) Optical arrangement
for moire interferometry.
(b) Four-beam arrangement for
N_x and N_y patterns.

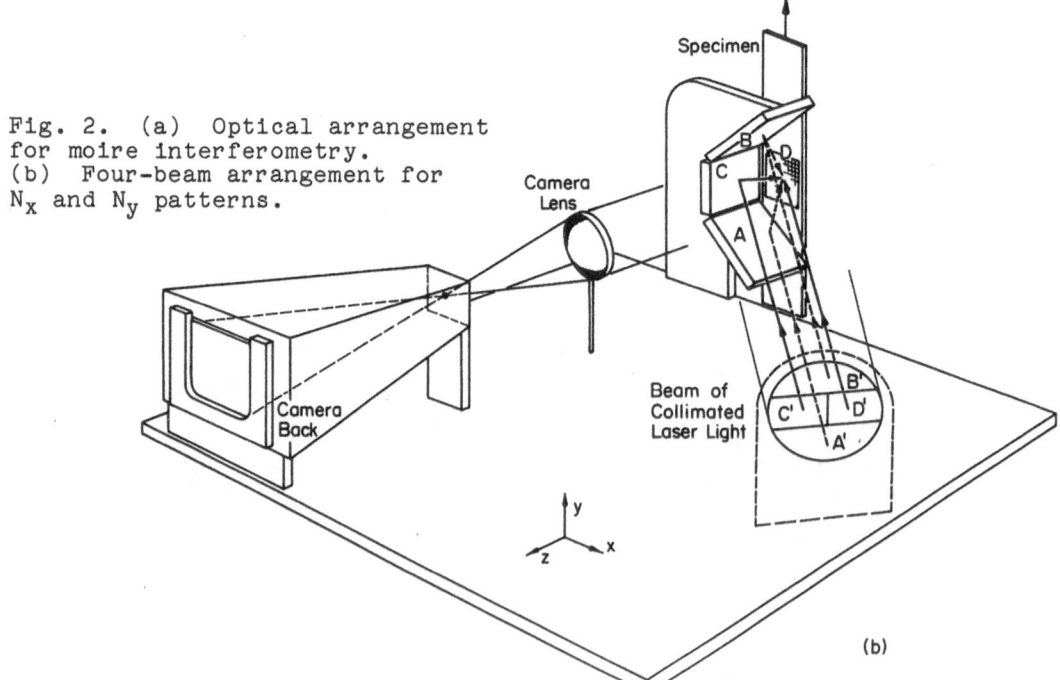

(b)

The virtual reference grating interacts with the deformed specimen grating to form the pattern recorded in the camera. A rigorous explanation considers diffraction of each incident beam by the specimen grating[1,2]. The initial frequency of the specimen grating is half that of the reference grating. With this condition the desired diffraction orders emerge perpendicular to the specimen.

Figure 2b illustrates a 4-beam optical system, suitable for analysis of both elements of the crossed-line specimen grating -- the superimposed X and Y specimen gratings. Two adjustable mirrors, A and B, are added to the system of Fig. 2a. Then light from sections A' and B' of the collimated laser beam is directed by mirrors A and B to intersect at an angle 2α in a vertical plane; this forms a virtual reference grating with its lines perpendicular to the y-axis. Light from sections C' and D' forms a virtual reference grating with its lines perpendicular to the x-axis.

The laser power required for this work depends upon the diffraction efficiency of the specimen grating and the magnification of the image in the camera. Laser power from 0.5 to 200 milliwatts has been used for various applications.

Corresponding specimen gratings and virtual reference gratings interact to produce the whole-field fringe patterns of U and V, where U and V are displacement components in the x and y-directions, respectively. The relationship between displacements and fringe orders is

$$U = (1/f)N_x; \quad V = (1/f)N_y \qquad (2)$$

where N_x is the moire fringe order at any point in the fringe pattern, for the fringe pattern obtained with the reference grating lines oriented perpendicular to the x-axis; N_y is the fringe order taken from the pattern obtained with the reference grating lines oriented perpendicular to the y-axis.

The sensitivity depends upon the frequency, f, of the virtual reference grating. High sensitivity corresponds to small displacement per fringe order and therefore to high reference grating frequency. By Eq. 1, the theoretical upper limit of frequency is 2/λ, which for visible light is about 4000 lines/mm (ℓ/mm). A frequency of f = 2400 ℓ/mm was used for the applications shown in this paper. The corresponding sensitivity is 1/2400 mm = 0.417 μm per fringe order.

APPLICATION I, CARRIER PATTERNS

Carrier patterns can be used for various purposes in interferometry. Here, it will be used to clarify fringe orders and gradients of fringe orders in the load-induced pattern.

Figure 3a illustrates the specimen and loading. The specimen was cut from a 48 ply quasi-isotropic laminate [45/0/-45/90]$_{6s}$ of graphite fibers and PEEK matrix. The specimen grating was applied to the edge of the laminate, where all 48 plies were exposed. Figure 3b shows the N_x fringe pattern for the portion of the specimen defined by the dashed box. Numerical values are fringe orders N_x.

With the applied 5-point loading, the bending moment at section A-A' is essentially zero, but the vertical shear force is large. The section is in a state of nearly pure shear stress. The U-displacements along A-A' are complicated, and they are not easily determined from Fig. 3b.

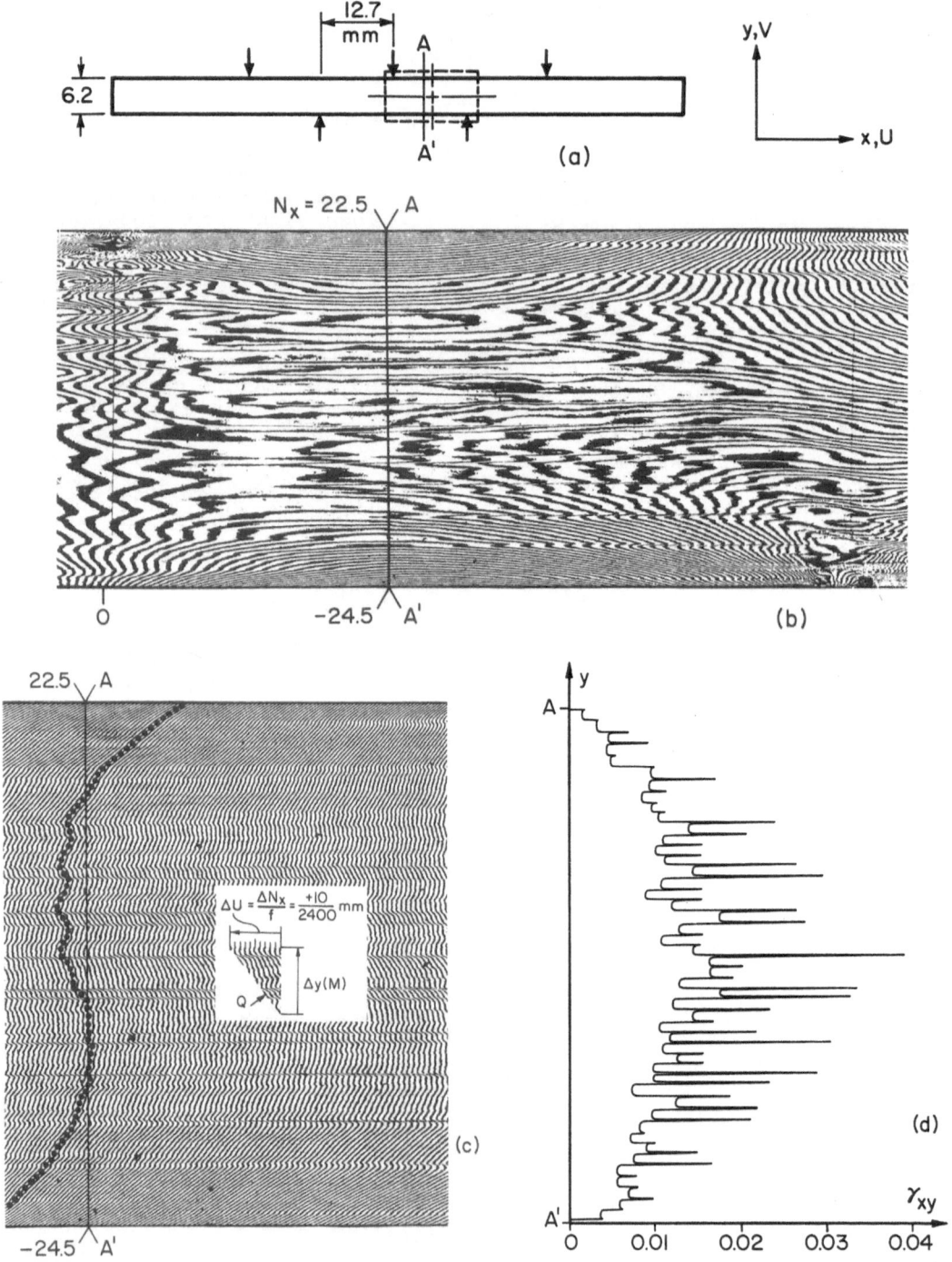

Fig. 3. (a) Quasi-isotropic composite specimen. (b) N_x or U-field for zone in dashed box. (c) Same pattern with carrier. (d) Shear strains on A-A'.

Figure 3c shows the same displacement field near A-A', but with a carrier pattern of extension added. This was done simply by changing the angle of incidence α (Fig. 2a) by a small amount. If the load was not applied, the pattern would be comprised of vertical fringes parallel to line A-A'. Therefore, the no-load fringe order, everywhere along A-A', is a constant. When under load, the contours of constant fringe orders change form to those of Fig. 3c. Now fringe orders along A-A' can be ascertained easily without ambiguity. Starting at A', where N_x is known from Fig. 3b, the fringe orders along A-A' change by one unit wherever a fringe crosses A-A'. This is true because the no-load fringes, being parallel to A-A', do not change the relative fringe orders along A-A'.

A special feature exists for cases such as this one, where the N_x fringes near A-A' are nearly horizontal (i.e., $\Delta U/\Delta x \approx 0$). Any continuous fringe near A-A' forms a graph of the variation of N_x along A-A'. For example, the fringe highlighted by dots in Fig. 3c is a graph of N_x consistent with the arbitrary selection of the zero fringe order in Fig. 3b.

Since N_x is proportional to U (by Eq. 2), the slope of the graph N_x vs. y gives $\partial U/\partial y$. Even when $\Delta U/\Delta x \neq 0$, the slopes of the fringes crossing A-A' in Fig. 3c gives $\partial U/\partial y$. To implement this technique, consider point Q in Fig. 3c. Draw a line tangent to the fringe at Q and form a triangle as shown. The vertical leg of the triangle is measured to determine $\Delta y(M)$ where M is the magnification of the fringe pattern relative to the specimen. The number of fringes crossed by the horizontal leg gives ΔU as shown. Then, at point Q, $\partial U/\partial y = \Delta U/\Delta y$ = $(1/f)\Delta N_x/\Delta y$. For consistency of signs, Δy is taken in the +y direction and ΔU is positive or negative depending upon whether the fringe orders in Fig. 3c are increasing or decreasing in the direction of the arrow. The validity of this technique stems from the condition that $\Delta U/\Delta y = 0$ for the carrier fringes.

The same argument applies for the N_y fringe pattern and $\partial V/\partial x$ can be determined from the slopes of fringes N_y as they cross A-A'. Again, a carrier pattern of extension is used. Since shear strains $\gamma_{xy} = \partial U/\partial y + \partial V/\partial x$, the shear strain along A-A' can be calculated using this technique. The result is shown in Fig. 3d. In this case, $\partial V/\partial x$ was nearly constant along A-A', and the strong variations of γ_{xy} are associated with the very strong variations of slope seen in Fig. 3c.

The envelope of the γ_{xy} graph is nearly parabolic, corresponding to an isotropic specimen. The severe undulations of γ_{xy}, however, are associated with strong interlaminar shear strains developed between the plies of the composite. The result applies to this material, with a PEEK matrix. Similar studies with graphite-epoxy composites did not show these large interlaminar shear strains[3].

The same technique can be used for normal strains along A-A', determined from $\partial U/\partial x$ and $\partial V/\partial y$, respectively. In this case, carrier patterns of rotation would be used, since $\partial U/\partial x$ and $\partial V/\partial y$ are zero for carrier patterns of rotation.

APPLICATION II, SUBTRACTION

Figure 4 shows the V-displacement field for a plastically deformed pure copper specimen. It shows only the nonuniform or anomalous part of the displacement field, since the uniform part was subtracted from the pattern. In this test, the specimen grating was replicated on the specimen in its virgin condition. The specimen was loaded into its small plasticity range and then unloaded; the unloaded plastically deformed specimen was installed in the optical system illustrated in Fig. 2. The N_y pattern exhibited closely spaced fringes, essentially perpendicular to the y-direction, but the fringes exhibited some irregularities or waviness that indicated a nonuniform strain field.

The angle of incidence α was adjusted slightly to change the frequency of the virtual reference grating. It was adjusted to reduce the number of fringes in the pattern and produce the most sparse pattern possible. With this adjustment, the frequency of the virtual reference grating matched (twice) the average frequency of the specimen grating in its plastically deformed condition, instead of in its initial undeformed condition. The result is a fringe pattern in which the average strain, or the uniform part of the strain, is subtracted from the total strain field. The pattern shows only the nonuniform part of the strain field.

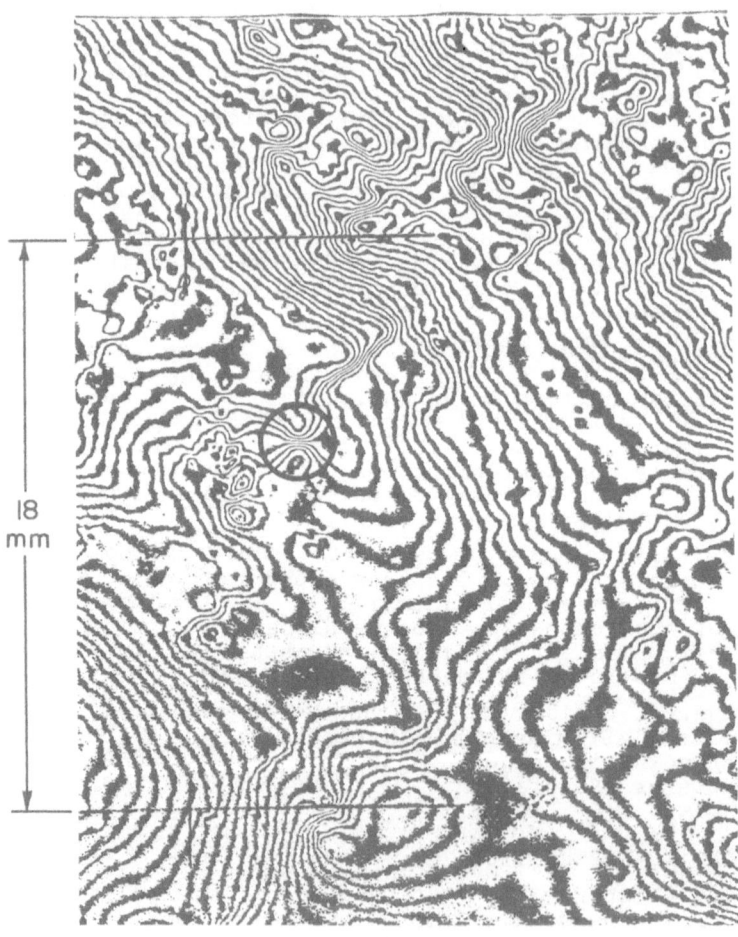

Fig. 4. Non-uniform or anomalous part of deformation field in plastically deformed copper specimen.

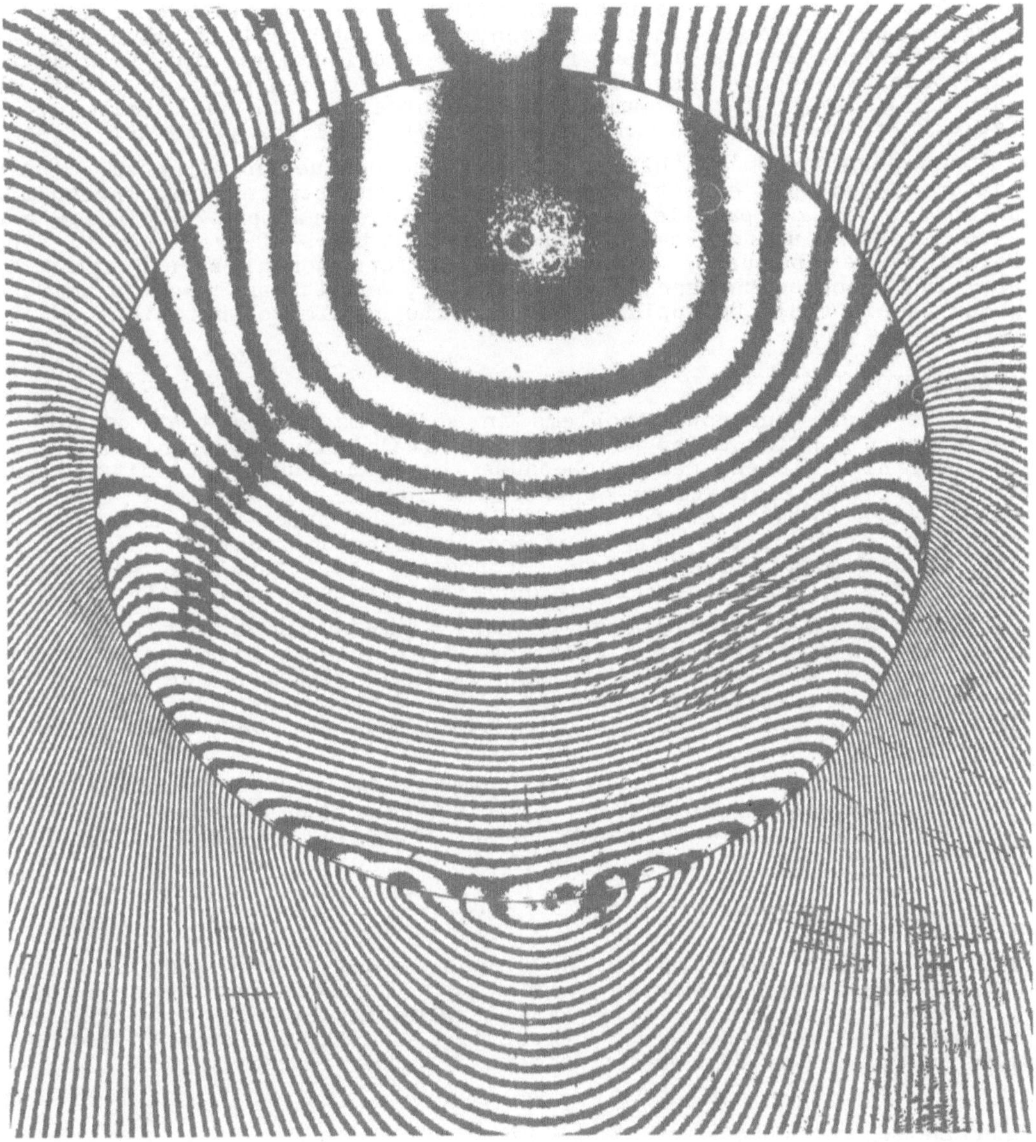

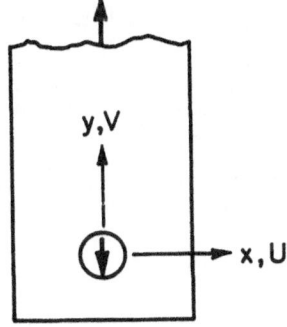

Fig. 5. V-displacement field for a pin joint. Experimentally determined displacements provide boundary data in contact region for finite element analysis.

This technique is valuable for finding and measuring anomalies in specimen performance. In the present case, the specimen was subjected to a uniform tensile stress field and a uniform strain field might have been expected. Instead, the cancelled uniform part of the strain was 0.85%, while the anomalous normal strain ε_y in the circled zone was 0.53%; the local maximum was the total, or 1.38% permanent strain.

One can think of this subtraction technique in terms of carrier patterns, too. In effect, the average uniform strain field produces a uniform array of parallel fringes of fringe gradient F. These are cancelled by a uniform carrier pattern of fringes of gradient -F. The technique can be used to subtract a uniform normal strain field by applying an opposite carrier of extension, and it can cancel a uniform shear strain field by applying an opposite carrier of rotation.

APPLICATION III, HYBRID ANALYSES

There are numerous problems in solid mechanics that cannot be solved with confidence by powerful numerical analysis techniques, such as the finite element method, because the true boundary conditions are not known. One such class of problems is all two-body contact problems, where frictional forces -- especially stick-slip friction -- cannot be prescribed. Hybrid experimental/theoretical methods can be used effecively, where the true boundary conditions are determined experimentally and used as input data for the numerical analysis.

Figure 5 illustrates such a case. The specimen is a pin joint, comprised of a plate with a hole loaded by a 19 mm diameter pin. The plate and pin were both made of high-strength aluminum and the pin had a slight clearance fit in the hole. The figure shows the V-displacement field in the vicinity of the joint, for the load applied in the y-direction.

It is clear that moire interferometry can give faithful data for the actual contact conditions. It provides a great abundance of data points on the boundary, more than typically used in numerical analyses.

ACKNOWLEDGEMENTS

Moire interferometry was developed in large part under sponsorship of the National Science Foundation. Applications shown in Figs. 3, 4 and 5 were sponsored by NASA Langley Research Center, NSF, and the Office of Naval Research, respectively. This sponsorship and permission to use the figures is gratefully acknowledged.

REFERENCES

1. D. Post, "Moire Interferometry at VPI & SU," Experimental Mechanics, 23(2), pp. 203-210 (June 1983).

2. D. Post, "Moire Interferometry," SEM Handbook on Experimental Mechanics, A. S. Kobayashi, Editor, Prentice-Hall, Englewood Cliffs, NJ (to be published 1986).

3. D. Post, R. Czarnek, D. Joh and J. Wood, "Deformation Measurements of Composite Multi-Span Shear Specimens by Moire Interferometry," NASA Contractor Report 3844 (Nov. 1984).

Scanning-Moire Method

Y. Morimoto and T. Hayashi

Faculty of Engineering Science, Osaka University, Toyonaka, Osaka, 560 Japan

INTRODUCTION

In the conventional moire method, a moire pattern appears as interference between a model grating and a master grating. However, a moire
pattern also appears when the model grating lines are sampled by a TV
camera. If the scanning lines of the TV camera are regarded as the
master grating lines, the geometric relations among the moire fringe
lines, the model grating lines and the scanning lines are obtained in
the same way as the conventional moire method. When the TV camera
scanning lines are thinned out, this method corresponds to a mismatch
method in the conventional moire method. We call this the "Scanning-
moire Method" (Morimoto 1984).

The analysis of strain or stress distribution from fringe patterns
obtained by photoelasticity, moire method, holographic interferometry,
etc. is laborious. However, in this scanning-moire method, the image
obtained by a TV camera is analyzed with an image processor system.
This system, which consists of a personal computer and an image memory
unit, is easy to operate and the cost is very low (Morimoto 1984, 1985).

In this paper, we will show the principle of the scanning-moire method
and its applications to strain analysis.

FUNDAMENTAL PHENOMENON OF SCANNING MOIRE METHOD

Let us show the phenomenon which moire fringes appear by thinning out
the TV camera scanning lines.

Figure 1 illustrates a schematic of the scanning-moire method. Figure
1(a) shows the TV camera scanning lines. Fig. 1(b) shows model grating
lines when the strain is small. If the scanning line is on a black line
of the model grating, we have a black line image. If it is on a white
line, we have a white line image. Thus, these black and white images
make a moire pattern as shown in Fig. 1(c). When the strain increases
as shown in Fig. 1(d), the image shown in Fig. 1(e) appears. In this
case, the pitch of the model grating is almost three times that of the
scanning lines. We cannot recognize a moire pattern. However, if the
line images from the alternate scanning lines are thinned out, as shown
in Fig. 1(f), the image is separated into white and black groups and
moire fringes appear. Conversely, Fig. 1(g) shows the image which
consists of the line images thinned out in Fig. 1(f). It differs from
Fig. 1(f) by π in the thinning out and moire fringe phases.

If every third scanning line is picked out from Fig. 1(e), the moire
pattern shown in Fig. 1(h) is obtained. It is a different moire
pattern from Fig. 1(f). In Fig. 1(h), we selected the first of the three

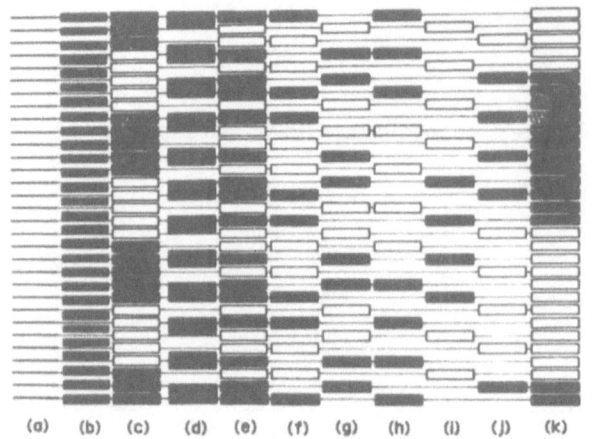

(a) TV camera scanning lines
(b) Deformed model grating
    lines (small strain)
(c) Sampled image by TV camera
(d) Deformed model grating
    lines (large strain)
(e) Sampled image by TV camera
(f) Thinned-out image
    (N=2, phase=0)
(g) Thinned-out image
    (N=2, phase=π)
(h) Thinned-out image
    (N=3, phase=0)
(i) Thinned-out image
    (N=3, phase=2π/3)
(j) Thinned-out image
    (N=3, phase=4π/3)
(k) Copied image

Fig. 1. Schematic of scanning-moire method

lines. However, if we select the second of the three lines, the image
in Fig. 1(i) is obtained.   Figure 1(j) shows the image obtained by
selecting the third of the three lines. If we consider that the fringes
in Fig. 1(h) have the continuous integer fringe orders, the fringes in
Fig. 1(i) have the order of (the integer fringe order + 1/3). Therefore
if we change the phase of the thinning out, we can obtain fractional
fringe orders.

Finally, if all of the scanning lines which are not picked out in
Fig. 1(j) are replaced by the foregoing lines which are picked out in
Fig. 1(j),   the moire fringes are obtained more clearly as shown in
Fig. 1(k). These thinning-out processes correspond to a mismatch in the
conventional moire method.

When the number of the scanning lines thinned out is N-1, we call N, the
thinning-out index.

We will now show an example of the changes of the moire patterns when
the thinning-out index N changes.   Figure 2(a) shows a model grating.
Its pitch is 12 times the scanning line pitch (i.e., 12 pixels) in the
vertical direction.   The angle from the horizontal direction to the
grating line is -45°.

The images thinned out by N=3, 5, 6, 8, 10, 12, and 14 are shown in

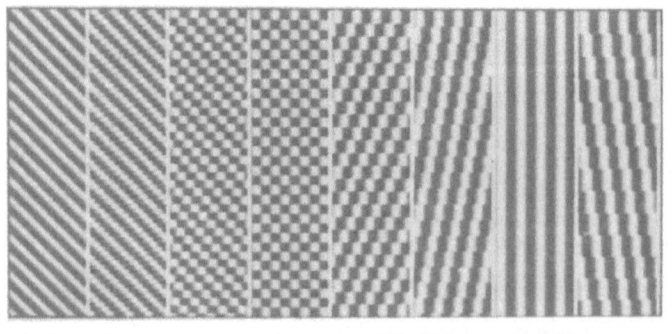

(a)   (b)   (c)   (d)   (e)   (f)   (g)   (h)
N=1   N=3   N=5   N=6   N=8   N=10 N=12 N=14

Fig. 2. Example of
moire fringe patterns
depending on thinning
out index N

Figs. 2(b) through 2(h). For N=3 and 5, the images show the original grating lines. However, when N is greater than 5, the images show moire patterns and the patterns differ by N.

Using this scanning-moire method, we can analyze all of the strain field from only one picture by computer image processing, even if the specimen contains a region of large strain, as in the case of a stress concentration.

This scanning-moire method has the following advantages:
(1) It is not necessary to prepare a master grating on a specimen and to develop and print a photographic film;
(2) The moire pattern obtained by this method has not an image of a master grating so the fringe pattern is very clear;
(3) It is easy to perform a mismatch method by changing the pitch ratio of the master grating to the model grating by using a zoom lens; and
(4) Instead of using a zoom lens, by systematically picking out the images from specific scanning lines and erasing the others with image processing, a mismatch method is performed easily.

MOIRE FRINGE - STRAIN RELATION

Let us show the geometric relation between the fringe spacing δy and the strain ε_y along the y axis. Figure 3 shows the geometric relation among the model grating lines, the TV camera scanning lines and the moire fringe lines in a sufficiently small part of a specimen after deformation.

Before deformation, the model grating lines are parallel to the x axis, the spacings of the grating lines are constant, and one of the model grating lines passes through the origin. Therefore, the equation of the model grating lines is expressed by

$$y = \ell \, p_0 \tag{1}$$

where p_0 is the pitch of the model grating lines before deformation and ℓ is the order number of each grating line increasing in the y direction.

Let us assume one-dimensional deformation in the y direction. When the pitch of the model grating lines after deformation is designated as p'

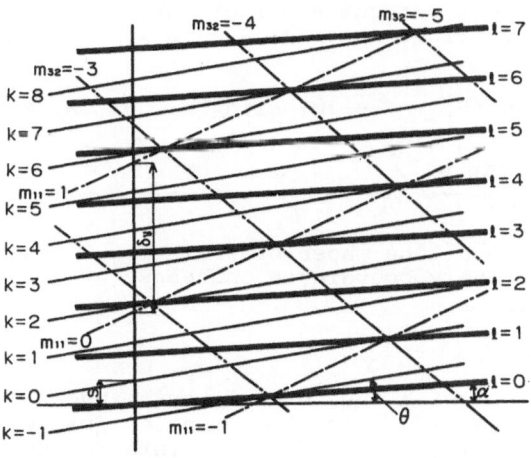

Fig. 3. Geometric relation among model grating lines, scanning lines, and moire fringe lines

and the angle from the x axis to the lines is designated as α , the Eulerian-normal strain ε_y in the y direction is

$$\varepsilon_y = \frac{p'/\cos \alpha + p_0}{p'/\cos \alpha} . \tag{2}$$

The equation of model grating lines is.

$$y = x \tan \alpha + \ell \, p'/\cos \alpha . \tag{3}$$

These deformed model grating lines are taken by a TV camera. The pitch of the TV camera scanning lines is designated as p and the angle from the x axis to the fixed scanning lines is designated as θ . This corresponds to the conventional moire method when the pitch of the model grating is p', the pitch of the master grating is p, and the angle of the misalignment is θ.

The equation of the scanning lines is

$$y = x \tan \theta + s + k \, p/\cos \theta \tag{4}$$

where k is the order number of the scanning lines increasing in the y direction and s is the y intercept of the scanning line whose order is zero.

Now, p is equal to λ multiplied by p_0, that is,

$$p = \lambda \, p_0 \tag{5}$$

where λ is called the mismatch coefficient.

Moire fringes are produced at the points in which the following general expression is satisfied:

$$j \, k - i \, \ell = m_{ij} \tag{6}$$

where j and i are prime to each other and m_{ij} is an integer (j is the model grating line thinning-out index, i is the scanning line thinning-out index, and m_{ij} is the order of the moire fringe lines generated when the indexes are i and j). That is, the moire fringe line whose order is m_{ij} passes through the intersection of every ith line of the model grating lines and every jth line of the scanning lines.

The moire fringe is obtained by connecting the intersections which have the same m_{ij}. However, we cannot recognize all of the fringes which satisfy Eq. (6), but those whose fringe spacings are large and contrast is high.

Substituting Eqs. (2) to (5) into (6) and eliminating p, p', k, and ℓ , the next relation is obtained.

$$y = \frac{\{j \sin \theta - i \lambda (1 - \varepsilon_y) \tan \alpha\} x + j \, s \cos \theta + \lambda \, p_0 \, m_{ij}}{j \cos \theta - i \lambda (1 - \varepsilon_y)} \tag{7}$$

This equation shows the moire fringe line whose order is m_{ij}. When the fringe order increase by Δm_{ij}, the y coordinate increases by

$$\Delta y = \frac{\partial y}{\partial m_{ij}} \Delta m_{ij} = \frac{\lambda \, p_0 \, \Delta m_{ij}}{j \cos \theta - i \lambda (1 - \varepsilon_y)} . \tag{8}$$

Let us now consider the case in which the thinning-out index is N. Since the increment of i is N in Eq. (6), the increment of fringe order is N. Therefore, substituting $\Delta m_{ij}=N$, i=N, and j=1 into Eq. (8), the y directional distance δ_y of the fringe is expressed by

$$\delta_y = \frac{\partial y}{\partial m_{ij}} N = \frac{N \lambda p_0}{\cos \theta - N \lambda (1 - \varepsilon_y)}.$$ (9)

From this,

$$\varepsilon_y = \frac{p_0}{\delta_y} - \frac{\cos \theta}{N \lambda} + 1.$$ (10)

For alignment, it becomes

$$\varepsilon_y = \frac{p_0}{\delta_y} - \frac{1}{N \lambda} + 1.$$ (11)

This equation gives the relation between the y directional distance δ_y between fringes and the y directional strain ε_y.

STRAIN ANALYSIS BY IMAGE PROCESSING

Since the number of the scanning lines in our image processing system is not so large, this scanning-moire method is adopted to a specimen which has large deformation. Now let us show some examples of the strain distribution obtained by this scanning-moire method.

Figure 4 shows a rubber plate with a circular hole. Before loading, the pitch of the model grating lines is 4 pixels. Figure 4(a) shows a deformed model grating under tension in the vertical direction. The hole is elongated to an ellipse. When the scanning lines are thinned out by N=4 and 6, moire patterns appear as shown in Fig. 4(b) and (c).

As shown in Fig. 4(b) and 4(c), the fringe spacings in the same small region differ according to N. If we use only the moire pattern obtained by one of N, there are few data of the fringe spacings in some regions like the upper and lower regions near the hole in Fig. 4(b). In the regions, the strain distribution obtained from interpolating the strain data might have much error. However, when we use many kinds of N, the density of data becomes higher in all the small region. For example, in the upper and lower regions near the hole in Fig. 4(c), the spacing is so small that the strain can be calculated at many points.

From the moire fringe patterns, the center lines of the fringes are obtained by the image processing thinning program. The strain is calculated using Eq. (11). Figure 4(d) shows the equi-strain lines interpolated from the strain data at about 1000 points in Fig. 4(b) and

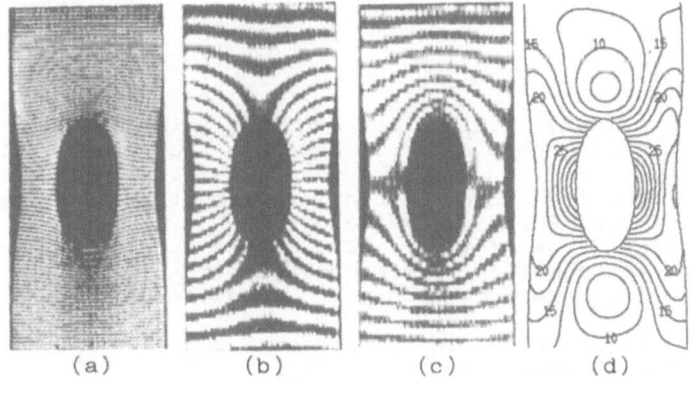

(a) Deformed grating
(b) Thinned-out image when N=4
(c) Thinned-out image when N=6
(d) Strain distribution (%)

Fig. 5 Rubber plate with hole

  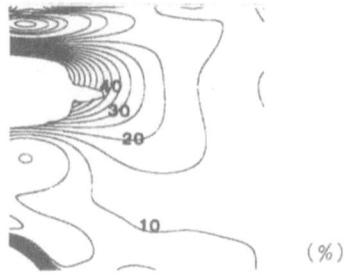

(a) Original image   (b) Thinned-out image   (c) Strain distribution

Fig. 5 Stainless steel plate with crack

 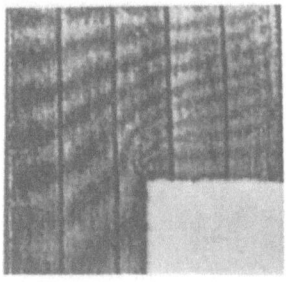 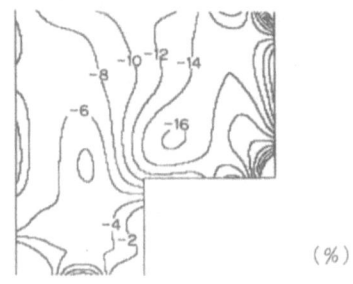

(a) Original image   (b) Thinned-out image   (c) Strain distribution

Fig. 6 Powder compact in T-shaped mold

(c). We select two-dimensional B-splines for interpolation.

When a stainless steel plate with a horizontal crack in the center is elongated vertically, the deformed model grating near the crack is as shown in Fig. 5(a). The scanning-moire pattern when N=4 is shown in Fig. 5(b). The strain distribution is shown in Fig. 5(c).

Figure 6(a) shows a corner part of the deformed model grating which is drawn on the powders compacted in a T-shaped mold. Figure 6(b) is the moire pattern when thinning out by N=2. Figure 6(c) is the strain distribution (Morimoto 1984).

Thus, we can easily obtain strain distribution by this scanning-moire method and image processing.

REFERENCES

Morimoto Y, Shiraishi T (1984) Fringe pattern analysis by image processing using personal computer. J Soc Mat Sci Jap (in Japanese) 33(367):495-500

Morimoto Y, Hayashi T, Yamaguchi N (1984) Strain measurement by scanning-moire method. Bulletin of JSME 27(233):2347-2352

Morimoto Y, Hayashi T (1984) Deformation measurement during powder compaction by a scanning-moire method. Exp Mech 24(2):112-116

Morimoto Y (1984) Gazoshori (Image Processing, in Japanese), Published by Baifukan

Morimoto Y, Kanoh K (1985) Image processing by personal computer and its programming (1). Pixel (in Japanese) No.36:166-174

Modification at the Reconstruction in Holographic Interferometry

W. Schumann

Laboratory of Photoelasticity, Swiss Federal Institute of Technology, Rämistrasse 101, CH 8092 Zurich, Switzerland

INTRODUCTION

Holography exhibits the two following features in addition to other properties:

Firstly a hologram can *store* at a recording a wave field emanating from an object, which may be reconstructed by diffraction at any subsequent time by means of a reference wave, thereby creating an image of the object in space.

Secondly it is possible to *modify* the reconstructed wave field by shifting the reference source, by deforming the hologram, or by changing the wavelength of light, thereby changing the image resulting in aberration and virtual deformation.

In *Holographic Interferometry*, two slightly different wave fields, which did not exist simultaneously at the recording, are made to interfere at the reconstruction. Analysis of the resulting macroscopic interference fringes constitutes the principal problem in this field of holography. Applications are mainly concerned with the determination of: (i) the surface deformation of an opaque body, (ii) changes of scalar index of refraction caused by temperature variations in fluids, or (iii) stresses from birefringence and interference in Photoelasticity.

Using the double exposure technique with one reference source and one hologram, the fringes are "frozen". Therefore, they cannot be modified as far as their order in each point is concerned, although it should be noted that their direction changes slightly when an optical modification is made. However, using *two* reference sources or *two* holograms, one for each exposure, we can act differently on each wave field. Thus, it is possible to alter considerably the fringes in addition to the image, and this modification can also be conveniently made. In particular, attention should be paid to the following techniques:

(a) In the *heterodyne method* [14-17,20,36], two reference sources are used (one for each state) and small wavelength changes are made at the reconstruction so that a modulation of the phase difference function occurs. This allows for precise measurements of fractions of fringe order. The method is also important in other fields of Optics [11,18, 19,34,46,47].

(b) In the *Sandwich method* [1-10,22-24,29,30,51], two holograms (again one for each state) are rigidly linked together. At the reconstruction, the sandwich is shifted or rotated so that the fringes may be conveniently modified.

(c) Similar to (b), *two rigidly linked reference sources*, or rather their images, may be shifted at the reconstruction by a mirror motion [12,13]. This technique is very similar to the sandwich method.

The essential feature of these different techniques is that the wavelength changes are small or that the holograms or reference sources are *close* to one another, so that an amplification effect occurs. For instance in (b) and (c), very small optical path differences are compensated for by relatively large motions, similar to what happens with a compensator of Babinet in Photoelasticity. Let us add (d) and (e):

(d) Instead of changing the wavelength of the reference source at the reconstruction as in (a), one may alternatively use two different wavelengths for the two exposures at the recording. When the object is undeformed and when a single reference wave is used at the reconstruction, the interference fringes give some information about the *contour lines* of the object [27,28,33,38,39,40,50].

(e) A modification is also possible upon a single exposure of one hologram and a single reference wave in the so-called *real time technique* [21,25,26,31,32,35,37,41,43,45,48,49,50].Here only the undeformed object is recorded on the hologram. Then, the diffracted wavefield for the undeformed state is superimposed at the recording,with the actual wavefield coming from the deformed object. A hologram motion then modifies the resulting fringes.

THE BASIC EQUATIONS FOR THE PRINCIPAL MODIFICATION TECHNIQUES

From the introduction it may be seen that the three first modification techniques are most relevant for analysis of deformation of an opaque body with fringe control and amplification:

(a) The principle of the *heterodyne method*, including a repositioning error of the hologram at the reconstruction, is shown in Fig. 1. (b) The *Sandwich technique* is outlined in Fig. 2, and (c) the *two reference sources method* is illustrated in Fig. 3.

All three figures show a number of corresponding quantities and, as we shall see, the basic equations also have a common structure. That is why we shall now lay out the basic concepts, so that step by step the reader may compare the three cases *in parallel*.

First, let us describe the recording and the reconstruction processes. At the recording, at any point $\hat{H}$ or $\hat{H}'$ of the hologram, we have the complex amplitudes

$$V = \frac{A_Q}{q}\exp\left(\frac{2\pi i}{\lambda}q\right) \qquad V' = \frac{A_{Q'}}{q'}\exp\left(\frac{2\pi i}{\lambda}q'\right) \tag{1}$$

of spherical waves coming from the point sources of the pinholes Q, Q'. The quantities q, q' denote optical paths $Q\hat{H}$, $Q'\hat{H}'$. A_Q , $A_{Q'}$ are amplitudes (with phases relative to Q, Q'), and λ is the wavelength of the laser light. At the same points $\hat{H}$, $\hat{H}'$ the so-called object waves also arise from the diffusely reflecting parts of the surface of the object in both its deformed and undeformed states. The corresponding complex amplitudes are

$$U = \iint\limits_A \frac{a_p}{p} \exp\left(\frac{2\pi i}{\lambda}p\right) dA \qquad U' = \iint\limits_{A'} \frac{a'_p}{p'} \exp\left(\frac{2\pi i}{\lambda}p'\right) dA'. \tag{2}$$

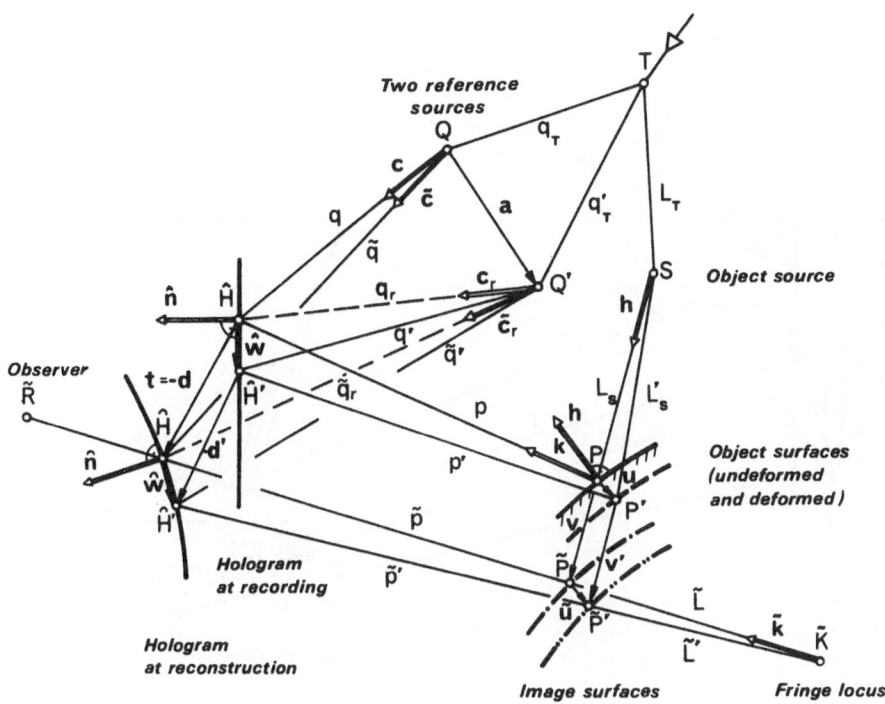

Fig. 1: The heterodyne method (a).

dA, dA' are surface elements, a_p, a'_p are specific statistical amplitudes (due to the surface roughness), and p, p' denote the distances
PĤ , P'Ĥ'. The reference waves and the object waves interfere and give
intensities

$$I = \frac{1}{2}(U + V)(U + V)* = \frac{1}{2}(|U|^2 + |V|^2 + UV* + U*V), \quad I' = \ldots , \quad (3)$$

where * indicates the complex conjugate. Therefore, after development
of the holograms, with equal exposure times $\tau/2$, we obtain the trans-
mittance

$$\bar{T} = \bar{T}_o - \frac{\beta\tau}{4}(UV* + U'V'* + U*V + U'*V') \quad (4a)$$

in the cases (a) and (c) (Fig. 1, 3) of a single hologram either in Ĥ
with U(Ĥ), U'(Ĥ) or also in Ĥ' with U(Ĥ'), U'(Ĥ'); whereas we have

$$T = T_o - \frac{\beta\tau}{4}(UV* + U*V) \qquad T' = T'_o - \frac{\beta\tau}{4}(U'V'* + U'*V') \quad (4b)$$

in case (b) of two separate holograms (Fig. 2). At the reconstruction,
the holograms are simultaneously illuminated by two waves

$$\tilde{v} = \tilde{V}\exp[- 2\pi i(\nu + \Delta\nu)t] \qquad \tilde{v}' = \tilde{V}'\exp[- 2\pi i(\nu - \Delta\nu)t] \quad (5a)$$

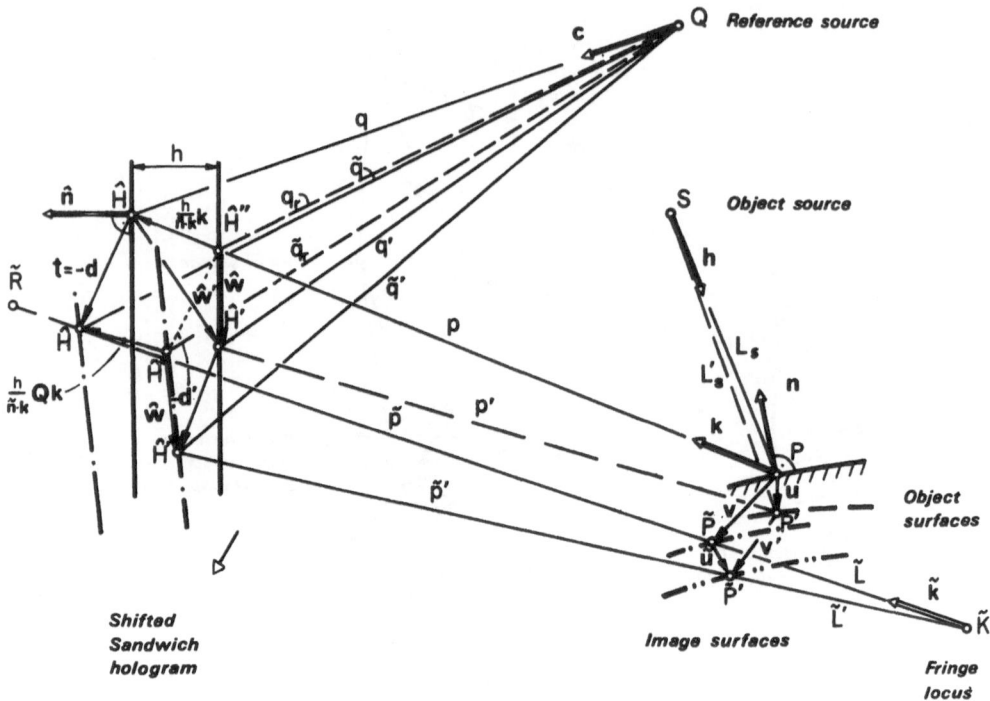

Fig. 2: The Sandwich method (b)

in case (a) of the heterodyne method and by

$$\tilde{v} = \tilde{V}\exp(-2\pi i\nu t) \qquad \tilde{v}' = \tilde{V}'\exp(-2\pi i\nu t) \qquad\qquad (5b)$$

in cases (b) and c) of the geometrical modification. Immediately behind the holograms, then we have the waves:

$$(\tilde{v} + \tilde{v}')\bar{T} \quad \text{(cases (a),(c))} \qquad \tilde{v}T, \tilde{v}'T' \quad \text{(case (b))} .$$

If we restrict our attention to the primary image wave fields and omit
the cross images in cases (a) and (c) (see f.i. left and right of Fig.
4), the important terms are then:

$$\tilde{v}UV* + \tilde{v}'U'V'* \qquad \text{and} \qquad \tilde{v}UV* , \tilde{v}'U'V'* .$$

Provisionally, we now use the concept of image points $\tilde{P}$, $\tilde{P}'$ when re-
constructing (with the modifications) the object points P, P' of the
undeformed and deformed configurations. If the rays through these
points were present alone (note: in this elementary consideration, we
drop visibility considerations concerning the neighboring rays through
the finite aperture), then at any point $\tilde{K}$ of distances $\tilde{L}$, $\tilde{L}'$ from $\tilde{P}$, $\tilde{P}'$
the terms $\tilde{v}UV*$, $\tilde{v}'U'V'*$ would give wave disturbances (for $\tilde{\lambda} = \lambda - \Delta\lambda$,
$\tilde{\lambda}' = \lambda + \Delta\lambda$ in case (a))

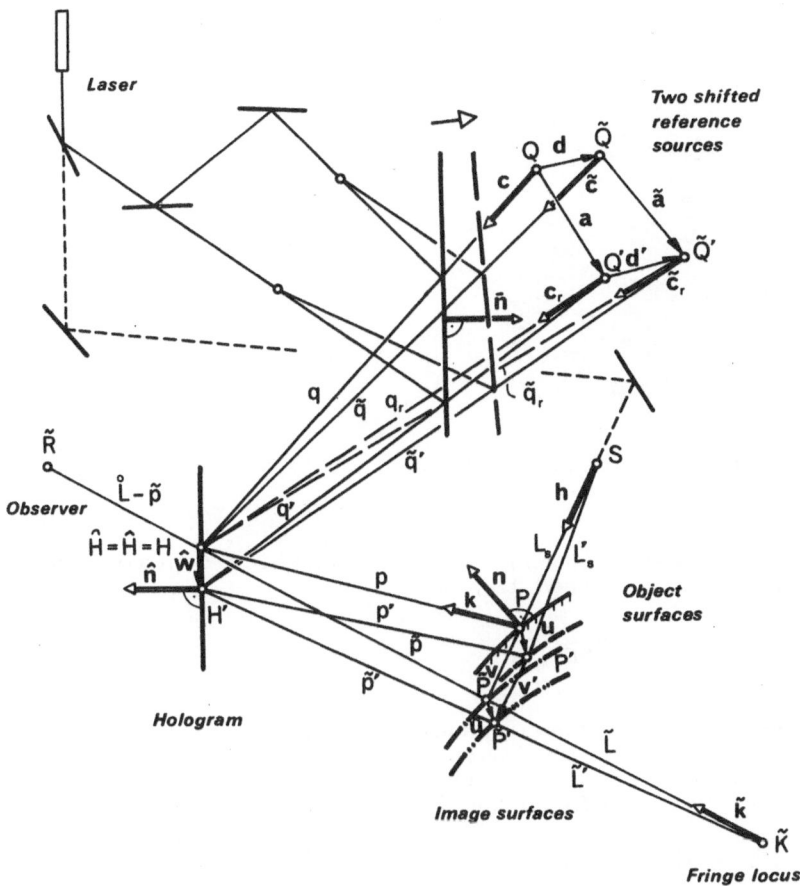

Fig. 3: The two reference sources method (c).

$$\tilde{u} = \frac{A_{\tilde{P}}}{\tilde{L}} \exp\left[i(\tilde{\phi} - \frac{2\pi}{\tilde{\lambda}}\tilde{L} - 2\pi\Delta\nu t)\right]\exp(-2\pi i\nu t)$$

$$\text{(6)}$$

$$\tilde{u}' = \frac{A_{\tilde{P}'}}{\tilde{L}'} \exp\left[i(\tilde{\phi}' - \frac{2\pi}{\tilde{\lambda}'}\tilde{L} + 2\pi\Delta\nu t)\right]\exp(-2\pi i\nu t)$$

where $\tilde{\phi}$, $\tilde{\phi}$ are phases at $\tilde{P}$, $\tilde{P}'$ and where $\Delta\nu \neq 0$ in case (a), $\Delta\nu = 0$ in cases (b) and (c). The resultant intensity at point K is:

$$J = \frac{1}{2}\left\{\frac{|A_{\tilde{P}}|^2}{\tilde{L}^2} + \frac{|A_{\tilde{P}'}|^2}{\tilde{L}'^2} + 2\frac{A_{\tilde{P}}A_{\tilde{P}'}}{\tilde{L}\tilde{L}'}\cos\left[(\tilde{\phi} - \frac{2\pi}{\tilde{\lambda}}\tilde{L}) - (\tilde{\phi}' - \frac{2\pi}{\tilde{\lambda}'}\tilde{L}') - 4\pi\Delta\nu t)\right]\right\}. \quad \text{(7)}$$

In the heterodyne method (a) the argument of the cos-function indicates that the fringe order oscillates with the beat frequency $2\Delta\nu$. Registration of this slow oscillation is one of the characteristics of this method since it permits accurate measurements of fractions of fringe order.

Fig. 4: Fringe pattern and cross images.

Relevant in all three cases in (7) is the phase which contains the optical path difference referred to λ:

$$D = \frac{\lambda}{2\pi}\left[(\tilde{\phi} - \frac{2\pi}{\tilde{\lambda}}\tilde{L}) - (\tilde{\phi}' - \frac{2\pi}{\tilde{\lambda}'}\tilde{L}')\right] . \tag{8}$$

In order to eliminate the unknown phases $\tilde{\phi}$, $\tilde{\phi}'$, we must have the conditions of *interference identity* which relate the proper interference at points $\hat{H}$, $\hat{H}'$ at the recording to the diffraction at points $\hat{H}$, $\hat{H}'$ at the reconstruction. This leads to the following equations:

$$\tilde{\phi} = \frac{2\pi}{\lambda}(L_S + L_T) + 2\pi\left[(\frac{p}{\lambda} - \frac{\tilde{p}}{\tilde{\lambda}}) - (\frac{q}{\lambda} - \frac{\tilde{q}}{\tilde{\lambda}}) - (\frac{1}{\lambda} - \frac{1}{\tilde{\lambda}})q_T\right] ,$$

$$\tilde{\phi}' = \frac{2\pi}{\lambda}(L_S' + L_T) + 2\pi\left[(\frac{p'}{\lambda} - \frac{\tilde{p}'}{\tilde{\lambda}'}) - (\frac{q'}{\lambda} - \frac{\tilde{q}'}{\tilde{\lambda}'}) - (\frac{1}{\lambda} - \frac{1}{\tilde{\lambda}'})q_T'\right] , \tag{9}$$

where paths L_T, q_T, q_T' between the sources S, Q, Q' and the bifurcation point T intervene. Inserting (9) into (8), we generally find:

$$D = (L_S - L_S') + (p - p') - (q - q') +$$

$$+ \lambda\left[\frac{\tilde{L}'+\tilde{p}'}{\tilde{\lambda}'} - \frac{\tilde{L}+\tilde{p}}{\tilde{\lambda}} + \frac{\tilde{q}}{\tilde{\lambda}} - \frac{\tilde{q}'}{\tilde{\lambda}'} - (\frac{1}{\lambda} - \frac{1}{\tilde{\lambda}})q_T + (\frac{1}{\lambda} - \frac{1}{\tilde{\lambda}'})q_T'\right] . \tag{10}$$

Using first order approximations for small wavelength changes $\Delta\lambda$, small displacement $\mathbf{u}$ and with a unit vector $\mathbf{h}$, we get:

$$D = - \mathbf{u} \cdot \mathbf{h} - (\tilde{L} + \tilde{p}) + (\tilde{L}' + \tilde{p}') + (p - p') - (q - q') + (\tilde{q} - \tilde{q}')$$

$$+ \frac{\Delta\lambda}{\lambda}\left[- (\tilde{L}' + \tilde{p}' + \tilde{L} + \tilde{p}) + (\tilde{q} + q_T + \tilde{q}' + q_T')\right] . \tag{11}$$

In case of the sandwich, where $\Delta\lambda = 0$, in addition to a unit vector $\tilde{k}$ on the line of sight, we have an orthogonal rotation tensor $\mathbf{Q}$, the thickness h and small in-plane vectors $\hat{\mathbf{w}}, \hat{\mathbf{w}}$, so that

$$p - p' = \mathbf{u}\cdot\mathbf{k} + \frac{h}{\hat{\mathbf{n}}\cdot\mathbf{k}} - \hat{\mathbf{w}}\cdot\mathbf{k} \quad, \quad - (\tilde{L} + \tilde{p}) + (\tilde{L}' + \tilde{p}') = -\frac{h}{\hat{\mathbf{n}}\cdot\mathbf{k}}\tilde{k}\mathbf{Q}\mathbf{k} + \hat{\mathbf{w}}\cdot\tilde{k} . \qquad (12)$$

Therefore, with $\tilde{k}\mathbf{Q}\mathbf{k} - 1 \simeq \mathbf{k}(\mathbf{Q} - \mathbf{I})\mathbf{k} = \mathbf{k}\mathbf{\Psi}\mathbf{k} = 0$ ($\mathbf{\Psi} = -\mathbf{\Psi}^T$ is antimetric), we get:

$$- (\tilde{L} + \tilde{p}) + (\tilde{L}' + \tilde{p}') + (p - p') = \hat{\mathbf{w}}\cdot\tilde{k} - \hat{\mathbf{w}}\cdot\mathbf{k} + \mathbf{u}\cdot\mathbf{k} . \qquad (13)$$

This auxiliary relationship holds not only for case (b), but trivially for cases (a) and (c) as well. Moreover, with reference distances q_r, $\tilde{q}_r$ to common points, either H or Q and a development $q' = q_r + \hat{\mathbf{w}}\cdot\mathbf{c}_r$ $\tilde{q}' = \tilde{q}_r + \hat{\mathbf{w}}\cdot\tilde{c}_r$ ($c_r, \tilde{c}_r$ are unit vectors) as well as with the sensitivity vector $\mathbf{k} - \mathbf{h} = \mathbf{g}$, for the optical path difference, we obtain:

$$D \simeq \mathbf{u}\cdot\mathbf{g} + \hat{\mathbf{w}}\cdot(\tilde{k} - \tilde{c}_r) - \hat{\mathbf{w}}\cdot(\mathbf{k} - \mathbf{c}_r) - (q - \tilde{q}) + (q_r - \tilde{q}_r)$$
$$+ \frac{\Delta\lambda}{\lambda}\left[- 2(\tilde{L} + \tilde{p}) + (q + q_T + q' + q_T')\right] \quad, \qquad (14)$$

where $\Delta\lambda \neq 0$ only in case (a). In practical application of the heterodyne method, $\Delta\lambda$ is very small so that the last term may be neglected in any case. Furthermore, the stationary behavior of the path difference $(q_r - \tilde{q}_r) - (p' - \tilde{p}')$ which, in fact, determines the approximate direction $\mathbf{k}$ of the image P' implies the aberration equation [44]

$$\hat{\mathbf{w}}\cdot(\tilde{k} - \tilde{c}_r) - \hat{\mathbf{w}}\cdot(\mathbf{k} - \mathbf{c}_r) = 0 . \qquad (15)$$

Therefore, the general expression of the optical path difference in all three cases is simply:

$$D = \mathbf{u}\cdot\mathbf{g} - (q - \tilde{q}) + (q_r - \tilde{q}_r). \qquad (16)$$

In this expression, $\mathbf{u}\cdot\mathbf{g}$ represents the well-known scalar product displacement vector times sensitivity vector of standard holographic interferometry, while the two remaining terms concern the modification.

We assume from now on that the shift $|\mathbf{d}|$ of the sources (c) or of the holograms (a,b) is small when compared with either the distance $|\mathbf{a}|$ of the sources or with the thickness h of the sandwich; $|\mathbf{d}|$ can, however, be moderately large relative to the mechanical displacement $\mathbf{u}$. In summary, we have $|\mathbf{u}| \ll |\mathbf{d}| \ll |\mathbf{a}|$, h $\ll \ell_o$, ℓ_o being some characteristic length. We therefore write a development

$$D = \mathbf{u}\cdot\mathbf{g} + \mathbf{d}'\cdot\mathbf{c}_r - \mathbf{d}\cdot\mathbf{c} - \frac{1}{2}\left[\frac{1}{q_r}\mathbf{d}'\cdot\mathbf{C}_r\mathbf{d}' - \frac{1}{q}\mathbf{d}\cdot\mathbf{C}\mathbf{d}\right] \qquad (17)$$

provisionally up to the second order terms in $\mathbf{d}, \mathbf{d}'$, and with projectors $\mathbf{C} = \mathbf{I} - \mathbf{c}\otimes\mathbf{c}$, $\mathbf{C}_r = \mathbf{I} - \mathbf{c}_r \otimes \mathbf{c}_r$, $\mathbf{I}$ denoting the identity tensor. As for $\mathbf{d}'$, we have the relation of a quadrangle (b, c)

$$d' = d + \tilde{a} - a \tag{18}$$

where a denotes either the vector separating the sources Q, Q' or where $a = (h/n \cdot k)k$ is the oblique vector separating the points $\hat{H}''$, $\hat{H}$ of the sandwich. Due to the rigid body motion of the modification at the reconstruction, we have $\tilde{a} = Qa$ with the orthogonal tensor Q ($Q^T Q = I$), so that

$$d' = d + (Q - I)a = \Psi a + d \tag{19}$$

where we recall that $\Psi = -\Psi^T$ is an antimetric tensor. In order to compare the different terms in (17), it is convenient to introduce some parameters. If $|u|/\ell_o = \varepsilon$, $|d|/\ell_o = \delta$ and $|a|/\ell_o = \eta$, then $\Psi a/\ell_o = 0(\eta\delta)$ and, therefore,

$$\frac{1}{2\ell_o}\left[\frac{1}{q_r}d' \cdot C_r d' - \frac{1}{q}d \cdot Cd\right] = 0(\eta\delta^2) \ . \tag{20}$$

Finally, developing

$$c_r = c - \frac{1}{q}Ca - \frac{1}{2q^2}aCa \tag{21}$$

where $C = c \otimes C + C \otimes c + C \otimes c)^T$ is a superprojector (the sign $)^T$ denoting partial transposition), the optical path difference becomes:

$$D = u \cdot g - a \cdot \left(\Psi c + \frac{1}{q}Cd - \frac{1}{q}\Psi Ca\right) - \frac{1}{2q^2}(aCa)d + 0(\eta\delta^2) \tag{22}$$

If the linear term in a is comparable to the standard term $u \cdot g$, then $\delta\eta = 0(\varepsilon)$. Therefore the quadratic terms are $0(\delta\eta^2)$ and in the case (c) of the two source method they preferably should not be neglected if these sources are some centimeters apart. In the case of the sandwich, however, h is usually only some millimeters, so that we have the good approximation [24]

$$D = u \cdot g - a \cdot \left(\Psi c + \frac{1}{q}Cd\right) = u \cdot g - \frac{h}{\hat{n} \cdot k}k \cdot \left(\Psi c + \frac{1}{q}Cd\right) \ . \tag{23}$$

Before considering the industrially relevant applications of eqs. (22) and (23), some remarks might be useful. A fringe field does not directly tell us something about the fringe order D/λ because the zero order is usually not known. Since, above all, it shows *fringe directions* and *interspaces*, it is a visible expression of the derivative of the fringe order. On the other hand, such a derivative must contain the deformation gradient $\nabla \otimes u$. Therefore, when the holograms are rotated until the fringes disappear in the sandwich technique, which means a vanishing gradient of D, then the basic idea of that operation is to somehow compensate strains and rotations in one point of the object by a motion of the sandwich. This concept is, therefore, very similar to the compensation technique used in Photoelasticity. On the other hand, modification in Holographic Interferometry gives an analoguous information about surface strains as does f.i. the interference fringe technique of Nisida and Saito 1964 in [42] for plane stresses.

Since the calculus of the derivative of D is rather cumbersome, we shall outline it here only briefly. The reader may find details about the so-called fringe vector in the literature. In case of the sandwich (b), for instance, we have

$$\frac{dD_R}{d\phi} = m \cdot \left\{ L_R M w - K u + \frac{h}{\hat{n} \cdot k} \hat{M} \left[(\Psi c + \frac{1}{q} C d) \right. \right.$$

$$\left. \left. + \frac{L_R - p}{q} (\Psi C - C \Psi) k - \frac{L_R - p}{q^2} k \cdot C d \right] \right\} , \tag{24}$$

whereas in case of the two reference sources (c) the analoguous equation is:

$$\frac{dD_R}{d\phi} = m \cdot \left\{ L_R M w - K u + \frac{L_R - p}{q} \hat{M} [-a\Psi C + \frac{1}{q} a C d - \frac{1}{q} a \Psi C a + \frac{1}{2q^2} (a \mathcal{C} a) d] \right\} \tag{25}$$

with a hyperprojector $\mathcal{C} = c \otimes C + {}^T (c \otimes C + C \otimes c)^T + C \otimes c - C \otimes C - (C \otimes c)^T - (C \otimes C)^T$ [44]. When the observer is located at some point R in front of the holograms, $dD_R/d\phi$ is the angular derivative of D in the direction of the unit vector ${}^R m$ (in the plane normal to the line of sight k). $L_R M w - K u = L_R M N w - K u = f_R$ is the fringe vector of Stetson for standard holography when no modification is performed at the reconstruction. This vector contains the distance $\overline{RP} = L_R$, an oblique projector $M = I - n \otimes k / n \cdot k$ (the transpose of it is sometimes called "shadow"), where n is the unit normal of the object surface. Furthermore, there are two normal projectors $N = I - n \otimes n$, $K = I - k \otimes k$ and a vector $N w$, which, in case of collimated object light $L_s \longrightarrow \infty$, is ($g = k - h$)

$$N w = (\Gamma + \Omega E + \omega \otimes n) g . \tag{26}$$

In eq. (26) Γ is the in-plane strain tensor (components ε_x, ε_y, $\frac{1}{2}\gamma_{xy}$) Ω describes the pivot rotation and ω (ω_x, ω_y) is the inclination vector of the surface element. Finally, $E \hat{=} \begin{bmatrix} 0 & 1 \\ -1 & 0 \end{bmatrix}$ denotes the two-dimensional permutation tensor ($E E = - N$).

The remaining terms in the brackets of (24) or (25) represent the modification of the fringe vector. Basically, they contain the shift d and the rotation Ψ of the sandwich, together with the small factor h or the distance $|a|$ of the sources, which involves the amplification effect mentioned in the introduction. Therefore, it follows that eqs. (24) or (25) must be applied in at least three directions: k_i (complete base g_i) and two directions m_α in order to get six equations for ε_x, ε_y, $\frac{1}{2}\gamma_{xy}$, Ω, ω_x, ω_y in addition to a rough determination of u which need not be precise if L_R is large. Fig. 5 shows an example of the two-sources method (c) where the inclined fringes are successively modified in a vertical and a horizontal position, thereby leading to measures of the modification term in eq. (25) [13].

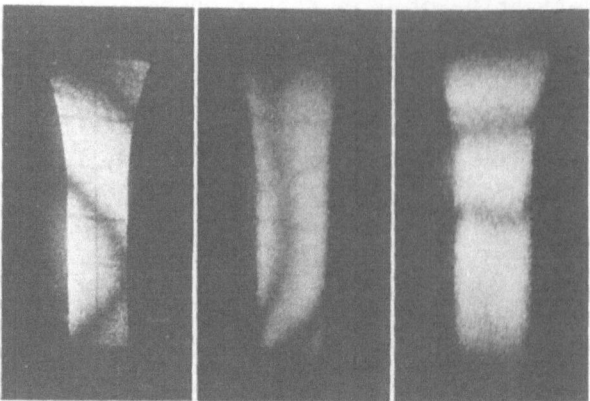

Fig. 5: Fringes and their modification.

In conclusion, we may state that a modification at the reconstruction in Holographic Interferometry represents a flexible tool for quantitative industrial applications.

BIBLIOGRAPHY

[1]   Abramson N. (1974): Sandwich hologram interferometry: A new di-
        mension in holographic comparison, Appl. Opt. 13: 2019-2025
[2]   Abramson N. (1975): Sandwich hologram interferometry 2: Some
        practical calculations, Appl. Opt. 14: 981-984
[3]   Abramson N. (1977): Sandwich hologram interferometry 4: Holo-
        graphic studies of two milling machines, Appl. Opt. 17:
        2521-2531
[4]   Abramson N., Bjelkhagen H. (1978): Pulsed sandwich holography
        2: Practical application, Appl. Opt. 17: 187-191
[5]   Abramson N., Bjelkhagen H. (1979): Sandwich hologram interfero-
        metry 5: Measurement of in-plane displacement and compensa-
        tion for rigid body motion, Appl. Opt. 18: 2870-2880
[6]   Abramson N., Bjelkhagen H., Skande P. (1979): Sandwich hologra-
        phy for storing information interferometrically with a high
        degree of security, Appl. Opt. 18: 2017-2021
[7]   Abramson N., Bjelkhagen H. (1980): Deformation, displacement
        and vibration investigations in manufacturing applications
        using a new hologram interferometra technique, Opt. Lasers
        Eng. 1: 51-68
[8]   Abramson N. (1981): *Making and Evaluation of Holograms*, Acade-
        mic Press, London
[9]   Amadesi S., D'Altorio A., Paoletti D. (1982): Sandwich hologra-
        phy for painting diagnostics, Appl. Opt. 21: 1889-1890
[10]   Bjelkhagen H. (1977): Pulsed sandwich holography, Appl. Opt.
        16: 1727-1731
[11]   Churnside J.H., Yura H.T. (1982): Laser vector velocimetry: A
        3-D measurement technique, Appl. Opt. 21: 845-850
[12]   Cuche D., Schumann W. (1983): Fringe modification with amplifi-
        cation in holographic interferometry and application of this
        to determine strain and rotation, SPIE Vol. 398: 35-45
[13]   Cuche D., (1984): "Modification des franges d'interférence en
        interférométrie holographique appliquée à la détermination
        des dilatations et des rotations" Thesis ETH Zürich No. 7459

[14]   Dändliker R., Ineichen B., Mottier F.M. (1973): High resolution
        hologram interferometry by electronic phase measurement, Opt.
        commun. 9: 412-416
[15]   Dändliker R., Ineichen B., Mottier F.M. (1974): Electric pro-
        cessing of holographic interferograms, in Digest of Papers,
        Int. Opt. Computing Conf. Zürich (IEEE, New York, 1974):
        69-72
[16]   Dändliker R. Marom E., Mottier F.M. (1976): Two-reference-beam
        holographic interferometry, J. Opt. Soc. Am. 66: 23-30
[17]   Dändliker R. (1980): "Heterodyne holographic Interferometry" in
        Progress in Optics, vol. XVII, chap. 1 (North-Holland, Am-
        sterdam)
[18]   Dändliker R., Willemin J.F. (1981): Measuring microvibrations
        by heterodyne speckle interferometry, Opt. Lett. 6: 165-167
[19]   Dändliker R. (1982): Measuring displacement, velocity and vi-
        bration by laser interferometry, in *Optoelectronics in Engi-
        neering,* ed. by W. Waidelich (Springer-Verlag, Berlin): 52-58
[20]   Decker A.J., Pao Y.H., Claspy P.C. (1978): Electronic heterody-
        ne recording and processing of optical holograms using phase
        modulated reference waves, Appl. Opt. 17: 917-921
[21]   De Larminat P.M., Wei R.P. (1976): A fringe-compensation tech-
        nique for stress analysis by reflection holographic interfe-
        rometry, Exp. Mech. 16: 241-248
[22]   Dirtoft I., Abramson N., Sandström U. (1979): Holographic mea-
        suring of deformations in complete upper dentures, SPIE Vol.
        211: 106-110
[23]   Doty J.L., Hildebrand B.P. (1982): The use of sandwich hologram
        interferometry for nondestructive testing of nuclear reactor
        components, Opt. Eng. 21: 542-547
[24]   Dubas M., Schumann W. (1977): Contribution à l'étude théorique
        des images et des franges produites par deux hologrammes en
        sandwich, Opt. Acta 24: 1193-1209
[25]   Dudderar T.D., Doerries E.M. (1979): Application of holographic
        interferometry to real-time studies of heat effects in multi-
        layer circuit boards, Mat. Evaluation 37: 41-50
[26]   Fischer B., Cronin-Golomb M., White J.O., Yariv A. (1981): Am-
        plified reflection, transmission, and self-oscillation in
        real-time holography, Opt. Lett. 6: 519-521
[27]   Friesem A.A., Levy U. (1976): Fringe formation in two-wave-
        length contour holography, Appl. Opt. 16: 3009-3020
[28]   Haines K.A., Hildebrand B.P. (1965): Contour generation by
        wavefront reconstruction, Phys. Lett. 19: 10-11
[29]   Hariharan P., Hegedus Z.S. (1976): Two-hologram interferometry:
        A simplified sandwich technique, Appl. Opt. 15: 848-849
[30]   Hariharan P. (1977): Hologram Interferometry: Identification of
        the sign of surface displacements, Opt. Acta 24: 989-990
[31]   Hariharan P., Oreb B.F., Brown N. (1983): A digital system for
        real-time holographic stress analysis, SPIE Vol. 370: 189-194
[32]   Hariharan P., Oreb B.F., Brown N. (1983): Real-time holographic
        interferometry: A microcomputer system for the measurement of
        vector displacements, Appl. Opt. 22: 876-880
[33]   Hildebrand B.P., Haines K.A. (1967): Multiple-wavelength and
        multiple-source holography applied to contour generation, J.
        Opt. Soc. Am. 57: 155-162
[34]   Hoffer T.M., Fischer W. (1977): Abnahme von Werkzeugmaschinen
        mit einem Laser-Messystem, Feinwerktechnik & Messtechnik 85,
        (I): 229-235, (II): 343-359
[35]   Hsu T.R. (1974): Large-deformation measurements by real-time
        holographic interferometry, Exp. Mech. 14: 408-411

[36]   Ineichen B., Dändliker R., Mottier F.M. (1977): Accuracy and
        reproducibility of heterodyne holographic interferometry, in
        Applications of Holography and Optical Data Processing, ed. by
        E. Marom, A.A. Friesem, E. Wiener-Avnear (Pergamon Press,
        Oxford: 207-212
[37]   Krepelkova H. (1980): The application of holographic interfero-
        metry to the analysis of composite material structure, Opt.
        Appl. X: 91-97
[38]   Küchel F.M., Tiziani H.J. (1981): Real-time contour holography
        using BSO crystals, Opt. Commun. 38: 17-20
[39]   Leung K.M., Lee T.C., Bernal E., Wyant J.C. (1979): Two-wave-
        length contouring with the automated thermoplastic hologra-
        phic camera, SPIE Vol. 192: 184-189
[40]   Menzel E. (1974): Comment to the methods of contour holography
        Optik 40: 557-559
[41]   Morizov N.V., Ostrovskii Y.I., Boeva L.M. (1982): Real-time
        holographic interferometry of moving objects in oppositely
        directed beams, Zh. Tekh. Fiz. 52: 1854-1858
[42]   Nisida M., Saito H. (1964): A new interferometric method of
        two-dimensional stress analysis, Exp. Mech. 4 (12): 366-376
[43]   Politch J. (1982): Real-time imaging and strain distribution of
        an angularly vibrating diffused plate, Opt. Acta 29: 485-492
[44]   Schumann W., Zürcher J.-P., Cuche D. (1985): *Holography and
        Deformation Analysis* (Springer, Heidelberg, Berlin, New York,
        Tokyo)
[45]   Sciammarella C.A., Rastogi P.K., Jacquot P., Narayanan R.
        (1982): Holographic moiré in real time Exp. Mech. 22: 52-63
[46]   Shapiro J.H., Capron B.A., Harney R.C. (1981): Imaging and tar-
        get detection with a heterodyne-reception optical radar,
        Appl. Opt. 20: 3292-3313
[47]   Sommargren G.E. (1981): Optical heterodyne profilometry, Appl.
        Opt. 20: 610-618
[48]   Tiziani H.J. (1982): Real-time metrology with BSO crystals,
        Opt. Acta 29: 463-470
[49]   Uyemura T., Yamamoto Y., Tenjimbayashi K., Yokoyama N. (1979):
        Real-time holographic interferometry with pulsed laser, SPIE
        Vol. 192: 190-195
[50]   Varner J.R. (1971): Simplified multiple-frequency holographic
        contouring, Appl. Opt. 10: 212-213
[51]   Vukicevic S., Vinter I., Vukicevic D. (1983): Sandwich hologram
        interferometry for determination of sacroiliac joint move-
        ments, SPIE Vol 370: 129-132

Strain Measurement by Laser-Speckle

Ichirou Yamaguchi

Optical Instrumentation Laboratory, The Institute of Physical and Chemical Research, Hirosawa, Wako, 351-01 Japan

INTRODUCTION

A laser-speckle pattern is a random granular pattern which arises in the light scattered from a laser-illuminated rough surface or in its image. It results from random interference of the light scattered from various surface points and can be used as distinct random marks attached to the surface. Applications of speckle to deformation measurements are especially useful because they are complementary to holographic interferometry in many respects (Ennos 1978; Yamaguchi 1985). In general, speckle methods are simpler in set-up and more suitable for measuring in-plane deformation that is difficult to determine by holographic interferometry.

The speckle methods for deformation measurement can be classified into two categories. The first, called speckle interferometry, provides contour lines of surface displacement components as a kind of moire fringes of speckle patterns formed by using interferometric set-ups (Leendertz 1970; Duffy 1972). The second utilizes speckle displacement caused by surface deformation. In speckle photography, the speckle displacement is detected by optical transformation of doubly exposed negatives, called specklegrams (Fink and Büger 1970; Köpf 1971; Archbold and Ennos 1972). This method provides vectorial distributions of in-plane displacement from which strain distribution can be derived.

Yamaguchi (1980) developed an optoelectronic method to detect speckle displacement by using a linear image sensor followed by a microcomputer. This technique was combined with a study on speckle displacement (Yamaguchi 1981a) to realize a laser-speckle strain gauge (Yamaguchi 1981b, 1982). Its advantages are: no need for surface preparation, automatic cancellation of rigid-body motion, simple optical set-up, quick time response, and full automation. In this paper the basic principles and improvements of the laser speckle strain gauge are presented with a lot of experimental results.

PRINCIPLES

Speckle displacement and decorrelation induced by displacement and deformation of a laser-illuminated diffuse surface have been derived for various optical systems by calculating the cross-correlation function of the intensity distributions before and after surface deformation (Yamaguchi 1981a). Figure 1 shows the coordinate system for deriving speckle displacement detected in the laser-speckle strain gauge.

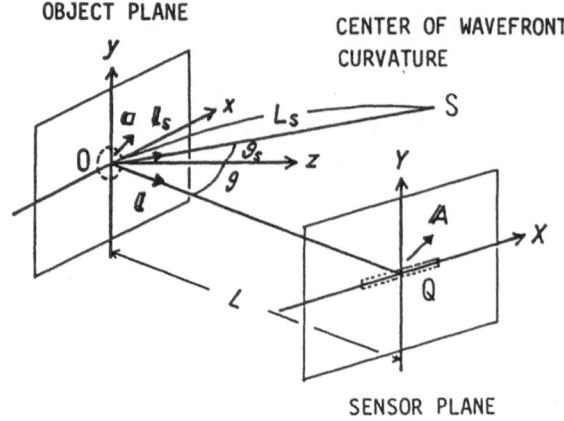

OBJECT PLANE

CENTER OF WAVEFRONT
CURVATURE

SENSOR PLANE

Fig. 1. Coordinate system for deriving speckle displacement

A laser beam is directly incident on a diffuse surface. Speckle displacement is derived as the peak position of the two-dimensional intensity cross-correlation function. Deformation of the region illuminated by a laser beam is represented by the displacement $a(x,y)$ of each object point. We assume the spot to be so small that the displacement within the illuminated object region can be approximated by a linear function of x and y. The speckle displacement at a point Q in the observation plane can then be represented by

$$A_X = - \frac{L}{\cos\theta} \left[\frac{\partial}{\partial x} (ls+l) \cdot a \right]_{x=y=0}, \quad A_Y = -L \left[\frac{\partial}{\partial y} (ls+l) \cdot a \right]_{x=y=0}, \quad (1)$$

where L is the distance from the center O of the spot to Q and denotes the angle between OQ and the surface normal. We set the x-axis of the object plane and the X-axis of the observation plane parallel to the plane of incidence. The vector $l_s(x,y)$ denotes the unit vector directed from (x,y) to the center S of wavefront curvature of the incident laser beam and vector $l(x,y)$ the unit vector directed from (x,y) to Q.

To derive the explicit representation of the components of Eq.(1), we introduce the translation vector $a_0 = (a_x, a_y, a_z)$, the rotation vector

$\Omega = [\partial a_z/\partial y, -\partial a_z/\partial x, (1/2)(\partial a_y/\partial x - \partial a_x/\partial y)]$, and the strain coefficients

$\varepsilon_x = \partial a_x/\partial x, \varepsilon_y = \partial a_y/\partial y, \gamma_{xy} = (1/2)(\partial a_y/\partial x + \partial a_x/\partial y)$, where the values in the

right-hand side are assigned at x=y=0. The final expressions for Eq. (1) are given by linear combinations of these quantities where their coefficients depend on θ_s (incident angle), θ, L_s, and L (Yamaguchi 1982). From A_X, surface strain ε_x can be separated by taking the difference between those for either symmetrical incident angles θ_s and $-\theta_s$ or symmetrical observation angles θ and $-\theta$. In the former (dual-beam) case, the differential speckle displacement is given by

$$\Delta A = A_X(\theta_s, \theta) - A_X(-\theta_s, \theta) = -2L\varepsilon_x \sin\theta_s \tag{2}$$

that is proportional to strain parallel to the plane formed by the beams. In the latter (double-sensor) system we have

$$\Delta A = A_X(\theta_s, \theta) - A_X(\theta_s, -\theta) = -2L\varepsilon_x \tan\theta - 2a_z \sin\theta \tag{3}$$

This system gives strain along the line connecting the detectors, but the difference also receives a contribution from out-of-plane translation a_z. Despite this shortcoming the double-sensor system is more practical because it is easier to implement.

DETECTION OF SPECKLE DISPLACEMENT

The component A_X is determined by two means. In the first we position a linear image sensor along the X-axis and calculate the peak position of the cross-correlation function between the sensor signals before and after surface deformation by a microcomputer. This method is limited in its response to a few tenths of Hertz by the time for cross-correlation computation. The second method, which is achieved in real time, uses a special type of detector having signal processing capability. It is called a spatial filtering detector with electronic scanning facility and delivers a voltage directly proportional to pattern displacement along a fixed direction up to several kHz (Yamaguchi, Furukawa, Ueda, Ogita 1985).

EXPERIMENTAL ARRANGEMENTS

Figure 2 shows the basic set-up of the double-sensor laser speckle strain gauge using the spatial filtering detectors.

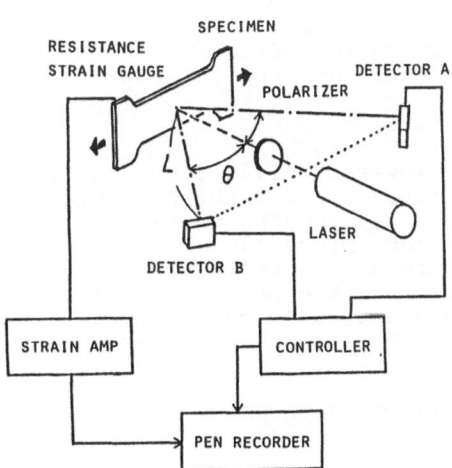

Fig. 2. Set-up of the laser speckle strain gauge

A direct beam from a 5-mW He-Ne laser is made incident on the specimen at nearly a right angle. A polarizer is inserted to adjust the beam

intensity to avoid saturating detectors in the case of a highly
reflective object. The gauge length of the present apparatus is given
by the beam diameter which can be made smaller than 1mm. A pair of
spatial filtering detectors, Si-photodiode arrays, are positioned at a
distance L = 300 mm and an angle θ = 45° to the laser beam.

The photodiode array consists of 128 Si-photocells each of which is 20
μm wide and 5.12 mm long and has a 40 μm pitch. The photocells are
connected as shown in Fig. 3, and the interconnections are changed
sequentially by electronic switching.

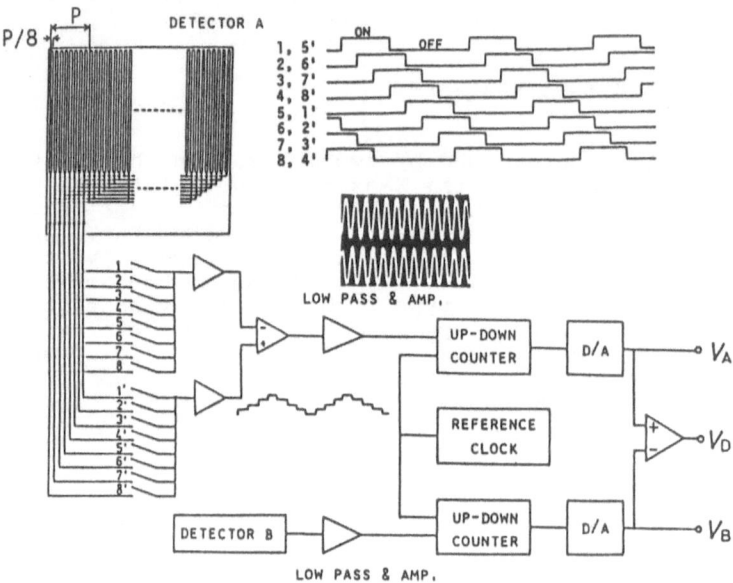

Fig. 3. Photodiode array and electronic interconnections with its
timing chart and the low-pass filtered signals from the both detectors

After low-pass filtering of the output from the differential
amplifier, a 62.5 kHz sinusoidal output is obtained when the pattern
on the array remains stationary. Displacement of the pattern across
the pitch produces a shift in the phase of the output which is
transformed to a voltage by an up-down counter followed by a D/A
converter. The processor delivers the voltages V_A and V_B for speckle
displacements at each detector (sensitivity 6.1 mV/μm) as well as
their difference V_D which is amplified about ten times.

CALIBRATION PROCEDURES

Considering Eq. (3), we can derive the strain sensitivity of the gauge
from the dependence of the differential voltage V_D on out-of-plane
translation a_z. A metal plate was placed in the specimen position in
Fig. 2 and moved by a manual stage. The result shown in Fig. 4 gives
the slope of the matched line equal to $dV_D/da_z = 6.7$ mV/μm.

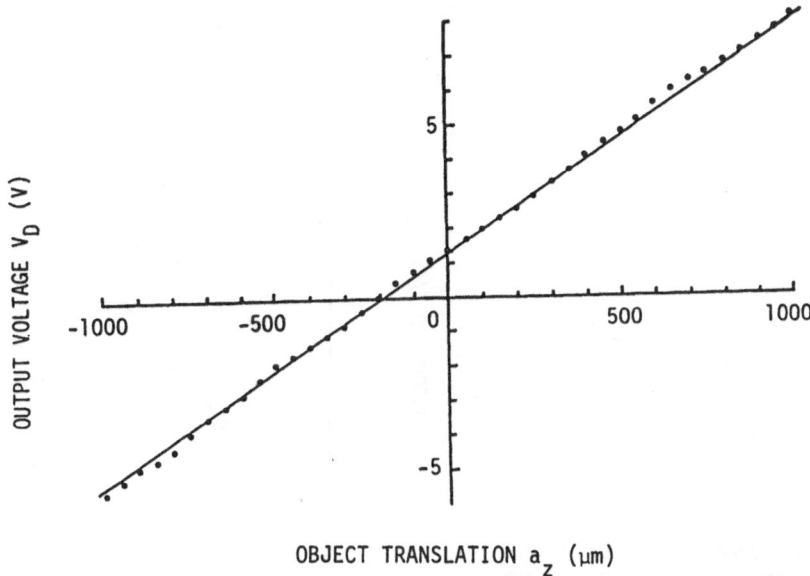

Fig. 4. Differential voltage versus out-of-plane translation.

Substituting this value into Eq. (3) provides the sensitivity to differential speckle displacement as

$$\frac{dV_D}{d\Delta A} = (\frac{1}{2\sin\theta}) \frac{dV_D}{da_z} = 4.73 \ (mV/\mu m) \tag{4}$$

This leads to the strain sensitivity of

$$\frac{dV_D}{d\varepsilon_x} = 2Lt\tan\theta \frac{dV_D}{d\Delta A} = 2.84 \ (mV/microstrain) \tag{5}$$

It has also been verified that V_D is insensitive to in-plane translation a_x.

COMPARISONS WITH RESISTANCE STRAIN GAUGE

We examined the strain sensitivity derived above by direct comparison with a resistance strain gauge. A brass plate specimen with a standard shape (100 mm long, 20 mm wide, 1 mm thick) was subject to tensile strain by mounting it in a test fixture and pulling on its edge by turning a screw. A resistance strain gauge with a 2 mm gauge length was attached at the rear of the spot position with its axis parallel to the detector arms. The differential voltage V_D was recorded against the output of the resistance strain gauge on a pen recorder. The result shown by Fig. 5 exhibits good linearity except in the region where the strain changes sign.

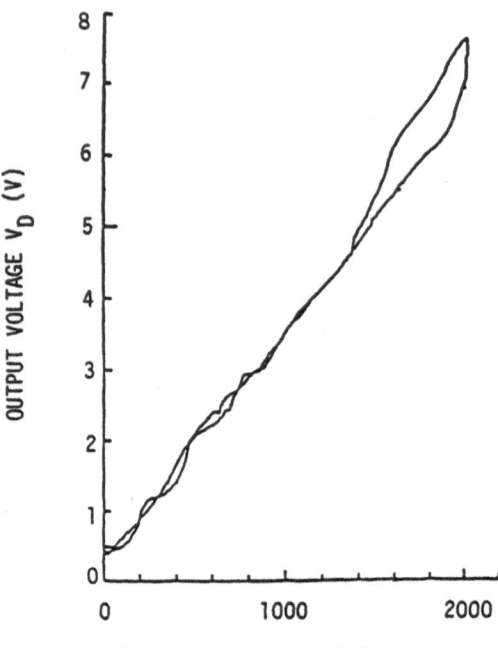

Fig. 5. Differential voltage versus output of a resistance strain gauge for uniform extension of a brass plate

The resultant strain sensitivity $dV_D/d\varepsilon_x$ equal to 3.1 mV/microstrain agrees well with Eq. (5) obtained from out-of-plane translation.

We also conducted the same experiment with specimens of mild steel subject to uniform extension as well as pure bending. Table 1 summarizes the results.

Table 1. Strain sensitivity resulting from uniform tension and pure bending of mild steel plate

		Tension	Bending		
Specimen length	(mm)	104	200	200	200
width	(mm)	6	30	30	30
thickness	(mm)	1	2	4	8
Sensitivity (mV/microstrain)		2.5	9.0	7.5	5.0

Although the strain sensitivity obtained from extension agrees well with Eq. (5), that for pure bending reveal larger discrepancy for thinner specimens. This tendency indicates the contributions from out-of-plane translation accompanying the strain. In Fig. 6, which resulted from bending of the 8 mm thick specimen, theoretical contributions to the differential voltage from the strain and the out-of-plane translation are shown. The out-of-plane displacement was measured by a dial gauge. The sum of each contribution agrees well with the experiment.

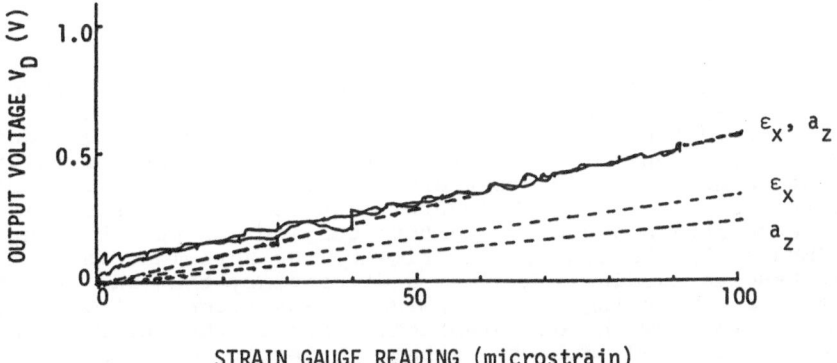

Fig. 6. Experimental and theoretical relations for bending of a mild steel specimen

MEASUREMENT OF DYNAMIC STRAIN

For examining time response of the present gauge we measured strain of a cantilever (150 mm long, 10 mm wide, 5 mm thick) subject to sinusoidal excitation at a point 60 mm distant from the fixed end and at a frequency of 220 Hz that corresponds to the fundamental resonance of the cantilever. Figure 7 shows the low-pass filtered outputs of each detector which show blurred phase shifts due to the repeated traces, outputs of a resistance strain gauge and the speckle strain gauge for a point near the fixed end, and the Lissajous figure between them. The half amplitude of strain is 8 microstrain. The Lissajous has a slope that agrees well with the theory. At 1 kHz we observed the slope to decrease down to about a half of this.

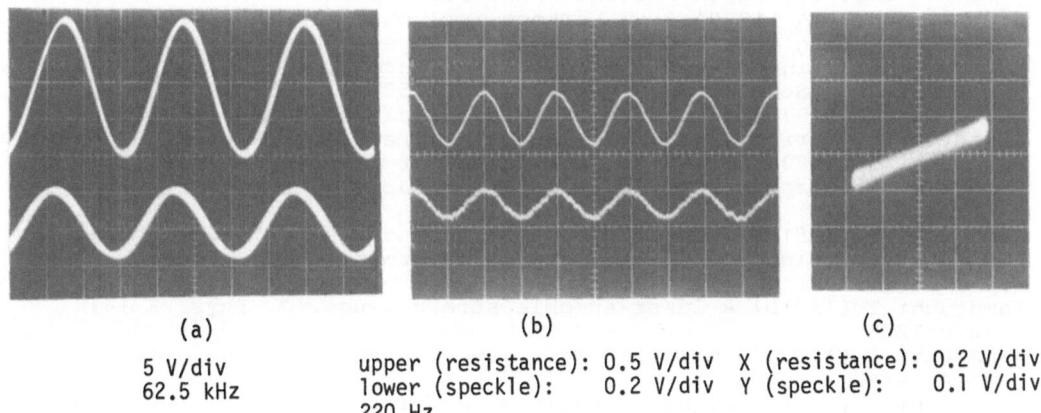

(a)	(b)	(c)
5 V/div	upper (resistance): 0.5 V/div	X (resistance): 0.2 V/div
62.5 kHz	lower (speckle): 0.2 V/div	Y (speckle): 0.1 V/div
	220 Hz	

Fig. 7. Results from strain measurement of a vibrating cantilever (a) Low-pass filtered outputs of each detector, (b) Outputs of a resistance and the speckle strain gauges, (c) their Lissajous

CONCLUSIONS

By introducing a new type of photodetector into a laser speckle strain gauge we were able to measure strain in real time. Its strain sensitivity depends only on the direction, distance, and pattern displacement sensitivity of the detectors. Strain sensitivity of the present gauge is 3 mV/microstrain with a gauge length of 1 mm and time response is at least 1 kilohertz. We compared this sensitivity with resistance strain gauge for uniform tension and pure bending of metal specimen. In the tension experiment, the agreement was satisfactory, while bending loading indicated a discrepancy that can be attributed to the accompanying out-of-plane translation. Solutions for this problem are now being examined. Since the present method is noncontacting and detects strain directly with a simple optical set-up, it will be especially useful where the conventional strain gauges are difficult to use, such as nonmetallic materials and thermal strain.

ACKNOWLEDGMENTS

The author wishes to thank T. Ueda and E. Ogita of Yokogawa-Hokushin Electric Corporation for producing and adjusting the detectors, T. Furukawa of the Institute of Physical and Chemical Research, M. Murata and S. Nishida of Mitsubishi Heavy Industories for discussions and experimental help.

REFERENCES

Archbold E, Ennos AE (1972) Displacement measurement from double-exposure laser photographs. Opt. Acta 19:253-271
Duffy DE (1970) Moire gauging of in-plane displacement using double aperture imaging. Appl. Opt. 11:1778-1781
Ennos AE (1978) Speckle interferometry. In: Wolf E (ed) Progress in Optics Vol. 16, North-Holland, Amsterdam, p 233
Fink W, Büger PA (1970) Eine Methode zur kontaktlosen Messung kleiner Verschiebungen rauher Oberflächen. Z. Angew. Phys. 30:176-178
Köpf U (1971) Ein kohärent-optisches Verfahren zur Messung mechanischer Schwingungen. Optik 33:517-521
Leendertz JA (1970) Interferometric displacement measurement on scattering surfaces utilizing speckle effect. J. Phys. E 3:214-218
Yamaguchi I (1980) Real-time measurement of in-plane translation and tilt by electronic speckle correlation. Jpn. J. Appl. Phys. 19:L133-136
Yamaguchi I (1981a) Speckle displacement and decorrelation in the diffraction fields for small object deformation. Opt. Acta 28:1359-1376
Yamaguchi I (1981b) A laser-speckle strain gauge. J. Phys. E 14: 1270-1273
Yamaguchi I (1982) Simplified laser-speckle strain gauge. Opt. Eng. 21:436-440
Yamaguchi I (1985) Fringe formations in deformation and vibration measurements using laser light. In: Wolf E (ed) Progress in Optics Vol. 22, North-Holland, Amsterdam, p 271
Yamaguchi I, Furukawa T, Ueda T, Ogita E (1985) Accelerated laser-speckle strain gauge. Proc. SPIE 556 (in the press)

Dynamic Stress Concentration Analysis in High Velocity Tension of Strips with Notches or Hole by Means of High Speed Photoelasticity

K. Kawata[1] and S. Hashimoto[2]

[1] Faculty of Science and Technology, Science University of Tokyo, Noda, Chiba, 278 Japan
[2] Institute of Interdisciplinary Research, University of Tokyo, Komaba, Meguro-ku, Tokyo, 153 Japan

ABSTRACT

The relations of dynamic stress concentration factor f_d varying with time vs static stress concentration factor f_s of very long strips with notches under dynamic tension are analyzed by high speed photo-elasticity by means of a multiple spark gap camera using hard epoxy resin specimens and by numerical analysis using FEM and Newmark's β method. As the shapes of the notches, central circular hole and U-shaped notches at both sides are chosen. The phenomenon of maximum stress position deviation towards incident side at the passage of stress wave front and the behaviour that f_d increases with time and then tends to f_s for all notches, are found. The results of corresponding FEM analyses well support the above mentioned behaviours obtained photoelastically.

INTRODUCTION

High speed photoelasticity is a powerful means of analysis for dynamic problems of long duration. Generally speaking, in dynamic problem analysis by FEM, numerical error which accumulates with time limits the accuracy for long time duration. In high speed photoelasticity, there is no such limiting factor. In the present paper, the results of experimental analysis by high speed photoelasticity using hard epoxy resin are reported, being compared with the ones by FEM. The relations of dynamic stress concentration factor f_d varying with time vs static concentration factor f_s of very long epoxy strips with notches under dynamic tension are analyzed. As the shapes of notches, central circular hole and U-shaped notches at the both sides are chosen.

HIGH SPEED PHOTOELASTIC ANALYSIS: SYSTEM AND TECHNIQUES

A high speed 16 frames multiple spark gap camera is used. Total experimental setup is shown in Fig. 1. Load is applied to impact block by a falling weight, the contact of the falling weight with the impact block being used as the trigger. The elastic longitudinal wave velocity obtained from the strain gage output: 1720 m/s well coin-cides with the value obtained from the velocity of fringes of the order 1.5 to 5.5: 1710 to 1780 m/s. Dynamic Young's modulus E_d is determined from these values as 370 Kgf/mm^2. The values of dynamic photoelastic coefficient α_d derived from dynamic strain obtained by strain gage, E_d, and dynamic photoelastic fringe order N_d are in the range of 0.97 to 1.1 mm/Kgf giving the mean value of 1.0 mm/Kgf. The specimen dimension are shown in Fig. 2 and Table 1. The input stress wave form measured using a flat strip specimen without notches is shown in Fig. 3. The length of stress wave plateau is about 320 mm and its duration is estimated as 320 mm/1720 m/s = 186 μsec. So, a stress wave with stress plateau of enough length can be given to the specimens with the notches as shown in Fig. 2, as the incident wave.

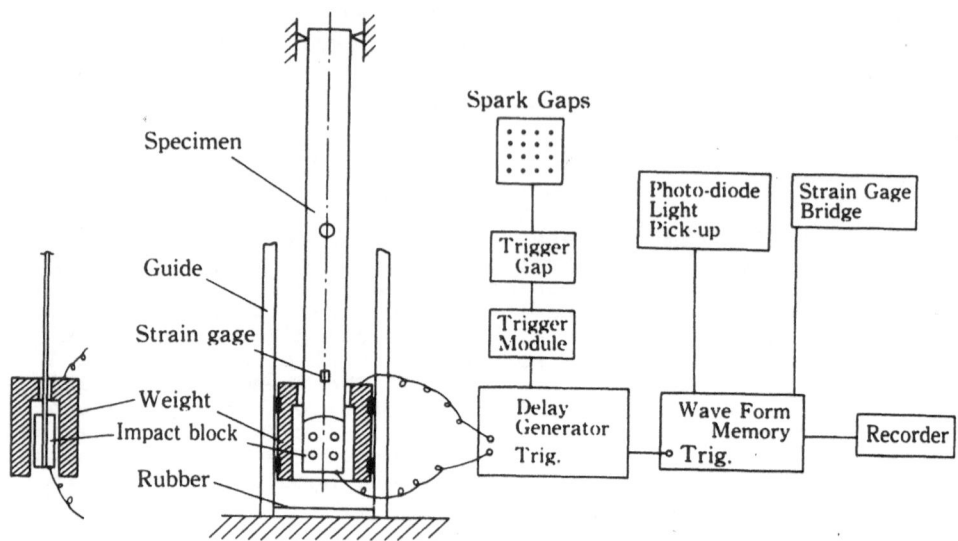

Fig. 1. Experimental arrangement

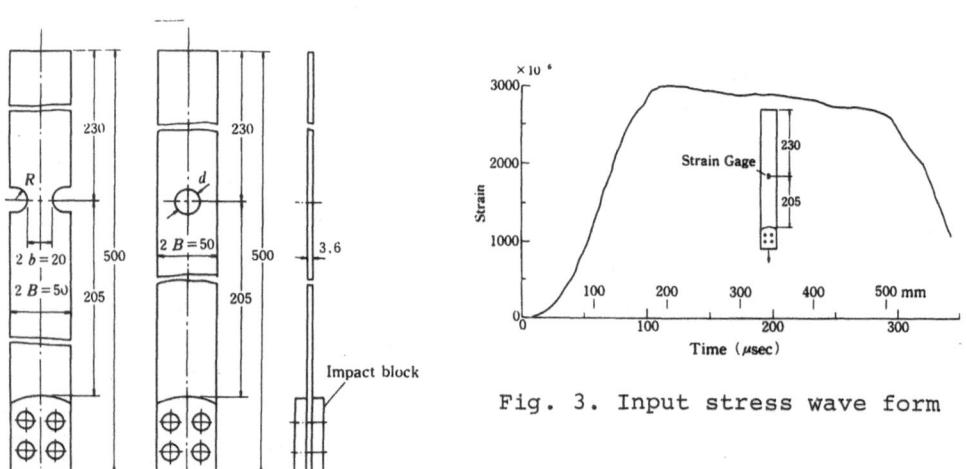

Fig. 2. Specimen dimension

Fig. 3. Input stress wave form

By the multiple spark gap camera, 16 frames of high speed photoelastic fringe pattern can be obtained with the sharpness the same with the statical photoelastic fringe patterns. Time durations between two adjacent frames can be determined precisely from a spark gap discharge record. In the series of the experiment, standard framing speed of 50,000 pps is adopted and clear high speed photoelastic fringe patterns as shown in Figs. 4 and 5, with time interval of 19 to 28 µsec are obtained.

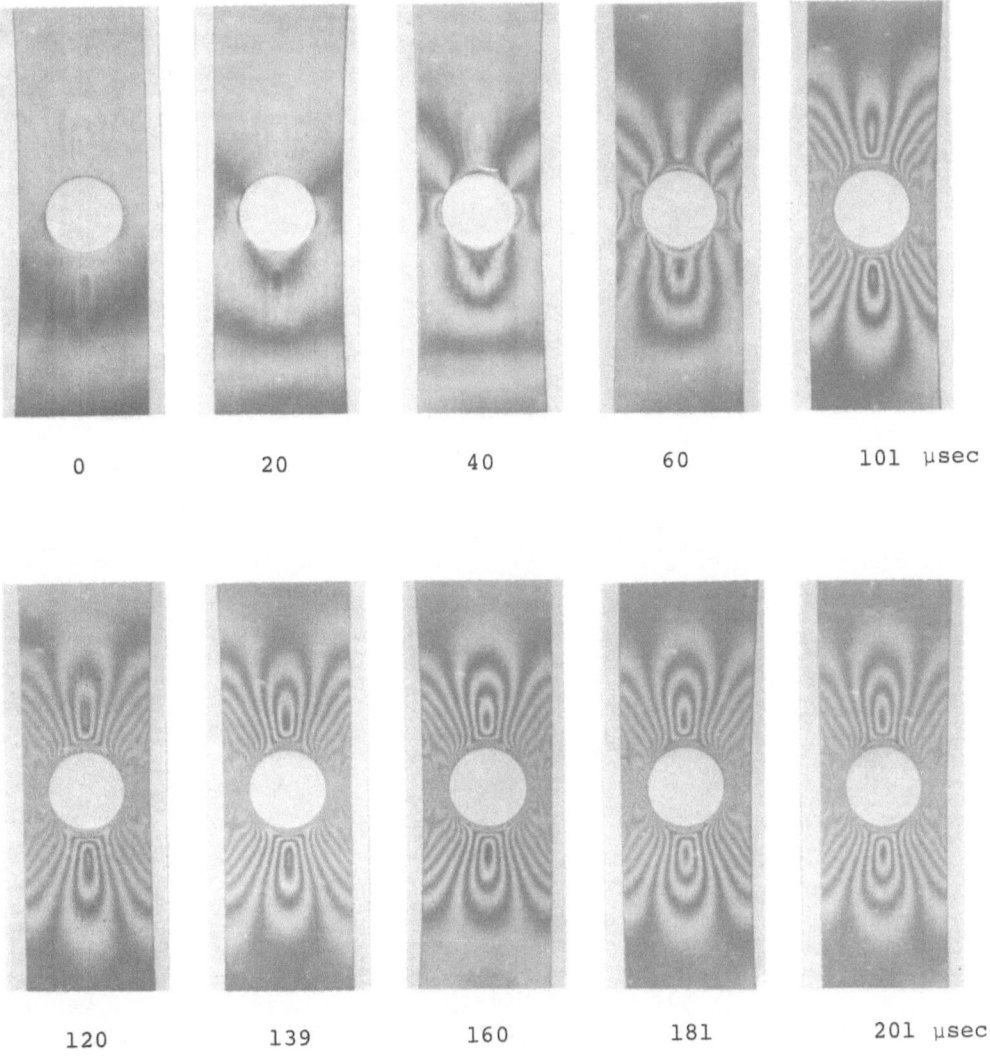

Fig. 4. High speed photoelastic fringe patterns for dynamic tension of epoxy strip specimen with a central circular hole   d/2B = 0.6 (d = 30 mm)

Until the present time, our group has conducted various high speed
photoelasticity experiments, such as the combination of polyurethane
rubber/HIMAC 16H — a high speed camera (10,000 pps), and polyurethane
or epoxy resin/Beckman-Whitley Model 192 (highest framing speed: 1.4
million pps, used range: 100,000 to 300,000 pps) also, and it is
concluded that the combination of epoxy resin/a high speed multiple
spark gap camera gives the sharpest high speed photoelastic fringe
pattern of large size.

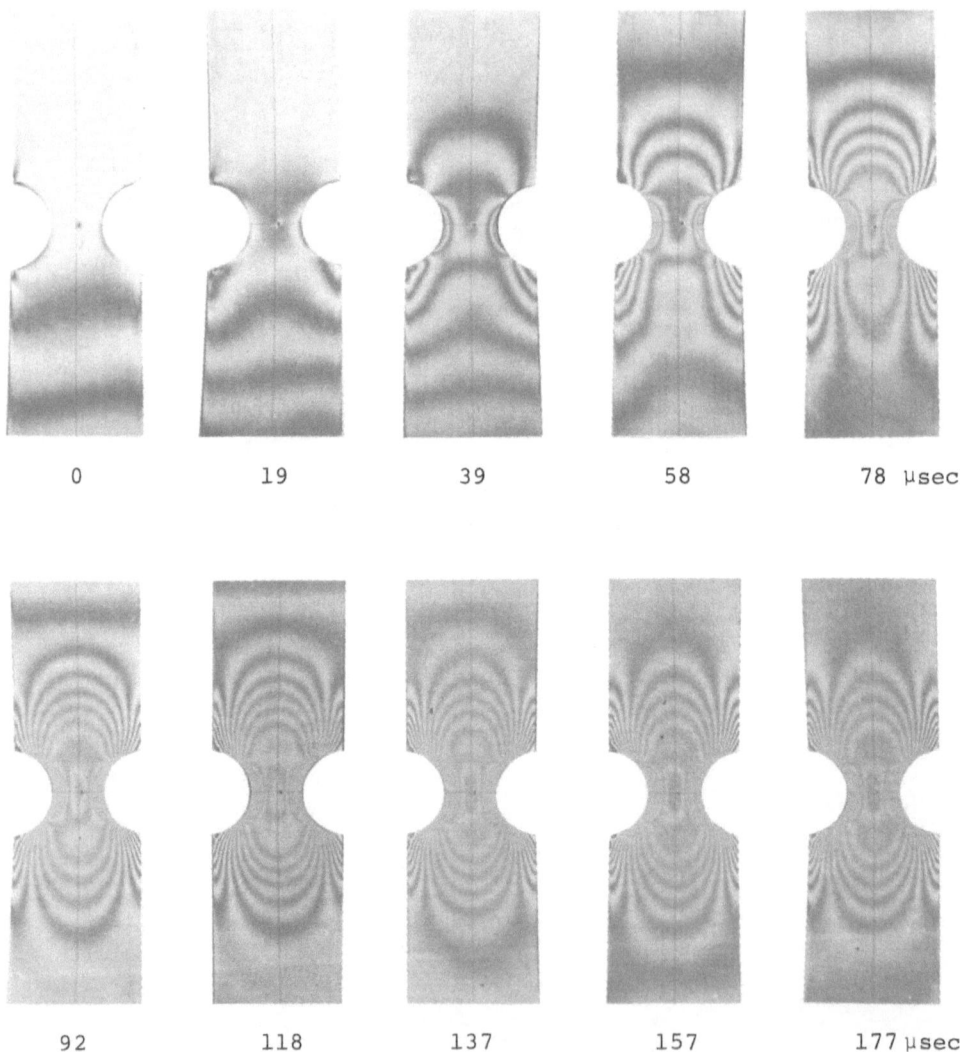

Fig. 5. High speed photoelastic fringe patterns for dynamic tension
of epoxy strip specimen with U-shaped notches at both sides   R = 15 mm

EXPERIMENTAL RESULTS BY HIGH SPEED PHOTOELASTICITY

Examples of obtained high speed photoelastic fringe patterns are shown
in Figs. 4 and 5. In these photographs, stress wave propagating
towards upside from lower side is shown. The behaviour of stress
concentration varying with time is clearly observed. The position of
maximum stress σ_{max} is located on the free boundary of notched area
and the phenomenon of deviation of the maximum stress point towards
the upper stream side at the passage of stress wave front, is found.
When time elapses, the position of maximum stress moves to the narrow-
est part of notch, that is, the same position with in static case.
The plots of dynamic stress concentration factor $f_d = \sigma_{max}/\sigma_0$ vs time,
compared with the corresponding static stress concentration factor f_s,
are shown in Figs. 6 and 7. f_d increases with time and then tends
to f_s. This tendency is observed for both of the cases of a central
circular hole and of U-shaped notches at both sides. The measured
cases are seven cases shown in Table 1.

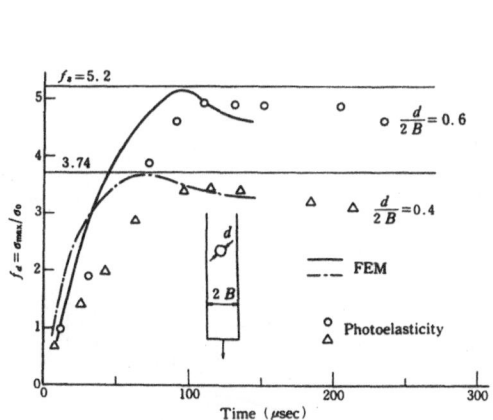

Fig. 6. Variation of dynamic
stress concentration factor with
time in dynamic tension of very
long strip with a central circular hole. Results by photoelasticity and FEM.

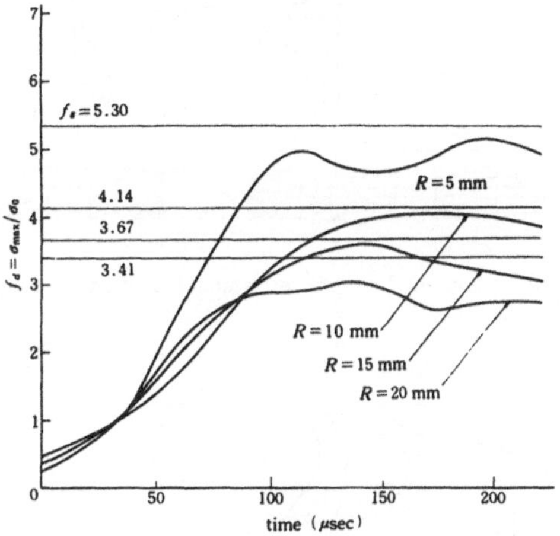

Fig. 7. Variation of dynamic
stress concentration factor with
time in dynamic tension of very
long strip with U-shaped notches.
Results by photoelasticity.

Table 1. Analyzed cases of dynamic tension (o : analyzed)

Notch		Static stress concentration factor	Dynamic analysis	
			Photo-elasticity	FEM
Central circular hole	d/2B 0.6 0.4	5.2 3.74	o o	o o
U-shaped notches at both sides	R(mm) R/2B 2.5 0.05 5 0.1 10 0.2 15 0.3 20 0.4	6.98 5.30 4.16 3.67 3.41	– o o o o	o o o – o

NUMERICAL RESULTS BY FEM

The results for two cases of d/B = 0.4 and 0.6 (for a circular hole), and for four cases of radius of curvature of notch R = 2.5, 5. 10 and 20 mm (for U-shaped notches at both sides), are shown in Figs. 6 and 10. The results of FEM analysis well support the experimental results by high speed photoelasticity. That is, the phenomenon of deviation of the maximum stress point towards the upper stream side at the passage of stress wave front and the tendency for f_d of increasing with time and tending to the corresponding f_s at large time region, are clearly obtained. The rate of tending of f_d to f_s seems faster in FEM results than in experimental results at first sight. This seems to be based upon the fact that the rise time of input wave is about 100 μsec in the experiment, and it is considered understandable.

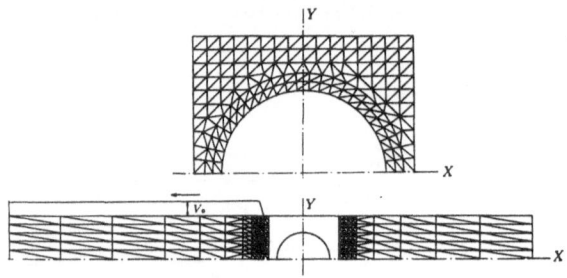

Fig. 8. Example of elements dividing

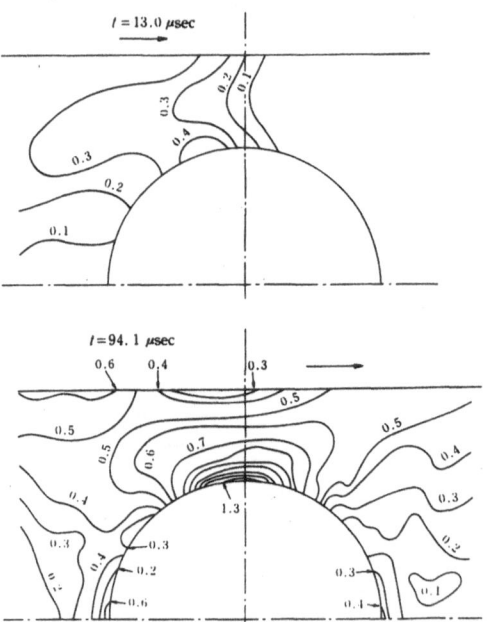

Fig. 9. Example of FEM results, showing calculated isochromatics. Maximum stress point varies with time on the free boundary: edge of hole.

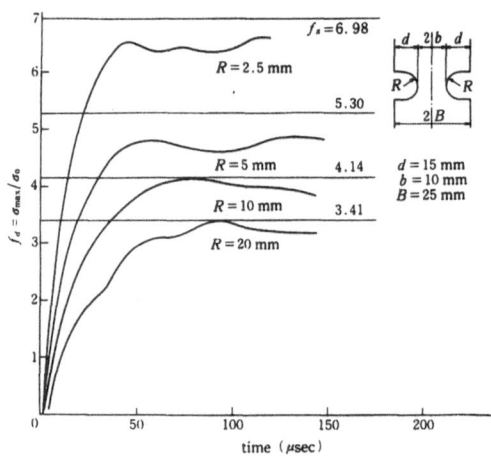

Fig. 10. Variation of dynamic stress concentration factor with time in dynamic tension of very long strip with U-shaped notches. Results by FEM.

CONCLUSION

Summarizing these results obtained, the following conclusions are obtained:

(1) Using a multiple spark gap camera, many series of 16 frames of large high speed photoelastic fringe patterns, having the same sharpness with static photoelastic fringe patterns, are obtained for dynamic uniaxial tension of very long hard epoxy strips with a central circular hole or U-shaped notches at both sides. This experimental method is the most suitable one to analyze dynamic stress patterns of the case in which elastic wave velocity is high as in hard epoxy resin.

(2) Using together measurements by wire strain gage, the materials constants necessary for high speed photoelastic stress analysis: elastic longitudinal wave velocity C, dynamic photoelastic sensitivity α_d, and dynamic elastic constant E_d are obtained.

(3) Dynamic uniaxial tension of very long epoxy strips with a central circular hole or U-shaped notches at the both sides (static stress concentration factor $f_s = 6.98$ to 3.41) is studied for the incidence of an approximate rectangular stress wave with time duration of 186 μsec of stress plateau (having some rise part). From the high speed photoelastic fringe patterns showing dynamic stress concentration for the above cases, the next behaviours are found:

    1) The phenomenon of the deviation of the maximum stress point on the free boundary of notched part towards upstream side, at the passage of incident stress wave front.

    2) Dynamic stress concentration factor increases with time and tends to the corresponding static stress concentration factor.

    3) The overshoot of f_d to above f_s at the initial part of f_d vs time relation may occur for the case with very short rise time of incident wave, but is not observed for the incident wave of Fig. 3.

(4) The numerical results by FEM and Newmark's β method well support the experimental results.

(5) The above mentioned results prove the appropriateness of the approximate theoretical analysis method by one-dimensionalization for dynamic stress concentration in dynamic tension of long strip with notches by Kawata.

REFERENCES

Kawata K, Hashimoto S, Hondo A, Ide T (1981) On the dynamic stress concentration of very long strips with notches under high velocity tension. Bull. Institute of Space and Aeronautical Sciences, University of Tokyo 17: 427-447

Kawata K, Hashimoto S (1965) On an analysis of dynamic stress concentration caused by an elastic wave with long stress plateau. ibid. 1: 69-102

Kawata K, Hashimoto S (1972) On the dynamic stress concentration in long elastic bars with notches under dynamic tension. ibid. 8: 377-384

Analysis of Impact Bending of Cantilever with Various Depth/Span Ratios by Means of High-Speed Photoelasticity

S. Hashimoto[1] and K. Kawata[2]

[1] Institute of Interdisciplinary Research, Faculty of Engineering, University of Tokyo, Hongo, Bunkyo-ku, Tokyo, 113 Japan
[2] Faculty of Science and Technology, Science University of Tokyo, Noda, Chiba, 278 Japan

ABSTRACT

Effect of Depth/Span ratio on bending stress wave propagation and on dynamic stress concentration factor in cantilever beam (dynamic load factor, DLF) under transverse impact load was studied by means of high-speed photoelasticity. The photoelastic isochromatics for the entire impact duration were obtained, and propagation of bending stress waves was investigated. It was found that the upper limit of depth/span ratio for generating bending waves was about 0.57. Positive stress produced at the lower edge of the fixed end by stress wave going ahead of bending wave decreases with increasing h/l and at h/l=0.55 and 0.91, this phenomenon is not observed. The results on DLF obtained from experiment were compared with theoretical solution by one dimensional equation for free vibrations of a beam in which transverse shear and rotary inertia were neglected. It was also shown that experimental results for DLF fell between theoretical solutions with the assumption of viscoelastic material and elastic material.

INTRODUCTION

Numerous investigators have studied elastic beams under impact load. There are mainly two types of theoretical approaches for elastic beam: the one based on the equation for forced vibrations (Timoshenko 1937, Lee 1940, Dengler 1951, Eringen 1953, Goland 1955, Barnhart 1957, Goldsmith 1960, Doyle 1984), and the other for the free vibrations (Timoshenko 1937, Goldsmith 1960, Suzuki 1970, Kida 1982). Recently, the studies by finite element method have also been performed. The basic equation for vibrations of beams used in these theoretical approaches is equation (1) in which transverse shear and rotary inertia are neglected,

$$EI\frac{\delta^4 W}{\delta x^4} + \rho A\frac{\delta^2 W}{\delta t^2} = F(t) \quad \text{----forced vibrations} \\ = 0 \quad \text{----free vibrations} \tag{1}$$

where E is the elastic constant, I the moment of inertia of cross section, ρ the density, A the beam cross sectional area, and W the transverse displacement. The equation in which transverse shear and rotary inertia are not neglected (known as the Timoshenko equation) has also been used (Timoshenko 1937, dengler 1951, goldsmith 1960). The procedure for solving the Timoshenko equation is the same as for basic equation (1), and it is indicated that these terms are not expected to seriously influence the motion in its early stages. For the present work, investigated time durations were comparable with or longer than the period of the fundamental harmonic of the beam.

Various experimental studies using strain gages have been performed to verify theoretical analysis. To compare these with numerical solutions by equations for forced vibrations, the experiments on a supported beam or elastically mounted beam subjected to transverse impact on its center have been investigated (Goland 1955, Cunningham 1956, Goldsmith 1956, 1958). For the solutions by the equation for free vibrations, the experiments for cantilever impacted at free end (Suzuki 1970) and for supported beam impacted at its center have been investigated. Elastic beam theories have also been applied to explain brittle fracture of beams (Menkes 1973, Freund 1976, Kida 1982).

Experimental studies using photoelasticity have also been performed to verify elastic beam theories (Tuzi 1935, Bester 1957, Goldsmith 1958, Clark 1970, 1972). The paper published by Tuzi and Nisida (1935) in which photoelastic experiment on elastic beam was performed with streak photography is the first study by high-speed photoelasticity. Since 1958 high-speed photoelastic experiments have been performed mainly by framing cameras. In dynamic case it is very difficult to obtain data that is directly applicable to practical construction as in the static case.

The purpose of the present study is to investigate the effect of depth/span ratio on dynamic stress distribution and on bending stress wave propagation of a cantilever under transverse impact by means of high-speed photoelasticity using a multiple spark gap camera.

EXPERIMENT

Specimens were prepared using a 6mm thick epoxy plate. The distance from the fixed end to the impact point was 165mm, and this length was held constant in all tested beams. Depths of beams h were 20, 30, 40, 70, 90, and 150mm, and the effects of depth/ span ratio on stress distribution and on DLF were investigated. Impacters were steel bars with square cross section and with 6mm diameter spherical tip. Impact speed was varied from 1m/s to 5m/s. In this range, the beam was not fractured. The epoxy plate used in the tests was dynamically characterized by the stress wave method (Sogabe 1981, Hashimoto 1984). Dynamic photoelastic sensitivity (or Dynamic stress-fringe value) was obtained by simultaneous measurements with strain gage output and photoelastic fringes. These values are shown in Table 1.

Table 1   Characterization of epoxy resin

E_s(Pa)	α_s(1/Pa·m) (λ=546nm)	E_d(Pa)	α_d(1/Pa·m) (λ=546nm)	ρ(kg/m3)
$3.4*10^9$	$\approx 1.0*10^{-4}$	$3.5*10^9$	$\approx 1.0*10^{-4}$	$1.22*10^3$

Three element model			
E(Pa)	K	η(Pa·s)	
$3.4*10^9$	0.15	$1.0*10^4$	

Notation
E Young's modulus α photoelastic sensitivity
ρ density   K,η shown in the attached figure
suffix s static   suffix d dynamic

The schematic diagram of the experimental arrangement is shown in Fig. 1. A 16-frame multiple spark gap camera was used. Data from a strain gage cemented 5mm from fixed end (X=5mm) was simultaneously recorded with high-speed photoelastic photography. Figuer 2 shows the high-speed photoelastic isochromatics for the full duration of impact of the beam of h/l = 0.12 and time variation plots of stress distribution along

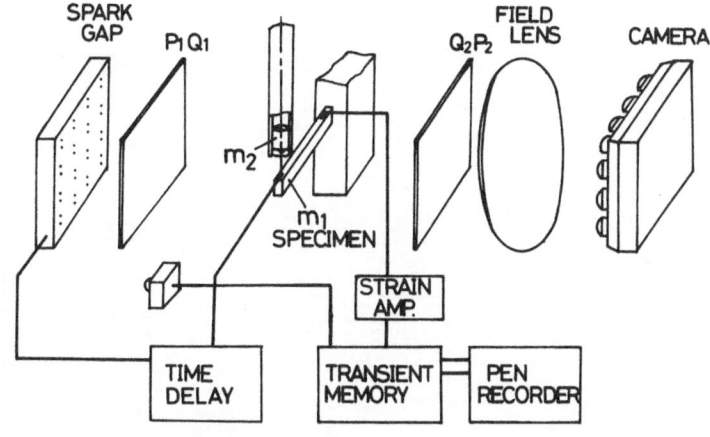

Fig. 1   Schematic diagram of experimental apparatus

upper and lower edges in the beam.   Figure 3 shows the photoelastic isochromatics obtained for systematically varied depth/span ratios.   Figure 4 shows the time variation plots of the stress distribution at the upper and lower edges of these beams.   Figure 5 shows the fringe order (principal stress difference $\sigma_1-\sigma_2=2\tau_m$, where τ_m is maximum shear stress) along the section at middle of span.   Figure 6 illustrates the bending wave pattern (left) and fringe order distribution along the section (right) where this bending pattern first appeared (with distance of "a" millimeter from the impact point).   Figure 7 shows the relation of maximum stress obtained from strain gage records vs. kinetic energy of impacters. Figure 8 shows the relation of DLF vs. mass ratio at x = 5mm (from strain   gage)   and   fig.   9 shows   the   same   for   x   =   0mm   (from photoelasticity).

DISCUSSION

An impressive change of stress distribution form with the variation of h/l is observed. It can be seen in Figs. 2 and 4 that positive stress is produced at the lower edge of the fixed end by stress waves going ahead of the bending wave.   This stress decreases with increasing h/l,   and   it can not be seen in the beams of h/l=0.55 and 0.91.   The bending   wave   pattern   illustrated   in   Fig.   6   first   appears at a distance of "a" mm from impact point. This distance "a" is nearly equal to "h", and "b" is about 1.5 times of "h". From the above investigation,   it   can be concluded that the lower limit of span for which the bending wave pattern can be produced is about 1.75 times of "h".   It is expected that if h/l becomes larger than 0.57 (consistent with   l=1.75h),   then   the   one   dimensional   beam   equation   is   not applicable.   The experimental results from the strain gage are shown in Fig. 8 and those for photoelastic isochromatics are shown in Fig. 9.   The   term DLF is the dynamic load factor, that is, the ratio of dynamic   maximum   moment to static maximum moment under static load of the impacter.

DLF = Md,max / Ms,max   (2)

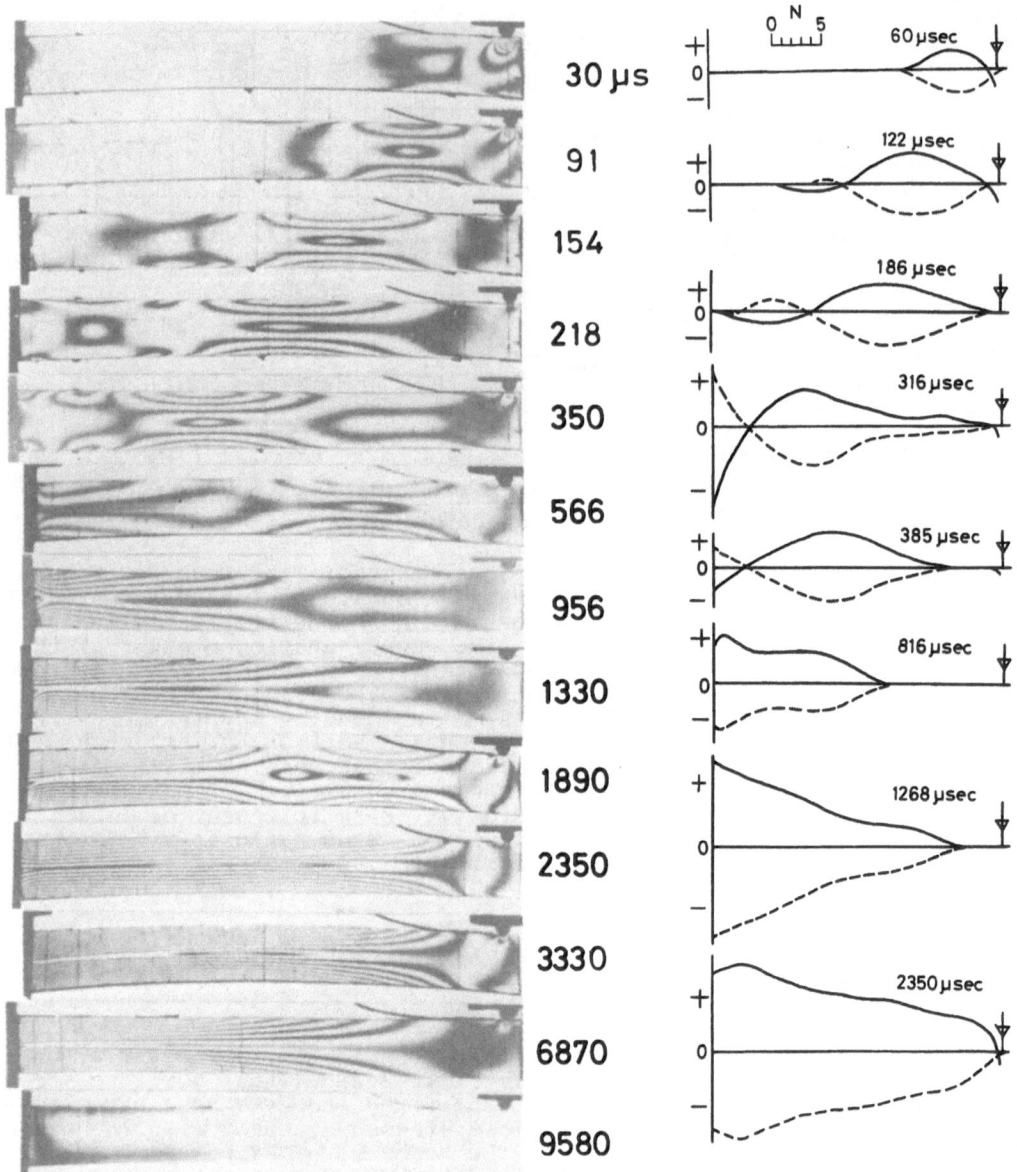

Fig. 2 Photoelastic fringe pattern of beams h/l=0.1,2, and
        stress distribution along upper (solid line) and
        lower (dotted line) edge

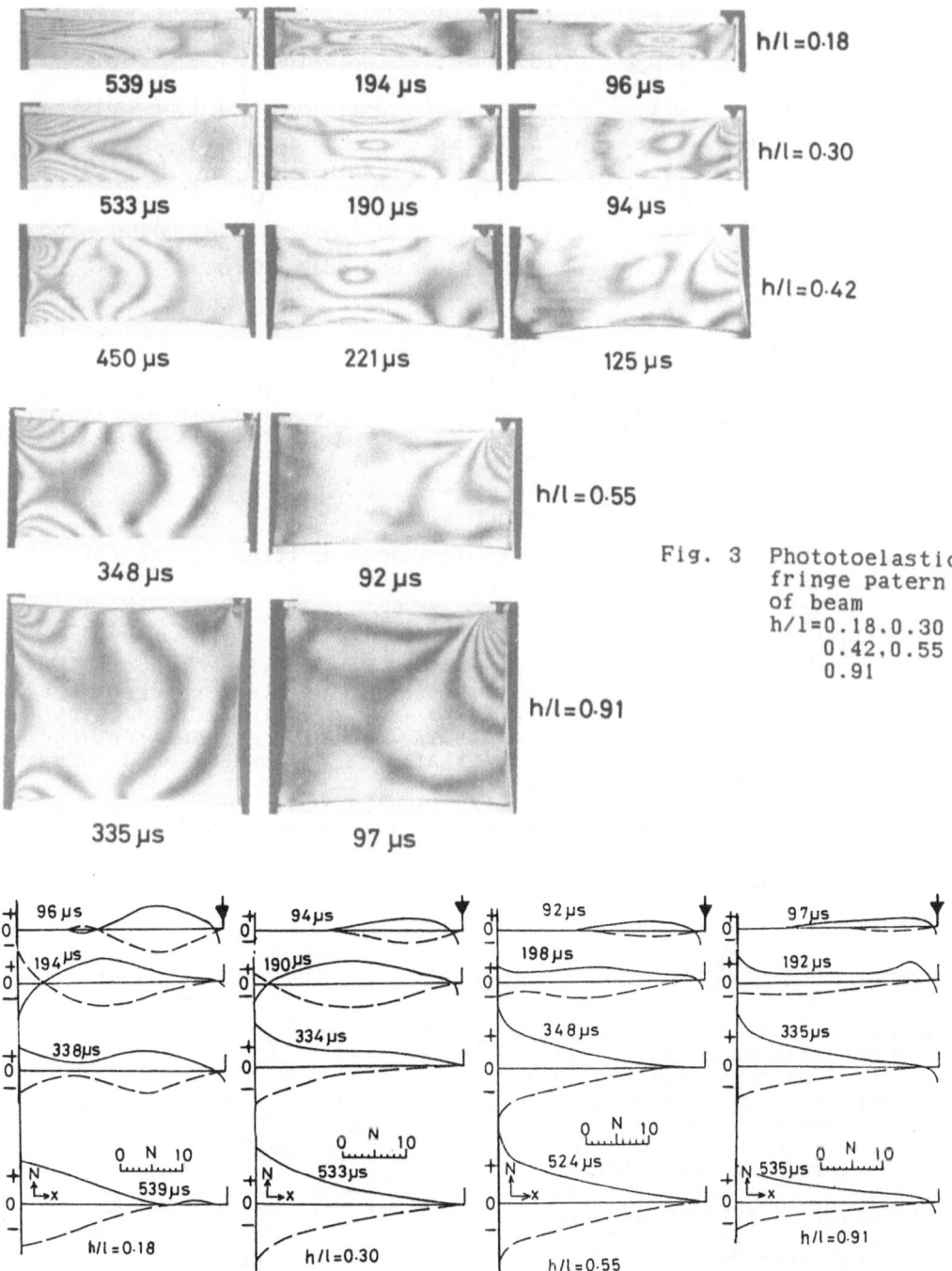

539 µs   194 µs   96 µs   h/l = 0.18

533 µs   190 µs   94 µs   h/l = 0.30

450 µs   221 µs   125 µs   h/l = 0.42

h/l = 0.55

348 µs   92 µs

Fig. 3   Phototoelastic
fringe patern
of beam
h/l=0.18,0.30
0.42,0.55
0.91

h/l=0.91

335 µs   97 µs

Fig. 4   Stress distribution along the upper (solid line) and the
lower (dotted line) edges of beams   h/l=0.18,0.30,0.55,0.91

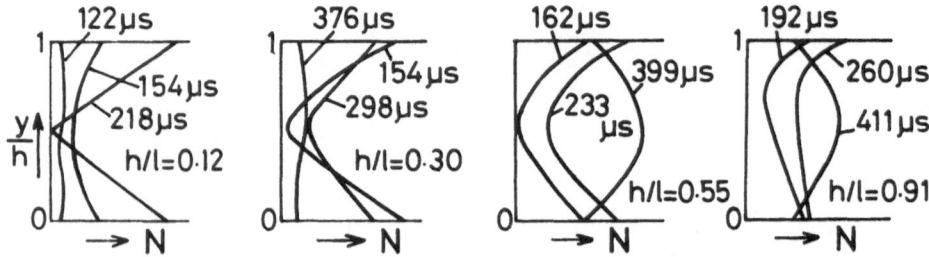

Fig. 5 Fringe order along the section at middle of span

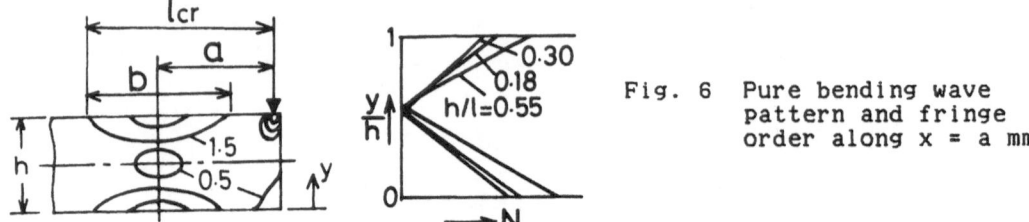

Fig. 6 Pure bending wave pattern and fringe order along x = a mm

In the elastic solution, the result can be written in the following expression:

$$F = DLF / \frac{v}{gl^2}\sqrt{\frac{EI}{\rho A}} = f(R) \qquad (3)$$

where $R=m_1/m_2$, m_1=mass of beam, m_2=mass of impacter, v=impact speed, g=acceleration of gravity, E=modulus of elasticity, I=moment of inertia of cross section, ρ=density, A=area of cross section, and l= span of beam. It is evident that F is dependent on R only and independent of the beam material.

In the viscoelastic solution, the result can be expressed as follow:

$$F = f(R,\beta,K) \qquad (4)$$

where

$$\beta = \frac{\eta}{EI^2}\sqrt{\frac{EI}{\rho A}} = (\frac{h}{l^2})\frac{\eta}{E}\sqrt{\frac{EI}{12}} . \qquad (5)$$

Then F is not only dependent on R but also on β and K. If the beam material and (h/l^2) are constant, F depends on R only.   When R is smaller than 0.5, the viscoelastic solution agrees well with experimental results.   The   result   of photoelastic   experiment   is applicable   to   common structural materials when R is nearly equal   or smaller than 0.5.   As R becomes larger than 0.5,   the   effect   of viscoelasticity   becomes significant.

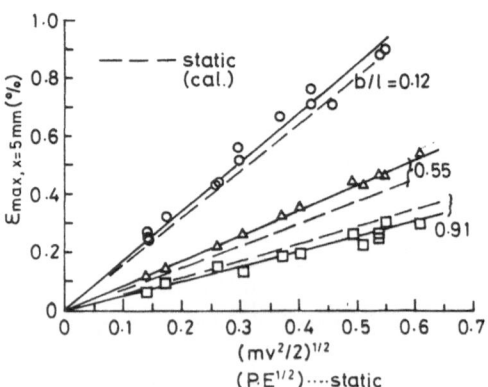

Fig. 7 Relation of maximum strain vs. kinetic energy of impacters

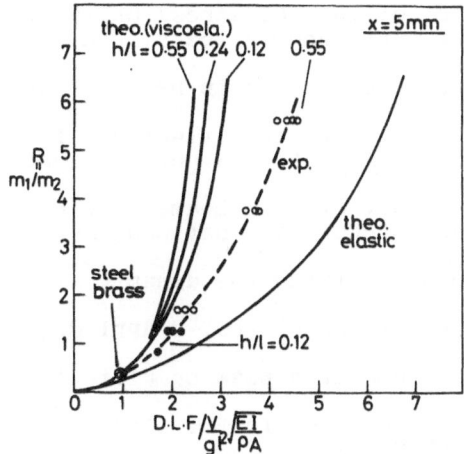

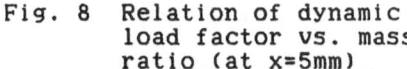

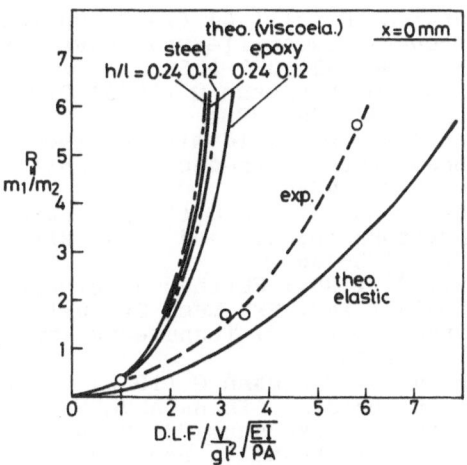

Fig. 8 Relation of dynamic
        load factor vs. mass
        ratio (at x=5mm)

Fig. 9 Relation of dynamic
        load factor vs. mass
        ratio (at x=0mm)

CONCLUSION

The photoelastic fringe pattern in a cantilever under transverse
impact that shows propagating and reflecting behavior and stress
concentration behavior was obtained for the full duration of impact.
A significant change of stress distribution with the variation of h/l
is observed. Positive stress produced at lower edge of the fixed end
by the stress wave preceding the bending wave decreases with
increasing h/l, and at h/l=0.55 and 0.91, this phenomenon is not
observed. It was found that the lower limit of depth/span ratio for
which the bending wave can be generated by transverse impact was about
0.57. It can be seen that the stress field in cantilever under impact
load varies according to depth/span ratio from the field where bending
stress is dominant to the field where shear stress is dominant. The
experimental result for DLF falls between the theoretical solution
for viscoelastic material with three element and for elastic
material. The relation σmax to root of (1/2)mv² was linear. These
proportional constants were not only dependent on beam dimension
(h/l²) but also beam materials. The results of dynamic photoelastic
experiments are applicable to common structural materials when R is
nearly equal to or smaller than 0.5.

REFERENCES

Barnhart KE, Goldsmith W (1957) Stresses in beams during transverse
    impact. J Appl Mech 24:440-446
Betser AA, Frocht MM (1957) A photoelastic study of maximum tensile
    stresses in simply supported short beams under central transverse
    impact. J Appl Mech 24:509-514
Clark JA. Durelli AJ (1970) Optical stress analysis of flexural waves
    in a bar. Trans ASME E37:331-338

Clark JA, Durelli AJ (1972) On the effect of initial stress on the propagation of flexural waves in elastic rectangular bars. J Acous Soc Amer 52:1077-1086

Colton LD, Herrmann G (1975) Dynamic fracture process in beams. Trans ASME E42:435-439

Cunningham DM, Goldsmith W (1956) An experimental investigation of beam stresses produced by oblique impact of a steel sphere. J Appl Mech 23:606-611

Denglar MA, Goland M (1951) Transverse impact of long beams, including rotatory inertia and shear effects. Proc first national cong Appl Mech 179-186

Doyle JF (1984) Further developments in determining the dynamic contact law. Exp Mech 24:10-16

Eringen AC (1953) Transverse impact on beams and plates. J Appl Mech 20:461-468

Freund LB, Herrmann G (1976) Dynamic fracture of a beam or plate in plane bending. Trans ASME E43:112-116

Goland M, Wickersham PD, Dengler MA (1955) Propagation of elastic impact in beams in bending. J Appl Mech 22:1-7

Goldsmith W, Cunningham DM (1956) An experimental investigation of the oblique impact of spheres upon simply supported steel beams. Proc SESA 14:171-180

Goldsmith W, Norris GW (1958) Stresses in curved beams due to transverse impact. Proc 3rd US National Cong Appl Mech 153-162

Goldsmith W (1960) Impact. London Edward Arnold Ltd P54-137

Hashimoto S, Kawata K (1984) On the effect of model material's viscosity in high-speed photoelasticity. Proc 6th Japan Soc Photoelasticity 97-100

kida s, Oda J (1982) On fracture behavior of brittle cantilever beam subjected to lateral impact load. Exp Mech 22:69-74

Lee EH (1940) The impact of a mass striking a beam. J Appl Mech 5:A129-A138

Menkes SB, Opat HJ (1973) Broken Beams. Exp Mech 13:480-486

Sogabe Y, Kishida K, Nakagawa K (1981) Study of damping characteristics of a high-damping alloy by means of stress wave propagation. Proc Japan Soc Mech Eng 47:748-756

Suzuki S (1970) Dynamic stresses of viscoelastic beams produced by rigid impacter of finite velocity. Proc Japan Soc Mech Eng 36:1405 -1412

Timoshenko SP (1937) Vibration problems in engineering. 2nd Edition New York D Van Nostrand Company Inc P331

Tuzi Z. Nisida M (1935) Photo-Elastic Study of Stress due to Impact. Scientific papers of the Institution of Physical and Chemical Research. Japan 26:277-309

Photoelastic Analysis of Dynamic Fracture Behavior

James W. Dally

Mechanical Engineering Department, University of Maryland, College Park, MD 20742, USA

INTRODUCTION

Since Tuzi (1928) initiated research in dynamic photoelasticity, the scientific community has witnessed a continuous development of this experimental method and an ever increasing range of application to major engineering problems. Dynamic fracture is one of the more contemporary engineering applications of photoelasticity. The development of this technical area has been delayed since Tuzi first showed that photoelasticity was an effective method for studying dynamic problems, for several reasons. Firstly, fracture mechanics did not evolve as a descipline with a suitable theoretical base until the 1950's. Secondly, the high-speed equipment, both electronic and photographic, necessary to conduct meaningful experiments was not available until the mid 1960's. Finally, the interest in the problem was deferred until the phenomena of crack initiation was well understood and engineers became concerned with structural integrity after crack initiation.

This paper describes the application of dynamic photoelasticity to the study of crack propagation with emphasis on crack arrest, crack branching and the relation between the stress intensity factor and the crack velocity. Since recording the isochromatic fringe pattern is paramount in any dynamic fracture experiment, the methods of high speed photography are covered in considerable detail. Next, a method for determining the stress intensity factor from the isochromatic fringe loops near the crack tip is described. This method utilizes the whole field potential of photoelasticity and a large number of data points are used to establish 4 to 10 unknown coefficients in the series representation of the stress field. The effect of the crack tip velocity $\dot{a}$ is taken into account by employing a dynamic solution (constant crack velocity) for the stress field. Results obtained from a series of experiments with Homalite 100 are presented in terms of the relation between K and $\dot{a}$. Two dynamic fracture properties which include the crack arrest toughness K_{Im} and the crack branching toughness K_{Ib} are defined. Finally, a discussion of different results and conclusions pertaining to the uniqueness of the K-a relation is presented.

HIGH SPEED PHOTOELASTIC RECORDING SYSTEMS

There are three different high speed photographic systems which can be utilized to record isochromatic fringes associated with dynamic events such as stress wave propagation or propagating cracks. These systems include the framing camera, the Cranz-Schardin multiple lens camera and the pulsed ruby laser. All three of these systems are capable of photographing high density fringe patterns propagating at velocities as high as 100,000 in/s (2500 m/s) but each system exhibits advantages and disadvantages. Flynn (1964 and 1966) demonstrated the use of a framing camera to simultaneously record dynamic isochromatic fringes in both normal and oblique incidence.

The multiple spark/lens camera developed originally by Cranz and Schardin (1929) has been widely used for many years in ballistics, shock wave studies and fracture observations. Christie (1955) adapted the Cranz Schardin camera to dynamic photoelasticity and showed high quality fringe photographs of both dilatational and distortional type stress waves. Wells and Post (1957) were the first to use this system in the U.S. in an early study of crack propagation. The Cranz-Schardin system is widely used today throughout the world and is the most popular recording system for dynamic photoelasticity, dynamic fracture and caustic studies. The system is moderate in cost, easy to operate, exhibits long life with little maintenance and can be closely synchronized with the dynamic event to be recorded. A typical dynamic fringe pattern recorded with this system presented in Fig. 1 shows the dilatational, distortional, von-Schmidt and Rayleigh waves propagating in a half-plane subjected to a load produced by a small explosive load on the boundary. The recording characteristics of the Cranz-Schardin system include: 1) framing rates from 20,000 to 800,000 frames/sec which can be varied between frames, 2) an adequate number of frames, 16 to 24, 3) dynamic resolution exceeding 10^6 lines/sec, 4) effective exposure times of 500 ns, 5) large field of view, 300 to 500 mm and 6) a large image size of 50 mm.

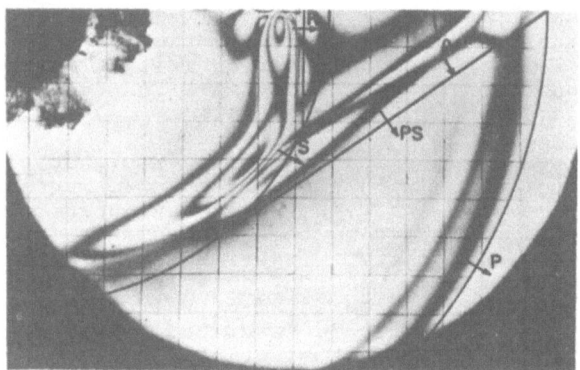

Fig. 1   Dynamic fringe pattern showing stress waves of several types propagating in an explosively loaded half-plane.

The third method of recording involves the use of a Q-switched laser to produce an extremely short duration but an intense pulse of monochromatic light. Taylor and associates (1968, 1975) have adapted a single sequentially pulsed ruby laser to a photoelastic bench. In one system developed by Taylor (1968) a streak camera was employed to record the image. In a second system Taylor (1975) used an acousto-optical deflector with a modified Cranz-Schardin optical bench to record the images. More recently Dally and Sanford (1982) developed a multiple ruby laser system where the traditional spark gaps were replaced with Q-switched ruby lasers. Fiber optic light guides were employed to transmit the light from the laser to the optical bench.

The quality of the dynamic isochromatic fringe patterns produced with a Q-switch ruby laser is outstanding as indicated by the photograph shown in Fig. 2. The very short exposure time, 30 ns and the narrow bandwidth of the light output provides a very high contrast ratio for the exposure. The use of the laser system for dynamic recording in photoelasticity is limited by the high cost of the optical elements and the associated electronics and the complexity of the systems.

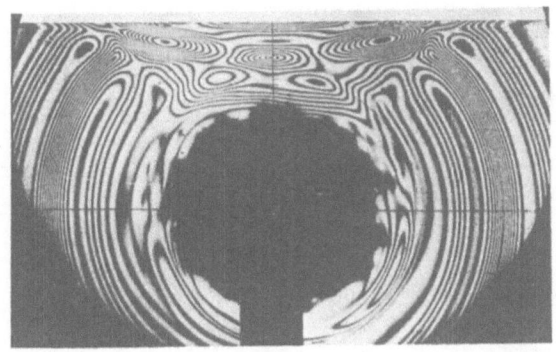

Fig. 2   Dynamic isochromatic fringe pattern recorded with a Q-switched ruby laser.

Modifications to the basic Cranz-Schardin system continue to be made to improve its performance. Dally and Sanford (1982) have used fiber optics for light transmission to the optical bench. The use of fiber optics permits the multiple light sources to be placed on close centers and eliminates the hazard of high voltage on the optical bench. More recently Kawata et al (1985) placed a 50 mm focal length lens near the light source to control the expansion of the beam of light to match the diameter of the field lens. This source lens enhanced the intensity of the light delivered through the optical arrangement to the film, and permitted the Cranz-Schardin system to be employed in the reflective mode.

STRESS INTENSITY FACTOR K_I FROM ISOCHROMATIC FRINGES

Irwin (1957) first described a method for determining the opening mode stress intensity factor K_I from the isochromatic fringe loops near the tip of a crack. Irwin used a two parameter approach, modifying the Westergaard (1939) solution, to describe the stress field, and showed equations for two parameters K_I and σ_{ox} in terms of the fringe order N and the coordinates r_m and θ_m of a single point on a given fringe loop located the maximum distance from the crack tip. Other investigators including Bradley and Kobayashi (1970), Schroedl and Smith (1973) and Etheridge and Dally (1970) have modified Irwin's method without significantly changing the basic approach of describing the stress field with only two or three parameters and determining K_I with deterministic methods involving only a small portion of the photoelastic data available.

A more complete approach is to express the stress field in terms of a higher order theory described by Sanford (1979) where

$$\sigma_{xx} = \sum_{n=0}^{N} A_n r^{n-1/2} \left[\cos(n-1/2)\theta - (n-1/2)\sin\theta\sin(n-3/2)\theta \right]$$

$$+ \sum_{m=0}^{M} B_m r^m \left[2\cos(m\theta) - m\sin\theta\sin(m-1)\theta \right] \tag{1}$$

$$\sigma_{yy} = \sum_{n=0}^{M} A_n r^{n-1/2} \left[\cos(n-1/2)\theta + (n-1/2)\sin\theta\sin(n-3/2)\theta \right]$$

$$+ \sum_{m=0}^{M} m B_m r^m \sin\theta\sin(m-1)\theta \tag{2}$$

$$\tau_{xy} = \sum_{n=0}^{N} -(n-1/2) A_n r^{n-1/2} \sin\theta\cos(n-3/2)\theta$$

$$- \sum_{m=0}^{M} B_m r^m \left[m\sin\theta\cos(m-1)\theta + \sin(m\theta) \right] \tag{3}$$

where r and θ are polar coordinates with the origin at the crack tip. These equations describe the stress field in terms of unknown coefficients A_O, A_1, ... A_N and B_O, B_1, ..., B_M and represent a multiple parameter theory which can be extended to more adequately match theoretical and experimental results.

Photoelastic data is taken from the local field about the crack tip to provide N_p, r_p, θ_p at : data points where $P > (N)(M)$. To utilize this data in determining the unknown coefficients the stress optic law is employed to relate the stress field to the fringe order.

$$\tau_m = \frac{\sigma_1 - \sigma_2}{2} = \frac{Nf_\sigma}{2h} \tag{4}$$

where f_σ is the material fringe value in tension
h is the model thickness.

Noting that

$$\left[(\sigma_1 - \sigma_2)/2 \right]^2 = \left[(\sigma_{yy} - \sigma_{xx})/2 \right]^2 + \tau_{xy}^2 \tag{5}$$

and then combining Eqns. (4) and (5) leads to

$$D^2 + T^2 = \left(\frac{Nf_\sigma}{2h} \right)^2 \tag{6}$$

where

$$D = (\sigma_{yy} - \sigma_{xx})/2 = \sum_{n=0}^{N} A_n (n-1/2) r^{n-1/2} \sin\theta\sin(n-3/2)\theta$$

$$+ \sum_{m=0}^{M} B_m r^m \left[m\sin\theta\sin(m-1)\theta - \cos(m\theta) \right] \tag{7}$$

$$T = \tau_{xy} = \sum_{n=0}^{N} - A_n(n-1/2)r^{n-1/2}\sin\theta\cos(n-3/2)\theta$$

$$- \sum_{m=0}^{M} B_m r^m[m\sin\theta\cos(m-1)\theta + \sin m\theta] \qquad (8)$$

Sanford and Dally (1979) introduced an overdeterministic method for the solution of Eqn. (6) which utilizes the full field data available from photoelasticity. This method involves least squares and the Newton-Raphson approach to give an iterative procedure which can be implemented with a simple computation on a computer.

Consider a set of functions g_k defined with respect to Eqn. (6) so that

$$g_k(A_o \cdots A_N, B_o \cdots B_N) = D^2 + T^2 - (\frac{Nf\sigma}{2h})^2 = 0 \qquad (9)$$

where $k = 1,2,\ldots P$ the total number of data points.

Next, consider a Taylor's series expansion of Eqn. (9) to obtain

$$(g_k)_{i+1} = (g_k)_i + [\frac{\partial g_k}{\partial A_o}]_i \Delta A_o + \cdots + [\frac{\partial g_k}{\partial A_N}]_i \Delta A_N$$

$$+ [\frac{\partial g_k}{\partial B_o}]_i \Delta B_o + \cdots + [\frac{\partial g_k}{\partial B_M}]_i \Delta B_M \qquad (10)$$

where i refers to the ith iteration step and ΔA_n and ΔB_m are corrections of the previous estimates of the unknown coefficients. The corrections ΔA_n and ΔB_m are determined by requiring $(g_k)_{i+1} = 0$ which leads to a set of equations which can be expressed in matrix notation as

$$[g] = [c][\delta] \qquad (11)$$

where

$$[g] = - \begin{bmatrix} g_1 \\ -- \\ g_P \end{bmatrix} \quad \delta = \begin{bmatrix} \Delta A_o \\ --- \\ \Delta A_N \\ \Delta B_o \\ --- \\ \Delta B_M \end{bmatrix} \quad [c] = \begin{bmatrix} \frac{\partial g_1}{\partial A_o} & \cdots & \frac{\partial g_1}{\partial A_N} & \frac{\partial g_1}{\partial B_o} & \cdots & \frac{\partial g_1}{\partial B_M} \\ \cdot & & \cdot & \cdot & & \cdot \\ \cdot & & \cdot & \cdot & & \cdot \\ \frac{\partial g_P}{\partial A_o} & \cdots & \frac{\partial g_P}{\partial A_N} & \frac{\partial g_P}{\partial B_o} & \cdots & \frac{\partial g_P}{\partial B_M} \end{bmatrix}$$

$$(12)$$

From Eqns. (6), (9) and (12), it is evident that the elements the of the matrix [c] are

$$\frac{\partial g_k}{\partial A_n} = 2D_k \left(\frac{\partial D}{\partial A_n}\right)_k + 2T_k \left(\frac{\partial T}{\partial B_m}\right)_k \qquad (13)$$

Since the matrix [c] will not be square as P > N+M the solution of Eqn. (11) requires the application of the least squares method. This is accomplished in matrix form by setting

$$[\delta] = [d]^{-1} [c]^T [g] \qquad (14)$$

where $[d] = [c]^T [c]$ $\qquad (15)$

The procedure for determining A_n, B_n from P data points involves an initial estimate of A_n, B_m and then the computation of the elements of [c] from Eqn. (13) and [g] from Eqn. (9). The correction terms are determined from Eqn. (14) and the initial estimates of A_n, B_m are revised according to

$$(A_n)_{i+1} = (A_n)_i + (\Delta A_n)_i$$

$$(B_m)_{i+1} = (B_m)_i + (\Delta B_m)_i \qquad (16)$$

The process is repeated until the column elements in $[\delta]$ become acceptably small. Convergence is usually rapid and the program required to conduct the numerical procedure is not complex. The number of parameters required depends on the geometry of the body containing the crack. Cottron and Lagarde (1982) have shown a systematic reduction of error in determining K_I by increasing the number of parameters from 1 to 6; however, if the crack tip is near a boundary or a loading point 10 or more terms in the series extension may be required. The adequacy of the match between the series representation and the isochromatic data may be checked by plotting the fringe pattern obtained from Eqn. (6). A comparison of the graph obtained using the coefficients A_n, B_n with the actual fringe pattern as illustrated in Fig. 3 clearly indicates the quality of the solution.

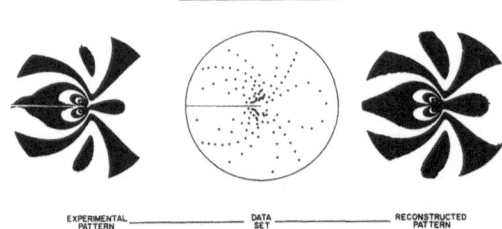

Fig. 3   Comparison of isochromatic pattern from the series solution and the experiment (after Chona and Sanford).

The value of K_I is determined from the coefficients A_o by:

$$K_I = -A_o/\sqrt{2\pi} \qquad (17)$$

Data from the ischromatic fringe pattern should be taken as close to the crack tip as possible to limit the number of terms required in the series representation. However, Rosakis and Ravi-Chander (1984) have shown that the state of stress in the zone (r/h) < 0.5 is three-dimensional and that the plane stress conditions assumed in developing the theory above do not apply. It is important to avoid the zone $(r/h) < 0.5$ in selecting data points p.

EFFECT OF CRACK VELOCITY ON K_I DETERMINATION

In dynamic photoelastic studies the crack is propagating with some velocity $\dot{a}$ and the method described above for determining K_I must be modified to account for the effect of a on the stress field. Irwin (1980) has provided a series solution for the stresses near a crack propagating at a constant velocity of the form

$$
\sigma_{xx} = \frac{1 + \lambda_2^2}{4\lambda_1\lambda_2 - (1+A_2)^2} \left\{ (1 + 2\lambda_1^2 - \lambda_2^2) \operatorname{Re} Z_1 - \frac{4\lambda_1\lambda_2}{1+A_2^2} \operatorname{Re} Z_2 \right\}
$$

$$
+ \frac{1}{\lambda_1^2 - \lambda_2^2} \left\{ (1 + 2\lambda_1^2 - \lambda_2^2) \operatorname{Re} Y_1 - (1 + \lambda_2^2) \operatorname{Re} Y_2 \right\}
\tag{18}
$$

$$
\sigma_{yy} = \frac{1 + \lambda_2^2}{4\lambda_1\lambda_2 - (1+\lambda_2^2)^2} \left\{ -(1+\lambda_1^2) \operatorname{Re} Z_1 \; \frac{4\lambda_1\lambda_2}{1+\lambda_2^2} \operatorname{Re} Z_2 \right\}
$$

$$
+ \frac{1 + \lambda_2^2}{\lambda_1^2 - \lambda_2^2} \left\{ \operatorname{Re} Y_2 - \operatorname{Re} Y_1 \right\}
\tag{19}
$$

$$
\tau_{xy} = \frac{2\lambda_1(1 + \lambda_2)}{4\lambda_1\lambda_2 - (1+\lambda_1^2)^2} (\operatorname{Im} Z_2 - \operatorname{Im} Z_1)
$$

$$
+ \frac{1}{\lambda_1^2 - \lambda_2^2} \left\{ \frac{(1 + \lambda_2)^2}{2\lambda_2} \operatorname{Im} Y_2 - 2\lambda_1 \operatorname{Im} Y_1 \right\}
\tag{20}
$$

where

$$
Z_j = Z(z_j) = \sum_{n=0}^{n=N} C_n z_j^{n-1/2}, \quad j = 1,2
\tag{21}
$$

$$
Y_j = Y(z_j) = \sum_{m=0}^{m=M} D_m z_j^m, \quad j = 1,2
\tag{22}
$$

$$
z_j = x + i\lambda_j y, \quad j = 1,2
\tag{23}
$$

$$
\lambda_j = 1 - (\dot{a}/c_j)^2, \quad j = 1,2
\tag{24}
$$

This solution for σ_{xx}, σ_{yy} and τ_{xy} is used in place of Eqns. (1), (2) and (3) in the overdeterministic method for determining the instantaneous stress intensity factor $K_I(t)$ from isochromatic fringe patterns recorded dynamically during the propagation period for the crack.

DYNAMIC FRACTURE BEHAVIOR OF HOMALITE 100

Photoelastic studies of fracture in Homalite 100 were conducted by T. Kobayashi and Dally (1979) using relatively large single edge notched (SEN) specimens. The loading of these specimens was with pins located along the center line to provide a K which increased with increasing crack length and with a split pin along the crack line to give a K which decreased with crack extension. Sixteen dynamic isochromatic fringe patterns were recorded during the crack propagation period with a Cranz-Schardin camera. The value of K(t) was determined using a two parameter deterministic method because at the time the overdeterministic method was not available. The position of the crack tip a was determined for each frame and the crack velocity å was obtained from the slope of the a(t) graph.

The results obtained from several different experiments with the two types of SEN specimens are shown in Fig. 4. The curve in this graph of K(t) as a function of a represents an average of about 100 data points. Examination of the K-å relation shows several characteristic features of dynamic fracture behavior which have been observed in several different materials by other investigators including Kalthoff (1983), Rosakis et al (1983) and Kanazawa et al (1981). Firstly, there is a minimum value of K below which the crack cannot propagate. This minimum value $K_{Im} = 0.418$ MPa$\sqrt{m}$ represents the arrest toughness of Homalite 100. The arrest toughness for Homalite 100 is slightly less than the initiation toughness and $K_{Im}/K_{Ic} = 0.94$.

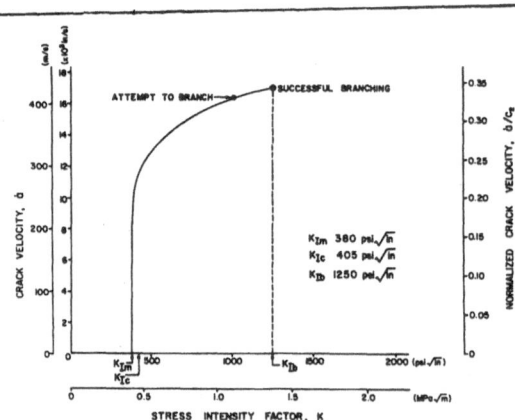

Fig. 4 Crack velocity å as a function of K(t) for Homalite 100.

Secondly, the K-å curve exhibits a nearly vertical stem which indicates that a value of K(t) slightly larger than K_{Im} results in relatively high crack velocities with a approaching 250 m/s. A transition region exists between 250 < å < 400 m/s where the value of K must be increased by a factor of more than 2 to increase the velocity from 250 to 400 m/s. The last portion of the K-å curve where 400 < å < 430 m/s is relatively flat and large increases in K are necessary to achieve small increases in å. In this third region of the K-å relation, the fracture surface is extremely rough indicating a large fracture process zone and the onset of crack instabilities associated with branching. Successful branching, illustrated in Fig. 5, occurs at $K = 1.38$ MPa$\sqrt{m}$ and a crack velocity a = 432 m/s. For the SEN specimen, these results indicate that the branching toughness of Homalite 100 is $K_{Ib} = 1.38$ MPa$\sqrt{m}$ and the ratio of the terminal velocity to the shear wave velocity $\dot{a}_T/c_2 = 0.35$.

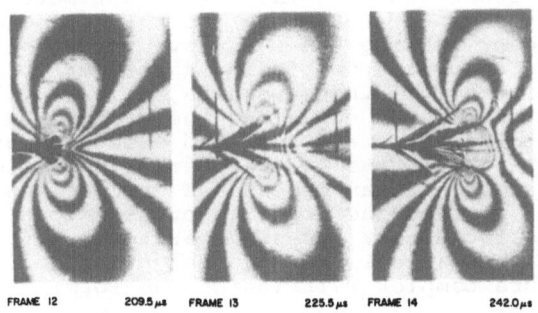

FRAME 12   209.5 μs   FRAME 13   225.5 μs   FRAME 14   242.0 μs

Fig. 5   Isochromatic patterns showing the instabilities in the crack front during the branching process.

Finally the ratio $K_{Im}/K_{Ib} = 3.3$ indicates the range of K(t) associated with crack propagation in this material. At low velocity å < 250 m/s, the crack requires only a relatively small energy G = 40.3 J/m^2 to propagate. However as a approaches $\dot{a}_T$ the energy required by the crack increases by more than an order of magnitude with $G_T = 435$ J/m^3.

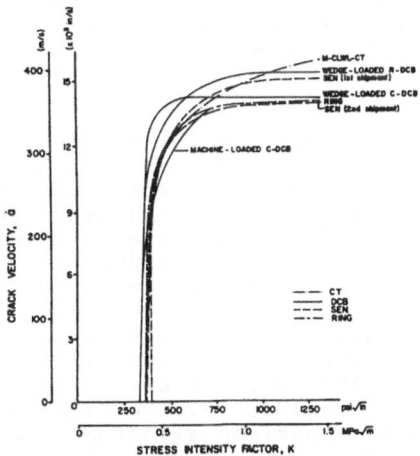

Fig. 6   Results of several experiments showing K-å for different types of fracture specimens.

Since these initial photoelastic studies were conducted with the relatively large SEN specimens, an extensive experimental program has been conducted at the University of Maryland to determine if the K-å relation in Homalite 100 depended upon specimen size, specimen type and the loading system. Five different specimens were employed which included in addition to the SEN a compact tension (CT), rectangular double-cantilever beam (RDCB), contoured double cantilever beam (CDCB) and a ring segment. The CDCB specimen was loaded with a wedge and by pin loading. In all cases a static load was applied with a stationary crack and the crack was initiated with a sharp knife edge drawn across a slightly blunt crack front. The results which were obtained are shown in Fig. 6 and represent measurements made with two different shipments of Homalite 100, over a three year period (1976-1979), and involved four different individuals. It is estimated that the error in individual measurements of K_I were about ±10 percent and the averaging obtained by drawing the K-$\overset{\bullet}{a}$ curve through 50 to 100 data points reduced the error in positioning the curve to ±5 percent. Based on the consistency of these results Dally (1979) concluded that the vertical stem of the K-$\overset{\bullet}{a}$ curve and the arrest toughness K_{Im} are independent of specimen size and type and also independent of the loading system. The K-$\overset{\bullet}{a}$ curve in the transition region and in the branching region shows a clear dependence on specimen type. In this region of the K-$\overset{\bullet}{a}$ relation the fracture process zone is relatively large and it is believed that non-singular terms affect the fracture to such an extent that it will be necessary to describe the crack velocity with a more complex relation of the form:

$$\overset{\bullet}{a} = f(A_n, B_m) \tag{25}$$

Experimental results achieved to date by Ramulu and Kobayashi (1983) and Anand (1983) appear inconsistent and a definition of Eqn. (25) remains to be resolved in the future.

DISCUSSION

The question of uniqueness of even a portion of the K-$\overset{\bullet}{a}$ relation has generated considerable interest and controversy during the past five years. Dally, Fourney and Irwin (1985) reviewed the major work conducted in determining K-$\overset{\bullet}{a}$ curves for polymers and steels. Differences exist between the results obtained in several different laboratories and these differences have lead several investigators to question the validity of expressing the crack velocity in terms of a single parameter K_I to describe the stress field.

A review of the experimental results for steel obtained by Rosakis (1983), T. Kobayashi and Dally (1979) and Kanazawa et al (1981) shows close correspondence in spite of differences in experimental methods and the difficulty of the experiments. Kanazawa et al clearly claim uniqueness and state that K_{Im} is a material property. Rosakis believes that the K-$\overset{\bullet}{a}$ relation for steel may be unique because of the influence of the plastic zone near the tip of the crack.

Results for brittle polymers show larger differences in spite of the fact that the experiments are less difficult. The differences between the results of Kalthoff (1983) for different sizes and shapes of fracture specimens in an epoxy Araldite B, are small (about 10 percent) and could be attributed to experimental error instead of the specimen shape. Errors which can occur in utilizing the method of caustics for measurements of K(t) are described in detail by Dally, Fourney and Irwin (1985).

The results of Ravi-Chandar and Knauss (1983), shown in Figure 7, were obtained by utilizing the method of caustics with Homalite 100. The general behavior of the results correspond to these reported by Dally (1979) as noted by the superimposed curve. However, the details of the results are markedly different, as the horizontal lines in Fig. 7

each represent the result of a single experiment. These horizontal
lines indicate that the crack velocity is independent of K(t) and the
velocity is a constant depending only on the value of K_O at crack initiation. It is difficult to explain these results since they differ
from all other measurements which show a K-å relationship with a common shape. The fact that Ravi-Chandar and Knauss used an electromagnetic loading device with a step loading of 150 µs in duration and
began the experiment with a stress free model may imply that the K-å
relation under transient and quasi-static loading may differ
significantly.

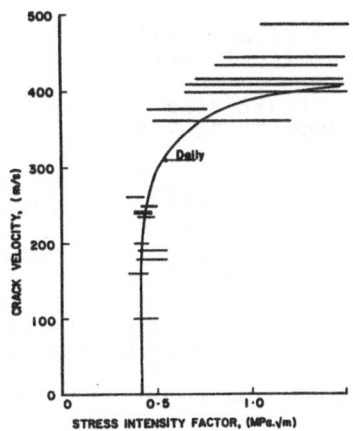

Fig. 7 K-å relation for Homalite 100 (after Ravi-Chandar and
        Knauss).

The subject of dynamic fracture is of major importance to engineers
designing large welded structures where cracks initiation can occur
unexpectedly at some time during the life of the structure. Fail-safe
design concepts require that the crack be arrested before it extends
far enough to destroy the integrity of the structure. Design of crack
arresting structures requires a knowledge of a K-å relation to be used
with numerical codes to predict crack length prior to arrest.
Photoelasticity represents an important experimental technique which
can be used with polymers to study fracture phenomena or with
birefringent coatings to directly measure fracture properties in engineering materials.

REFERENCES

Anand S, (1983) High speed crack propagation and branching under
    uniaxial and biaxial loading. MS Thesis Univ. of Rhode Island:1-82.

Bradley WB, Kobayashi AS (1970) An investigation of propagating crack
    by dynamic photoelasticity. Expl Mech 10-3:106-113.

Christie DG, (1955) A multiple spark camera for dynamic stress analy-
    sis. Jrnl Phot Sci 3:153-159.

Cottron J, Lagarde A (1982) A farfield method for the determination of
    mixed-mode stress-intensity factors from isochromatic fringe pat-
    terns. SM Arch 7-1:1-18.

Cranz C, Schardin H (1929) Kinematographic auf ru hendem film und mit extrem hoher bildfrequenz. Zeits Phys 56:147.

Dally JW (1979) Dynamic photoelastic studies of fracture, Expl Mech 19-10:349-361.

Dally JW, Fourney WL and Irwin GR (1985) On the uniqueness of the stress intensity factor-crack velocity relationship. Intl Jrnl Fract 27:159-168.

Dally JW, Sanford RJ (1982) Multiple ruby laser system for high-speed photography. Opt Eng 21-4:704-708.

Etheridge JM, Dally JW (1978) A simplified three parameter method for determining stress intensity factor. Mech Res Comm 5-1:21-26.

Flynn PD (1966) Photoelastic studies of dynamic stresses in high modulus materials. Jrnl SMPTE 75:729-734.

Flynn PD, Gilbert JT, and Roll AA (1964) Some recent developments in dynamic photoelasticity. Jrnl SPIE 2-4:128-131.

Hendley DR, Turner JL, Taylor CE (1975) A hybrid system for dynamic photoelasticity. Expl Mech 15-8:289-294.

Irwin GR (1980) Series representation of the stress field around constant speed cracks. Univ. MD Lect. Notes.

Irwin GR (1957) Discussion of Wells and Post paper. Proc. of SESA 16-1:93-96.

Kalthoff JF (1983) On some current problems in experimental fracture dynamics. Workshop on Dynamic Fracture. Cal Inst Tech:11-25.

Kanazawa T, Machida S, Teramoto T, Yoshmari H (1981) Study on fast fracture and crack arrest. Expl Mech 21-2:78-88.

Kawata K, Takeda N, Hashimoto S (1985) Photoelastic-coating analysis of dynamic stress concentration in composite strips. Expl Mech 24-4:316-327.

Kobayashi T, Dally JW (1977) The relation between crack velocity and the stress intensity factor in birefringent polymers. ASTM STP 726:257-273.

Ramulu M, Kobayashi AS (1983) Dynamic crack curving - a photoelastic evaluation. Expl Mech 23-1:1-9.

Ravi-Chandar K, Knauss WG (1983) Processes controlling the dynamic fracture of brittle solids. Workshop on Dynamic Fracture. Cal Inst Tech:119-128.

Rosakis AJ, Duffy J, Freund LB (1983) Dynamic crack growth in structural metals. Workshop on Dynamic Fracture. Cal Inst Tech:110-118.

Rosakis AJ, Ravi-Chandar K (1984) On crack tip stress state: an experimental evaluation of three dimensional effect. Cal Inst Tech Report:SM 84-2.

Rowlands RE, Taylor CE, Daniel IM (1968) Ultra-high speed framing photography employing a multiple pulsed ruby laser and a smear type camera; application to dynamic photoelasticity. 8th Intn'l Cong High-Speed Photography.

Sanford RJ (1979) A critical re-examination of the Westergaard method
for solving opening-mode crack problems. Mech Res Com 6:289-294.

Sanford RJ, Dally JW (1979) A general method for determining mixed-
mode stress intensity factors from isochromatic fringe patterns. Eng
Fract Mech 11:621-633.

Schroedl MA, Smith CW (1973) Local stresses near deep surface flaws
under cylindrical bending fields. Progess in flaw growth and fracture
toughness testing. ASTN STP 536:45-63.

Tuzi Z (1928) Photographic and kinematographic study of photoelasticity. Jrnl Soc Mech Eng 31-136:334-339.

Wells AA, Post D (1957) The dynamic stress distribution surrounding a
running crack - a photoelastic analysis. Proc SESA 16-1:69-92.

Westergaard HM (1939) Bearing pressures and cracks. Tran ASME
61:A49-A53.

Visualization of Pulsed Ultrasound
Using a Combined Photoelastic-Schlieren System

H. Shimada and K. Date*

Department of Metal Processing and Mechanical Metallurgy, Faculty of Engineering,
Tohoku University, Sendai, 980 Japan

INTRODUCTION

There are many applications of visualization methods of pulsed ultrasound
in the field of ultrasonic nondestructive testing: checking and improving
ultrasonic probes, aiding ultrasonic tests on components of complex
shape, studying the behavior of ultrasonic pulses, tools for training
ultrasonic operators, etc.

In recent years, the study of ultrasonic wave propagation in transparent
media has been made possible by the introduction of photoelastic and
schlieren system. (Wyatt 1972; Baborovsky et al. 1973; Hanstead 1974;
Hall 1978) The advantages of the photoelastic system are its relative
compactness, the optical simplicity, and simple relation between optical
phase retardation and ultrasonic wave stress. The major merit of the
schlieren system, on the other hand, lies in its ability to visualize
the ultrasonic wave in water as well as solids, because liquid coupling
is used in most ultrasonic nondestructive systems. A combined schlieren-
photoelastic system, therefore, has been successfully used by Marsh
(1973) to observe ultrasound in both solids and water, but the optical
equipment is expensive and large, because the system is basically a "Z"
schlieren layout.

The equipment described in this paper was a simple combined photoelastic-
schlieren system, developed as a tool to improve ultrasonic nondestructive
testing. It provides a simple means of studying the behavior of pulsed
ultrasound in both solids and water. The paper describes the usefulness
of this system by showing some typical examples of ultrasonic waves at
free boundary, artificial defects and liquid/solid interface.

EQUIPMENT DESCRIPTION

The combined photoelastic-schlieren system developed in this study is
basically a photoelastic optical layout as shown in Fig. 1 but having
a knife edge for schlieren optics. A commercial stroboscope light source,
with a very short duration light flash (less than 1 μs), is positioned
at the focal point of a collimating lens, whose effective diameter of
field is 200 mm. The collimated beam of light so produced illuminates
the transparent visualizing medium in which ultrasonic pulses are to be
imaged. The light source is imaged by the viewing lens on a knife edge.

Electrically, the stroboscope light source is synchronized with the
ultrasonic flaw detector, whose repetitional output is adequately
powerful to drive the ultrasonic probe used in most visualization
experiments. A smoothly variable electronic delay determines the exact

* Miyagi National College of Technology, Natori, Miyagi-ken, 981-12/JAPAN

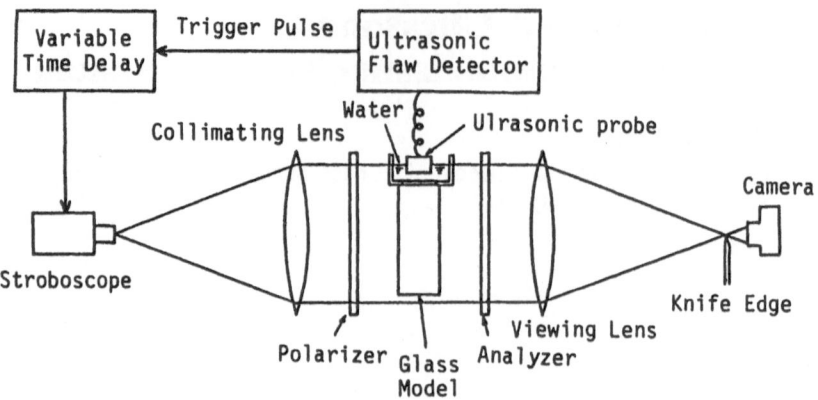

Fig. 1 Basic Layout of Combined Photoelastic-Schlieren Apparatus

timing of light flashes. By varying the flash delay time, an observer
varies the position at which the pulsed ultrasound is stroboscopically
"frozen" and imaged.

All of the photographic examples given in this paper were obtained using
Pyrex glass, 20-mm thick, as a visualizing medium in solid. Pyrex glass
has a better stress-optical coefficient than most inorganic glasses.
Its shear wave velocity (3420 m/s) is very close to that in steel, but
its compressional wave (5490 m/s) is some 7% lower. The velocities of
ultrasound in Pyrex glass and steel are similar so refraction direction,
when using shear wave angle probes or water coupling, are fairly well
simulated. Machine oil was used as a coupling agent for direct contact
of ultrasonic probe to glass model. When water coupling was used for
the study of liquid/solid interfaces in water immersion testing, the
model could carry a small water cell as shown in Fig. 1 or the whole
model could be immersed in a water tank. The cell and tank were made
of 1-mm thick acrylic resin plate.

Polarizer-analyzer orientation and knife edge setting are very important
for observing ultrasonic pulses in glass model and water using this
combined photoelastic-schlieren system. Imaging ultrasonic pulses in
glass model, the polarizer and analyzer were maintained in a "crossed"
condition and the knife edge was not used. In this condition, the
brightness at any point depends on the ultrasonic stress and the
orientation of the principal stress axes relative to the pass axes of
the crossed polarizers. When considering compressional waves, sensitivity
is maximum for wave propagation at 45° to the polarizers axes and zero
for propagation parallel to the polarizer axes. For shear waves, the
propagation direction corresponding to maximum and zero sensitivity are
at 45° to those for compressional waves. The variation of sensitivity
with propagation direction could be avoided by using circularly polarized
light, but this appears to lead to a loss of contrast and sensitivity.

When ultrasonic pulses are observed in water, the polarizer and analyzer
were set parallel and the knife edge was used as shown in Fig. 1. In
this simple schlieren system, an observer could only view the images of
ultrasound in water and not in the glass model. Observation of ultrasonic
pulses in both water and the glass model at the same time was possible
without knife edge by using photoelastic and shadowgraph methods, if
crossed condition of polarizer and analyzer is slightly disturbed (cross

angle of about 85°) to provide a bright background display. In practice, this appears to lead to a loss of contrast and is only to be recommended if simultaneous imaging is essential.

The visualization procedures used in this combined photoelastic-schlieren system are summarized in Table 1.

Table 1. Visualization Procedures of Ultrasonic Wave by Using Combined Photoelastic-Schlieren System

No.	Visualization Method	Polarizer Orientation	Knife Edge	Visualized Wave in
1	Photoelastic	Cross	Not used	Solid
2	Photoelastic & Shadowgraph	Almost Cross (about 85°)	Not used	Solid and Water
3	Schlieren	Parallel	Used	Water

ULTRASONIC PULSE IMAGES

An important feature of ultrasonic wave behavior in solids is the effect of reflection and mode conversion at free surfaces. This feature is strikingly evident when using the ultrasonic visualization method.

Figures 2a through d show the corner reflection of a 2MHz shear wave. In Fig. 2a the incident shear wave is recorded. In Figs. 2b through d the reflection of shear wave at the corner is displayed for the reflected shear wave propagating in the incident direction. Figures 3a through d show the corner reflection of a 2MHz compressional wave. The incident compressional wave is almost mode converted to a shear wave (T) and surface wave (S) by the reflection of the corner so that the reflected compressional wave cannot be observed. It is evident that the corner reflection of ultrasonic wave depends on the ultrasonic wave mode: compressional or shear.

Other examples of reflection and mode conversion at free surfaces are shown in Figs. 4 and 5. A compressional wave, shown in Fig. 4a, incident to a 3-mm diameter cylindrical hole produces both a cylindrical compressional wave (RL) and shear wave (RT) as shown in Figs. 4b and c. An internal crack-like defect is displayed in Figs. 5a through d. A 2MHz shear wave is incident in Fig. 5a and the reflection and mode conversion are shown in Figs. 5b through d. In Fig. 5b RL is the mode converted compressional wave produced by the reflection of incident shear wave at the defect tip. In Fig. 5c, the surface wave (S) propagating along the defect side surface can be observed. The reflected shear wave at defect tip is viewed as a cylindrical wave front shape, and part of its wave propagates in the incident direction (RT1).

Figures 6a through d present typical examples of ultrasonic pulses in both the glass model and water obtained by No.2 procedure in Table 1. Figures 6a and b show the incident compressional wave propagating in water. Figures 6c and d are the reflected compressional wave at the model surface and transmitted wave to the glass model. The propagation distance of the compressional wave in the glass model is larger than that in water because of the ultrasound velocity difference in glass and water.

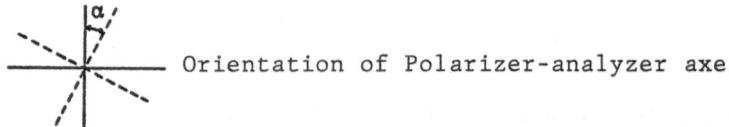

Orientation of Polarizer-analyzer axes

Dt Relative Time Delay

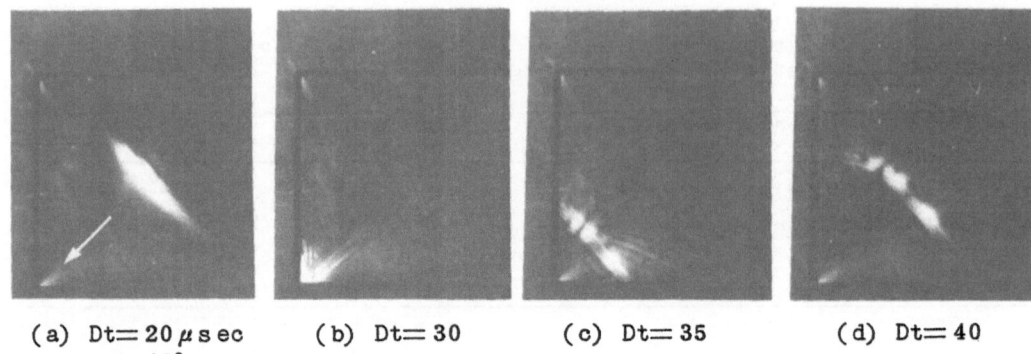

(a) Dt=20μsec   (b) Dt=30   (c) Dt=35   (d) Dt=40
α=45°

Fig. 2 Corner Reflection of 2MHz Shear Wave

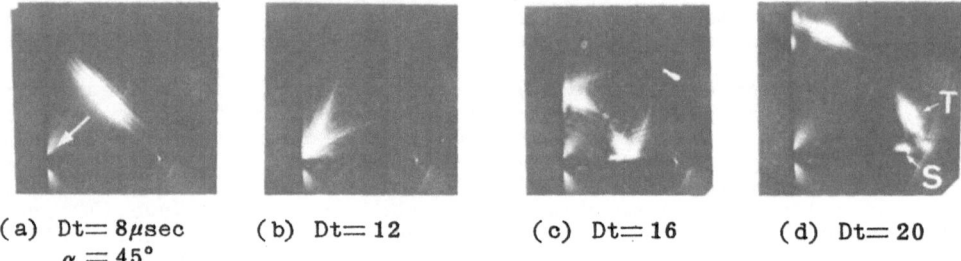

(a) Dt=8μsec   (b) Dt=12   (c) Dt=16   (d) Dt=20
α=45°

Fig. 3 Corner Reflection of 2MHz Compressional Wave

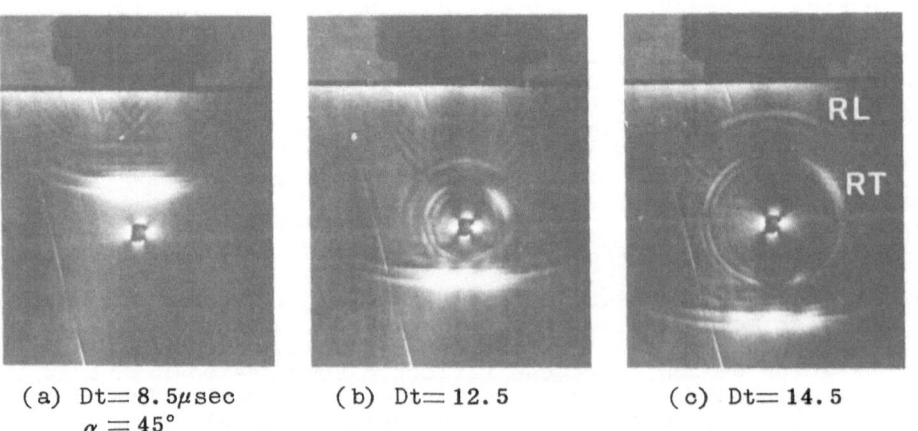

(a) Dt=8.5μsec   (b) Dt=12.5   (c) Dt=14.5
α=45°

Fig. 4 Reflection of Compressional Wave from a 3-mm Diameter
Cylindrical Hole

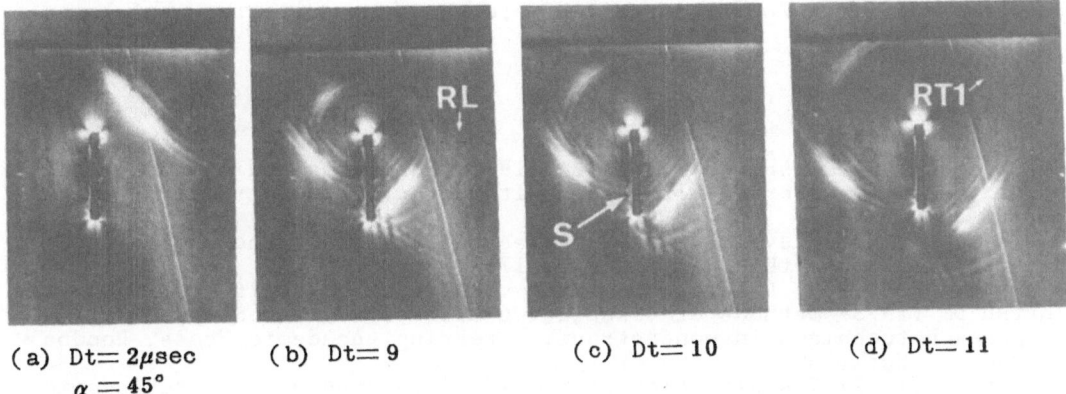

(a) Dt= 2μsec   (b) Dt= 9   (c) Dt= 10   (d) Dt= 11
    α = 45°

Fig. 5 Reflection of Shear Wave from an Internal Crack-Like Defect

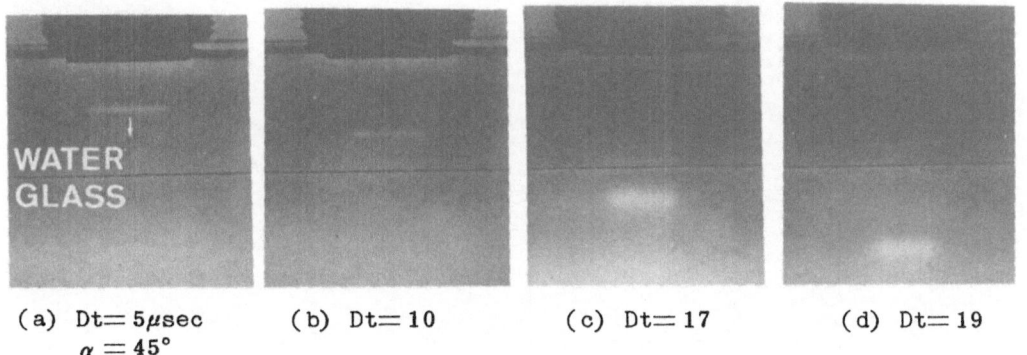

(a) Dt= 5μsec   (b) Dt= 10   (c) Dt= 17   (d) Dt= 19
    α = 45°

Fig. 6 Incidence of Compressional Wave to Glass Model through Water

Figures 7 and 8 demonstrate the effect of radius of focus lens. The incident focused wave, as shown in Figs. 7a through c, is passed through the acrylic focus lens of radius 15-mm and produces both compressional waves (L) and shear waves (T) at the incidence og glass model as shown in Figs. 7d through f. The incident wave shown in Figs. 8a and b is passed through the 52-mm radius acrylic lens and produces only a compressional wave as shown in Figs. 8c through e.

These results suggest that the behavior of ultrasonic waves in solids and solid/liquid interfaces is very much more complicated than had been anticipated. For the study of such ultrasound behavior, an ultrasonic visualization system, such as the combined photoelastic-schlieren system developed in this study, appears to be convenient and effective.

CONCLUSIONS

A simple combined photoelastic-schlieren system was developed as a tool for improving ultrasonic nondestructive testing. It provides a simple means of studying the behavior of pulsed ultrasound in both solid and water. Typical examples of ultrasonic waves at free surfaces, internal

defects and liquid/solid interfaces are shown to prove the usefulness
of this system.

REFERENCES

Baborovsky VM, Marsh DM, Slater EA (1973) Schlieren and computer studies
    of the interaction of ultrasound with defect. Nondestructive Testing:
    200-207
Hall KG (1978) An evaluation of ultrasonic probes by the photoelastic
    visualization method. Brit J NDT: 171-184
Hanstead PD (1974) Direct ultrasonic visualization. Brit J NDT: 34-44
Marsh DM (1973) Methods of visualizing ultrasound. In: Sharpe RS (ed)
    Reseach technique in nondestructive testing, Academic Press, London
    and New York, p333
Wyatt RC (1972) Visualization of pulsed ultrasound using stroboscopic
    photoelasticity. Nondestructive Testing 5: 354-358

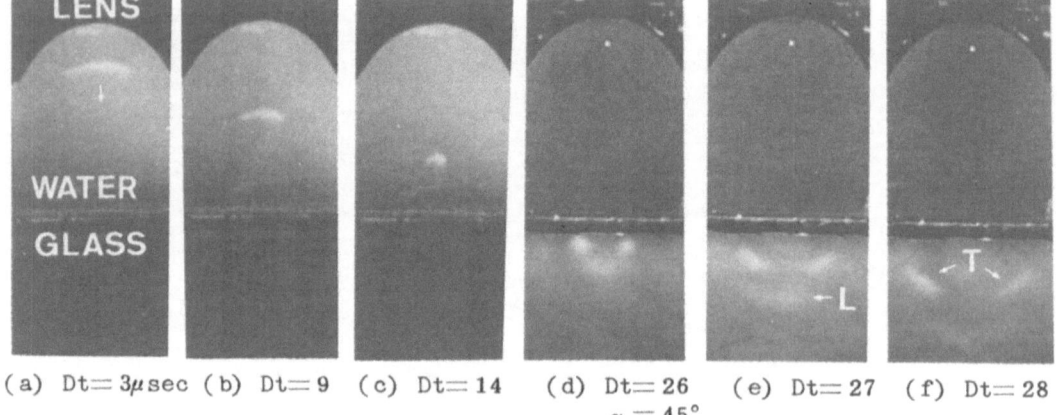

(a) Dt= 3μsec (b) Dt= 9 (c) Dt= 14 (d) Dt= 26 (e) Dt= 27 (f) Dt= 28

$\alpha = 45°$

Fig. 7 Incidence of Focused Compressional Wave to Glass Model through
       Water (Lens Radius, 15-mm)

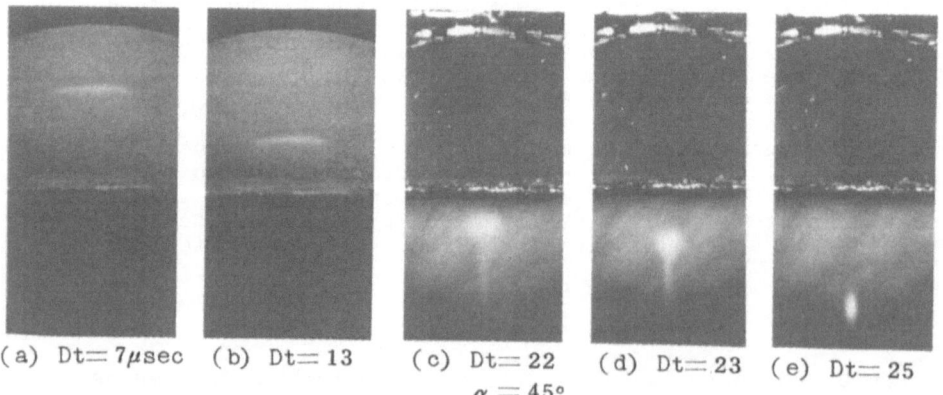

(a) Dt= 7μsec (b) Dt= 13 (c) Dt= 22 (d) Dt= 23 (e) Dt= 25

$\alpha = 45°$

Fig. 8 Incidence of Focused Compressional Wave to Glass Model through
       Water (Lens Radius, 53-mm)

The Shadow Optical Method of Caustics
An Overview on its Applications in Stress Concentration Problems

J.F. Kalthoff

Fraunhofer-Instiut für Werkstoffmechanik, Wöhlerstrasse 11, D-7800 Freiburg/Brsg., Federal Republic of Germany

INTRODUCTION

The shadow optical method of caustics is a relatively new experimental technique in stress strain analysis. It was originally introduced by Manogg (1964) for investigating crack tip stress intensifications. The method is sensitive to stress gradients and therefore is an appropriate tool for quantifying stress concentration problems. The technique was extended later by Theocaris (1971-1981), Rosakis (1982, 1983) and the author and his colleagues (1976-1986) to different conditions of loading, material behavior, in static as well as dynamic situations. This summarizing article reviews the basic physical principles and the mathematical analyses of the method and gives applications to various problems of practical relevance.

PHYSICAL PRINCIPLE AND THEORETICAL BACKGROUND

Stresses alter the optical properties of a solid, i.e. the thickness of the body (due to Poisson's effects) and the refractive index of the material. These changes in the optical properties are utilized in the shadow optical method of caustics to make stress distributions in the solid visible.

The physical principle of the method is illustrated in Fig. 1. A specimen with a notch as a stress riser is subjected to tensile loading. The specimen shall be of a transparent material for the moment. It is illuminated by a parallel light beam. Due to the tensile stress concentration around the notch tip the thickness of the specimen and the refractive index of the material are steadily reduced when the notch tip is

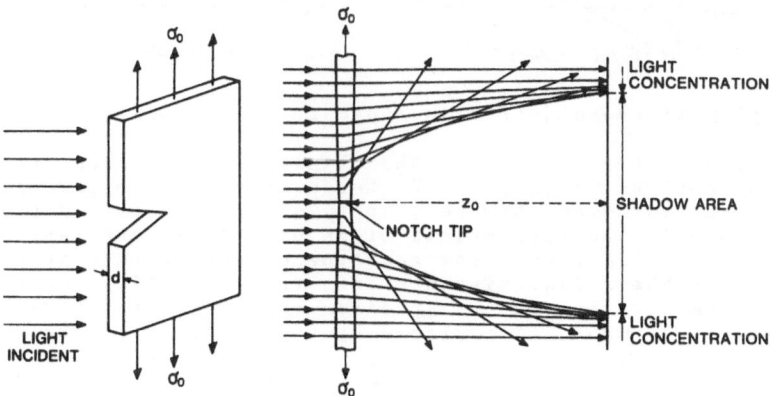

Fig. 1. Physical principle of the shadow optical method

approached. Thus, a light ray traversing the specimen near the notch
tip is deflected in a direction away from the notch tip. With that
respect the area surrounding the notch tip acts similar to a divergent
lens. But, the closer the light ray to the notch tip the larger the
deflection angle. Consequently, on a screen (image or reference plane)
at a distance z_0 from the specimen the notch tip appears as a shadow
area which is surrounded by a region of light concentration, i.e. the
caustic. This shadow pattern represents a quantitative description of
the stress distribution around the notch tip in the plate.

Shadow optical light patterns are obtained for tensile as well as
compressive stress concentrations. They can be observed with transparent specimens or in reflection with non-transparent specimens, as
real or as virtual images. Figure 1 represents the most simple case of
a tensile stress concentration in a transparent specimen with the observation of a real shadow optical image. The virtual image is obtained
on the opposite side of the specimen where the real image is observed.
A consideration of the light rays which are reflected at the front side
the specimen, i.e. the side facing the light source, leads to the formation of shadow images as well, which again can be real or virtual
depending on the position of the image plane.

For the mathematical description of light ray deflection the following
sign assignments and definitions are made: Tensile stresses are
positive. The observation direction defines the sign of distances. The
distance z_0 between the reference plane and the specimen is negative
(positive) if the reference plane with regard to the observation
direction is located ahead of (behind) the specimen. In transmission
(reflection) arrangements the observation direction is opposite to
(in) the direction of the illuminating light beam.

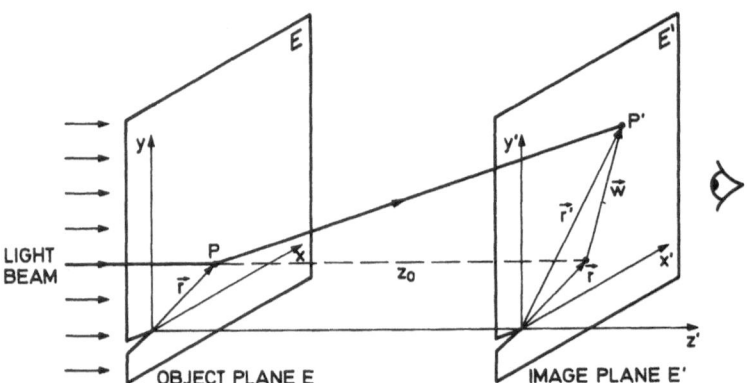

Fig. 2. Mapping of the object plane onto the image plane

Figure 2 considers the mapping of the object plane E (specimen) onto
a real shadow optical image plane E' for a transmission arrangement,
$z_0 < 0$. The given formulas, however, apply quite generally for any
observation mode if the appropriate signs of the representative distances are used. A light ray traverses the object plane E at the point
$P(\vec{r})$, where $\vec{r}$ is the radial distance from the notch tip. Due to the
influence of the stresses in the specimen this light ray is deflected
and hits the image plane E' displaced by the vector $\vec{w}$ at the point
$P'(\vec{r'})$ with

$$\vec{r'} = \vec{r} + \vec{w} \tag{1}$$

Quantitatively the displacement vector $\vec{w}$ is determined by the changes in the optical path length Δs in the object plane,

$$\vec{w} = -z_0 \ \overrightarrow{grad} \ \Delta s(r, \varphi) \tag{2}$$

with $z_0 < 0$ for real images in transmission or reflection and $z_0 > 0$ for virtual images in transmission or reflection.

Changes Δs in the optical path length for a plane parallel plate of thickness d are given by

$$\Delta s = (n-1) \ \Delta d_{eff} + d_{eff} \ \Delta n \tag{3}$$

with   n = refractive index, $d_{eff} = d$ for transmission, and   n = -1   , $d_{eff} = d/2*$ for reflection.
Furthermore, changes Δn in the refractive index due to the principal stresses σ_1, σ_2, σ_3 are described by Maxwell-Neumann's law

$$\Delta n_1 = A\sigma_1 + B(\sigma_2+\sigma_3), \quad \Delta n_2 = A\sigma_2 + B(\sigma_1+\sigma_3) \tag{4}$$

with A, B = material constants. For optically isotropic, non-birefringent materials A = B and for reflection A = B = 0. Changes Δd_{eff} due to the prevailing stresses are described by Hooke's law

$$\Delta d_{eff} = \left[\frac{1}{E}\sigma_3 - \frac{\nu}{E}(\sigma_1+\sigma_2) \right] d_{eff} \tag{5}$$

with σ_3 = 0 for plane stress, and Δd_{eff} = 0 for plane strain. With eqs. (4,5) the eq. (3) then can be rearranged to

$$\Delta s_{1/2} = c \ d_{eff} \left[(\sigma_1+\sigma_2) \pm \lambda(\sigma_1-\sigma_2) \right] \tag{6}$$

with

$$c = \frac{A+B}{2} - (n-1)\nu/E, \quad \lambda = \frac{A-B}{A+B-2(n-1)\nu/E} \quad \text{for plane stress}$$

$$c = \frac{A+B}{2} + \nu B, \quad \lambda = \frac{A-B}{A+B+2\nu B} \quad \text{for plane strain.}$$

Numerical values for the constants used in eqs. (3) - (6), in particular for the deduced shadow optical contant c and the anisotropy coefficient λ are given for different materials in Table 1.

TABLE 1 - Constants for Caustic Evaluation

Material	Elastic Constants		General Optical Constants			Shadow Optical Constants				Effective Thickness
						for Plane Stress		for Plane Strain		
	Young's Modulus MN/m^2	Poisson's Ratio	Refractive Index	A m^2/N	B m^2/N	c m^2/N	λ	c m^2/N	λ	d_{eff}
TRANSMISSION:										
Optically Anisotropic:										
Araldite B	3660**	0.392**	1.592	-0.056×10^{-10}	-0.620×10^{-10}	-0.970×10^{-10}	-0.288	-0.580×10^{-10}	-0.482	d
CR-39	2580	0.443	1.504	-0.160×10^{-10}	-0.520×10^{-10}	-1.200×10^{-10}	-0.148	-0.560×10^{-10}	-0.317	d
Plate Glass	73900	0.231	1.517	$+0.0032 \times 10^{-10}$	-0.025×10^{-10}	-0.027×10^{-10}	-0.519	-0.017×10^{-10}	-0.849	d
Homalite 100	4820**	0.310**	1.561	-0.444×10^{-10}	-0.672×10^{-10}	-0.920×10^{-10}	-0.121	-0.767×10^{-10}	-0.149	d
Optically Isotropic:										
PMMA	3240	0.350	1.491	-0.530×10^{-10}	-0.570×10^{-10}	-1.080×10^{-10}	~ 0	-0.750×10^{-10}	~ 0	d
REFLECTION:										
All materials	E	ν	-1	0	0	$2\nu/E$	0	$-$	$-$	d/2

* Only half of the specimen thickness d contributes to the surface deformation on one side   ** Dynamic values

The complete family of light rays which are deflected according to
eqs. (1,2,6) forms a shadow space behind the object plane (see also
Fig. 1). Its surface is an envelope to the light rays and is called the
caustic surface. The intersection of this surface with the image plane
forms the caustic curve. The caustic is a multivalued, singular solution of the mapping equations, i.e. the mapping of points along the
caustic is not reversible. Thus, a necessary and sufficient condition
for the existence of the caustic curve is obtained if the Jacobian of
eqs. (1,2,6) becomes zero, i.e.

$$\frac{\partial x'}{\partial r} \frac{\partial y'}{\partial \varphi} - \frac{\partial x'}{\partial \varphi} \frac{\partial y'}{\partial r} = 0 \qquad (7)$$

The coordinates r, φ of points P which fulfill eq. (7) form the so-
called initial curve in the object plane. The mapping of this initial
curve onto the image plane is the caustic.

MAPPING EUQATIONS AND CAUSTICS FOR SPECIFIC EXAMPLES

The specific mapping equation for a special stress concentration prob-
blem is obtained by inserting into the general equation (6) the individual stress distribution formula. Three stress concentration problems (see Fig. 3) are discussed simultaneously in a comparative manner.
These are: a) a compressive edge load P acting on a half plane b) a
circular hole in a plate subjected to a biaxial stress field p,q, and
c) a crack in a plate under tensile load with mode I stress intensity
factor K_I*.

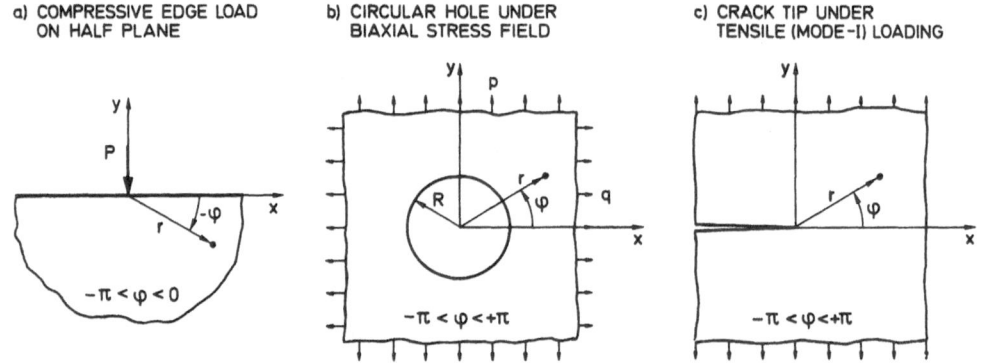

Fig. 3. Typical stress concentration problems

The linear elastic stress concentration fields for these three examples
are given by eqs. (8), Table 2. With these stress distributions and
eq. (6) for λ = o (for simplicity only the isotropic case shall be considered in this context) the mapping equations (1,2,6) for the considered problems result in eqs. (9), Table 2. The necessary and sufficient condition for the existence of the caustic curve is obtained by
applying eq. (7) to the mapping equation (9). The resulting equation
of the initial curve is given in Table 2, eq. (10). For all three cases
the initial curves are circles around the center point of stress concentration with fixed radii r_o. The caustic curves are finally obtained
as images of the initial curves, and are given in Table 2, eq. (11).

* For the definition of the stress intensity factor K see textbooks on
  fracture mechanics, e.g. Broek (1982)

TABLE 2 – Shadow Optical Equations

	a) EDGE LOAD	b) CIRCULAR HOLE	c) MODE I CRACK	
STRESSES	$\sigma_r = \frac{2P}{\pi}\frac{\sin\varphi}{r}$ $\sigma_\Theta = 0$ $\tau_{r\varphi} = 0$	$\sigma_r = \frac{p+q}{2}\left(1-\frac{R^2}{r^2}\right) - \frac{p-q}{2}\left(1-\frac{R^2}{r^2}+3\frac{R^4}{r^4}\right)\cos 2\varphi$ $\sigma_\varphi = \frac{p+q}{2}\left(1+\frac{R^2}{r^2}\right) + \frac{p-q}{2}\left(1+3\frac{R^4}{r^4}\right)\sin 2\varphi$ $\tau_{r\varphi} = -\frac{p-q}{2}\left(1+2\frac{R^2}{r^2}-3\frac{R^4}{r^4}\right)\sin 2\varphi$	$\sigma_r = \frac{K_I}{\sqrt{2\pi r}}\frac{1}{4}\left(5\cos\frac{1}{2}\varphi - \cos\frac{3}{2}\varphi\right)$ $\sigma_\varphi = \frac{K_I}{\sqrt{2\pi r}}\frac{1}{4}\left(3\cos\frac{1}{2}\varphi + \cos\frac{3}{2}\varphi\right)$ $\tau_{r\varphi} = \frac{K_I}{\sqrt{2\pi r}}\frac{1}{4}\left(\sin\frac{1}{2}\varphi + \sin\frac{3}{2}\varphi\right)$	(8)
MAP. EQ.	$x' = r\cos\varphi + \frac{2P}{\pi} z_0 c\, d_{eff}\, r^{-2}\sin 2\varphi$ $y' = r\sin\varphi - \frac{2P}{\pi} z_0 c\, d_{eff}\, r^{-2}\cos 2\varphi$	$x' = r\cos\varphi + 4 z_0 c\, d_{eff}\, R^2(p-q)\, r^{-3}\cos 3\varphi$ $y' = r\sin\varphi + 4 z_0 c\, d_{eff}\, R^2(p-q)\, r^{-3}\sin 3\varphi$	$x' = r\cos\varphi + \frac{K_I}{\sqrt{2\pi}} z_0 c\, d_{eff}\, r^{-3/2}\cos\frac{3}{2}\varphi$ $y' = r\sin\varphi + \frac{K_I}{\sqrt{2\pi}} z_0 c\, d_{eff}\, r^{-3/2}\sin\frac{3}{2}\varphi$	(9)
INIT.C.	$r = \left[\frac{4}{\pi}|z_0||c|\, d_{eff}\, P\right]^{1/3} \equiv r_0$	$r = \left[12|z_0||c|\, d_{eff}\, R^2(p-q)\right]^{1/4} \equiv r_0$	$r = \left[\frac{3}{2}\frac{K_I}{\sqrt{2\pi}}|z_0||c|\, d_{eff}\right]^{2/5} \equiv r_0$	(10)
CAUSTIC	$x' = r_0\left(\cos\varphi + \mathrm{sgn}(z_0 c)\frac{1}{2}\sin 2\varphi\right)$ $y' = r_0\left(\sin\varphi - \mathrm{sgn}(z_0 c)\frac{1}{2}\cos 2\varphi\right)$	$x' = r_0\left(\cos\varphi + \mathrm{sgn}(z_0 c)\frac{1}{3}\cos 3\varphi\right)$ $y' = r_0\left(\sin\varphi + \mathrm{sgn}(z_0 c)\frac{1}{3}\sin 3\varphi\right)$	$x' = r_0\left(\cos\varphi + \mathrm{sgn}(z_0 c)\frac{2}{3}\cos\frac{3}{2}\varphi\right)$ $y' = r_0\left(\sin\varphi + \mathrm{sgn}(z_0 c)\frac{2}{3}\sin\frac{3}{2}\varphi\right)$	(11)
EVA.F.	$D = 2.6\, r_0$	$D = 2.67\, r_0$	$D = 3.17\, r_0$	(12)
EVA.F.	$P = \frac{\pi}{4(2.6)^3 z_0 c\, d_{eff}} D^3$	$p-q = \frac{1}{12(2.67)^4 z_0 c\, d_{eff}\, R^2} D^4$	$K_I = \frac{2\sqrt{2\pi}}{3(3.17)^{5/2} z_0 c\, d_{eff}} D^{5/2}$	(13)

114

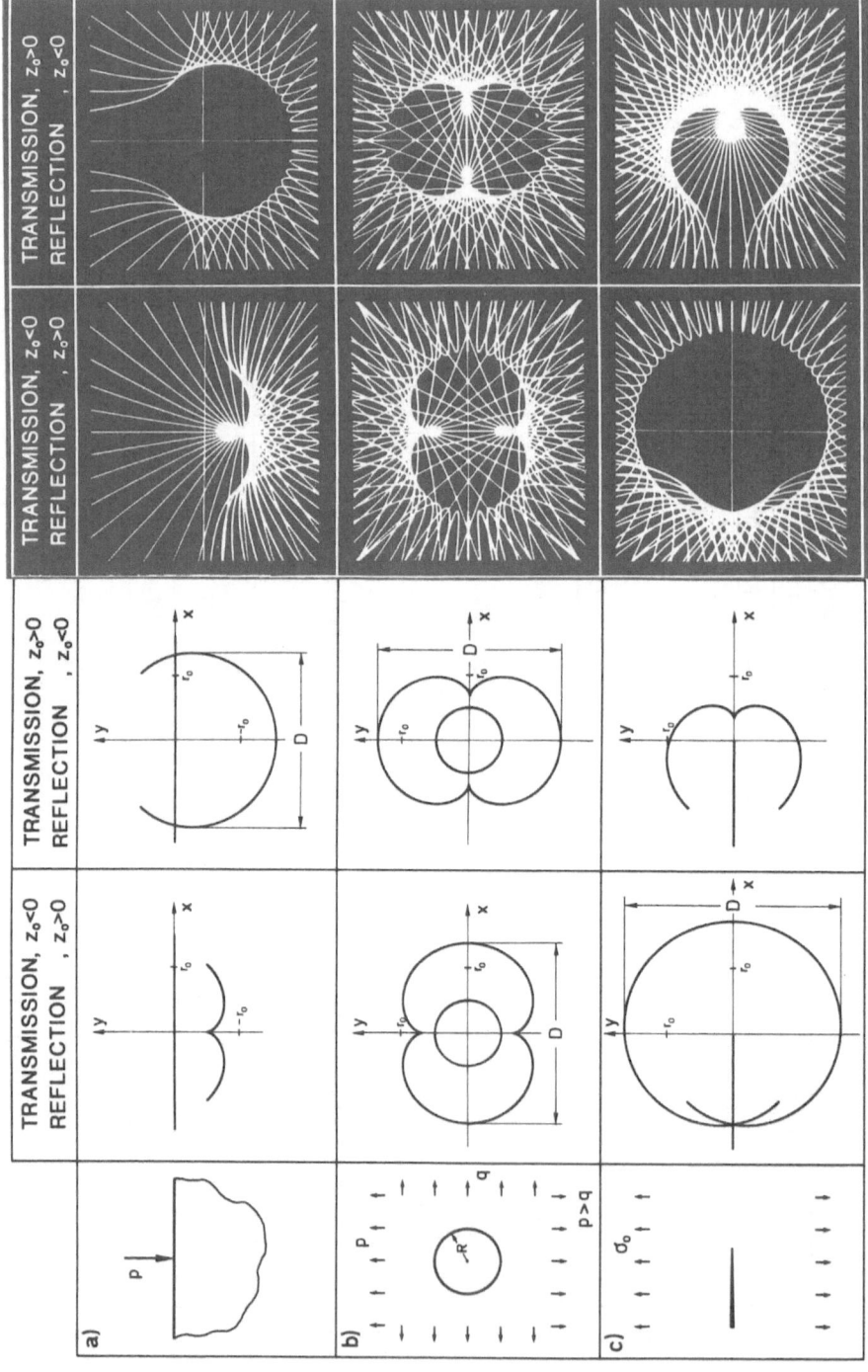

Fig. 4. Caustics and shadow patterns for typical stress concentration problems

Mathematically the caustic curves are generalised epicycloids. The caustics are graphically shown in the left diagrams of Fig. 4 for different observation modes, i.e. transmission and reflection arrangements and positive and negative distances z_0. Illustrative pictures of the complete light distributions in the image planes are given by the patterns at the right side of Fig. 4. The presented lines are the images of light rays which traverse the specimen along straight lines φ = const. The caustic curves appear as envelopes to the obtained families of image lines.

For the quantitative evaluation of caustics a length parameter between characteristic points on the caustic curve is defined, e.g. the distances D given in Fig. 4, left diagrams. These distances are related to the radii of the initial curves by eqs. (12), Table 2. With eqs. (10) and (12) a quantitative formula is then obtained in each case relating the size of the shadow optical pattern with the generating load parameter, given by eq. (13), Table 2. Thus, from the distance D measured in an experimentally observed caustic the generating load parameter for the specific stress concentration problem can be determined, i.e. the magnitude of the compressive edge load P, or the difference of the biaxial stresses p-q, or the crack tip stress intensity factor K_I.

With materials that are optically anisotropic the analysis becomes somewhat more complex. As a result the single caustic curve splits up into a double caustic (see also next Chapter a, b). The evaluation formulas for each of the two caustics are the same as given by eqs. (13) except for slightly different values of the numerical factors. For details see Kalthoff (1986).

Since shadow optical techniques have most extensively been used in fracture mechanics the theory of caustics around crack tips has been developed to the greatest degree. Some of the results that are of greater interest shall be presented.

Figure 5 shows crack tip caustics for mode II (in-plane shear) and mode III (anti-plane shear) loading of cracks, which are shown in

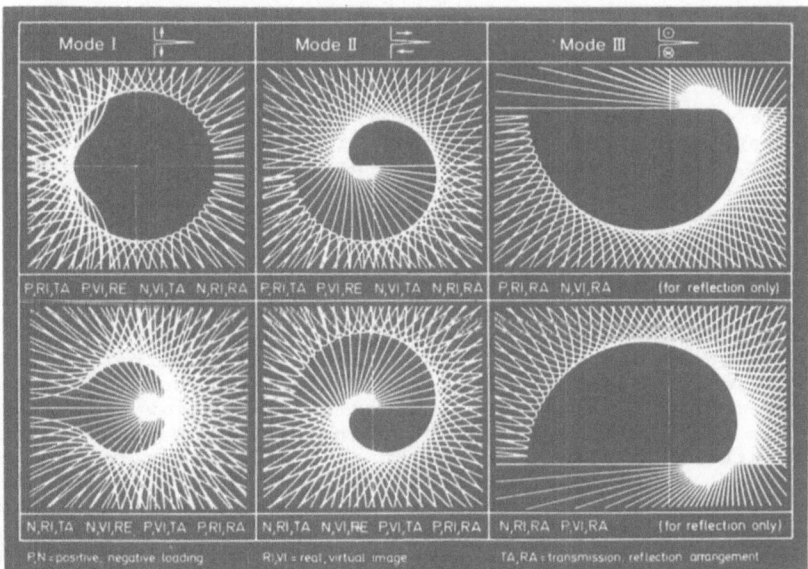

Fig. 5. Crack tip shadow patterns for different modes of loading

addition to the previously derived mode I (tensile) caustic. The
mode II and the mode III caustics become asymmetric. But as discussed
before a characteristic length parameter of the caustics determines the
stress intensity factors K_{II} or K_{III} respectively. For cracks subjected to a combined mode I mode II loading both stress intensity factors
K_I and K_{II} are determined by two characteristic length parameters
taken from the resulting mixed mode caustic. Quantitative formulas are
given by Theocaris (1971, 1980), Kalthoff (1986).

With materials that do not show a linear elastic but an elastic-plastic
behavior (e.g. structural steels) crack tip caustics can be observed as
well. Figure 6 shows numerically calculated (Rosakis et al. 1983)
mode I shadow patterns of a crack in a power-law hardening material for
different hardening exponents n. With increasing influence of plasticity effects the caustic changes its shape from the limiting case n = 1,
i.e. a linear elastic material, to the case n = ∞, i.e. an elastic-
perfectly-plastic material. Analogous to the previous discussions, the
diameter of the elastic-plastic caustic now determines the elastic-
plastic fracture mechanics parameter, i.e. the J-integral value. For
quantitative formulas see Rosakis et al. (1982, 1983), also Kalthoff
(1986).

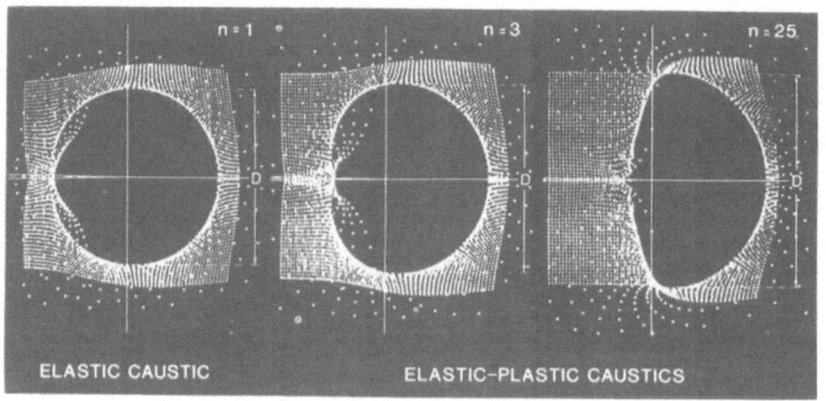

Fig. 6. Elastic-plastic crack tip shadow patterns

For further details, e.g. shadow optical analyses for cracks under
dynamic loading conditions, i.e. propagating cracks and cracks under
impact loading, the influence of higher order terms of the stress distribution on the caustic, and the influence of local plasticity, state-
of-stress etc., see Beinert, Kalthoff (1981), Kalthoff (1986).

APPLICATIONS

The application of the caustic technique for investigating stress concentration problems does not require a sophisticated experimental
equipment. The only essentials are a suitable light beam for illuminating the specimen and a device for recording the shadow pattern. The
light beam has to fulfill only one but very stringent requirement, it
must be generated by a point-like light source in order to produce high
quality caustics. Conventional cameras can be utilised for the recording of shadow patterns. The camera is simply focused onto the real or
virtual image plane ahead of or behind the specimen, depending on the
chosen observation mode. For further details on experimental techniques, e.g. specimen preparation, use of non-parallel light beams for
illumination of the specimen etc., see Kalthoff (1986).

The shadow optical images presented in the previous Chapter are
characterised by very simple geometric patterns which can easily be
evaluated. The shadow optical method, therefore, is very well suited
for investigating complex phenomena, for example in dynamic loading
situations. Figure 7 summarizes applications of the caustic technique
to various transient problems of practical interest. The schematic
drawings on the left side of the figure illustrate the physical problem
to be investigated; in particular the experimental set-up and the
shadow optical recording arrangements are shown. The obtained shadow
patterns are presented on the right side of the figure. They were photographed with a Cranz-Schardin 24 spark high-speed-camera.

a) The history of the load input into a specimen by a falling knife
edge is considered (see also Kalthoff 1986). A specimen made from the
optically anisotropic model material Araldite B is utilised. The shadow optical technique is applied in transmission. Virtual caustics
are photographed in a reference plane $z_o > 0$. The experimentally observed caustics are in good agreement with the theoretically predicted
shape shown in Fig. 4a. Since an optically anisotropic material is used
a double caustic instead of a single caustic is obtained. The high speed
series of shadow patterns indicates an oscillating load increase with
time.

b) The stress history is investigated in a specimen which is dynamically loaded by a projectile impinging on its edge (see also Kalthoff,
Winkler 1983). The stresses in the specimen are made visible by the
shadow patterns generated around a row of holes drilled into the
specimen (compare Fig. 4b). The size of the caustic is a measure of
the magnitude of the stress at the location of the hole. Furthermore,
for a fixed observation mode, the direction of the generated shadow
pattern is an indication whether the applied stress is tensile or
compressive. The specimen for this investigation is again made from
the material Araldite B. The shadow optical technique is applied in
transmission. Real shadow patterns are recorded in a reference plane
$z_o < 0$. The high speed series of shadow photographs shows the propagation of the compressive stress wave into the specimen. The last photograph, taken at much later times, indicates the change of the compressive stress field into a tensile stress field due to the reflection of the compressive waves as tensile waves at the free ends of
the specimen.

c) The stress intensity factor behavior of a crack in a three point
bend specimen under impact loading is investigated (see also Kalthoff
et al. 1980, 1982). A specimen made from a high strength steel is
utilised for this investigation. Virtual shadow patterns are photographed in a reflection arrangement with $z_o > 0$. The experimentally
observed caustics are in good agreement with the theoretically predicted pattern given in Fig. 4c. A quantitative evaluation of the
measured caustics according to eq. (14c) demonstrates that the actual
dynamic stress intensity factor at the crack tip can be very different
from those values which are derived from load values P_{II} registered
at the tup of the striking hammer utilising the conventional static
stress intensity formula.

d) An example of a crack under dynamic mode II loading is given last.
A specimen with two parallel edge notches is impacted at the edge
between the two notches by a projectile with a diameter equal to the
distance of the two notches. Polymethylmethacrylat (PMMA) is used as
specimen material. The shadow optical technique is applied in transmission. Real shadow patterns are recorded in a reference plane $z_o < 0$
at a time 60 µs after impact. The caustics represent an ideal shear
loading at the notch tip (see Fig. 5 for comparison). The loading

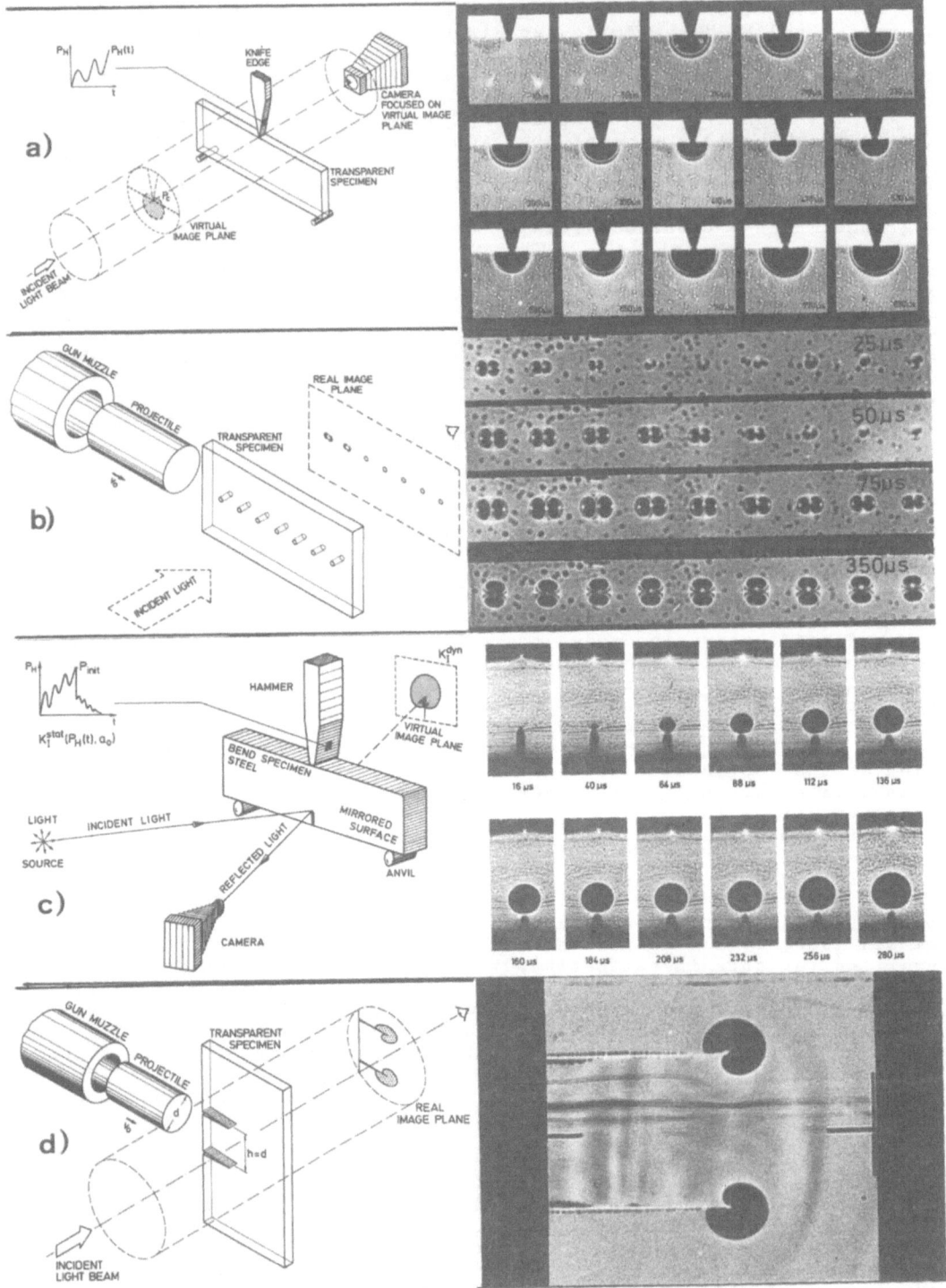

Fig. 7. Applications

technique, therefore, is very well suited for investigating the shear failure of materials at high rates of strain.

SUMMARY AND DISCUSSION

The physical principle and the mathematical description of the shadow optical imaging process has been presented. It has been shown that the shadow optical effect is based on the deflection of light rays due to stress gradients. For typical examples of stress concentration problems the shadow optical formulation is presented and caustic evaluation formulas have been derived. The applicability of the method to problems of practical relevance has been demonstrated by several examples. For many other stress concentration problems the shadow optical method of caustics lends itself as an appropriate tool for investigation.

Advantages and disadvantages of the shadow optical technique in comparison to other techniques of experimental stress analysis become evident by a discussion of the shadow optical and the photoelastic picture of a crack tip stress distribution, Fig. 8. Due to the large number of iso-

SHADOWOPTICAL PICTURE   **PHOTOELASTIC PICTURE**

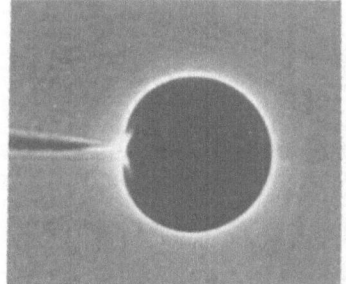 

OF CRACK TIP STRESS DISTRIBUTION

Fig. 8. Shadow optical and photoelastic crack tip patterns

chromatic fringes the photoelastic pattern is rather complicated. In the near field region around the crack tip, i.e. the region of interest, the isochromatic fringes are not resolved any more. Thus, only by extrapolation of far away data towards the center of stress concentration information on the real crack tip loading condition can be obtained. The shadow optical picture, on the other hand, is much simpler. Only one characteristic line, the caustic curve, is obtained which is directly related to the crack tip loading condition. This simplicity results from the fact that the shadow optical effect is caused by stress gradients. Thus, for any kind of stress concentration problem with large stress gradients the shadow optical technique is the appropriate measuring technique. The dependence on the stress gradient, of course, implies also certain disadvantages. The far field of the crack tip stress distribution, for example, does not become visible in the shadow optical picture, since the variations in stresses are too small. The photoelastic pattern, however, yields accurate information in particular in this region.

The shadow optical method of caustics and the photoeleastic method of isochromatic fringes, therefore, are not to be considered as competitive experimental tools which can be used with the same success and efficiency in obtaining the same kind of data. Depending on the specific property of interest and problem to be investigated the one or the other measuring technique is more appropriate. Both techniques have their specific ranges of applicability and complement each other.

REFERENCES

Beinert J, Kalthoff JF (1981) Experimental Determination of Dynamic Stress Intensity Factors by Shadow Patterns. In: Sih GC (ed) Mechanics of Fracture 7, Experimental Fracture Mechanics, Martinus Nijhoff Publ, The Hague Boston London, pp 280 - 330
Broek D (1982) Elementary Engineering Fracture Mechanics. Martinus Nijhoff Publ, The Hague Boston London
Kalthoff JF, Beinert J, Winkler S (1977) Measurements of Dynamic Stress Intensity Factors for Fast Running and Arresting Cracks in Double-Cantilever-Beam Specimens. In: Hahn GT, Kanninen MF (eds) Fast Fracture and Crack Arrest, ASTM STP 627: pp 167 - 176
Kalthoff JF (1980) Measurements of Dynamic Stress Intensity Factors in Impacted Bend Specimens. In: Proc. CSNI Specialists Meeting on Instrumented Precracked Charpy Testing, EPRI, Palo Alto, Calif.
Kalthoff JF, Winkler S, Böhme W, Klemm W (1981) Determination of Dynamic Fracture Toughness K_{Id} in Impact Tests by Means of Response Curves. In: Francois D (ed) Advances in Fracture Research, Pergamon, Oxford New York, pp 363 - 372
Kalthoff JF, Böhme W, Winkler S (1982) Analysis of Impact Fracture Phenomena by Means of the Shadow Optical Method of Caustics. In: Proc VIIth Int Conf Experimental Stress Analysis, Haifa, pp 148 - 160
Kalthoff JF, Winkler S (1983) Fracture Behavior under Impact. IWM-Reports W 8/82 and W 10/82 prepared for US ARO, European Research Office, Fraunhofer-Institut für Werkstoffmechanik, Freiburg
Kalthoff JF (1986) The Shadow Optical Method of Caustics. In: Kobayashi AS (ed) Handbook on Experimental Mechanics, Prentice Hall, Englewood Cliffs, New Jersey
Manogg P (1964) Anwendung der Schattenoptik zur Untersuchung des Zerreißvorgangs von Platten. Dissertation, Universität Freiburg
Manogg P (1964) Schattenoptische Messung der spezifischen Bruchenergie während des Bruchvorgangs bei Plexiglas. In: Proc Int Conf Physics Non-Crystalline Solids, Delft, pp 481 - 490
Rosakis AJ, Freund LB (1982) Optical Measurement of the Plastic Strain Concentration of a Tip in a Ductile Steel Plate. J Eng Mat Tech 104: 115 - 125
Rosakis AJ, Ma CC, Freund LB (1983) Analysis of the Optical Shadow Spot Method for a Tensile Crack in a Power-Law Hardening Material. J Appl Mech 50: 777 - 782
Theocaris PS, Joakimides N (1971) Some Properties of Generalised Epicycloids Applied to Fracture Mechanics. J Appl Mech 22: 876 - 890
Theocaris PS (1973) Stress Concentrations at Concentrated Loads. J Exp Mech 13: 511 - 528
Theocaris PS (1981) The Reflected Caustic Method for the Evaluation of Mode III Stress Intensity Factor. Int J Mech Sci 23: 105 - 117

Some New Trials on the Technique
of the Method of Caustics

K. Shimizu[1], S. Takahashi[1] and H.T. Danyluk[2]

[1] Department of Mechanical Engineering, Faculty of Engineering, Kanto Gakuin University, Yokohama, 236 Japan
[2] Department of Mechanical Engineering, University of Saskatchewan, Saskatoon, S7N 0W0, Canada

ABSTRACT

The correlation between the deviation of light rays and the method of caustics is discussed, and the magnitude of this deviation is shown in diagrams. The application of caustics to stress-frozen models is studied. Lastly, the caustics method is used to determine the stress intensity factor of a crack in a rotating disk.

INTRODUCTION

The method of caustics (Manogg 1966; Theocaris 1970), a recently developed experimental method for stress analysis, has many advantages particularly in measuring the stress intensity factor. Lately, it has applied to various problems (Shimizu et al. 1985; Theocaris 1982; Shockey et al. 1983). This method, however, is not fully established and there still remain some technical points to be clarified.

In this paper, we discuss the correlation between the deviation of a light ray and the method of caustics, which is the fundamental principle used in caustics. Next, the possibility of application of this method to a stress-frozen model is studied. Lastly, we apply the caustics method to determine the stress intensity factor of cracks in rotating disks and the results obtained by caustics are compared with those obtained by the conventional stress-freezing technique of photoelasticity.

THE DEVIATION OF A LIGHT RAY AND THE CAUSTIC PATTERN

A light ray which passes through a medium is deviated if there is a change in the refractive index. The caustics method utilizes this deviation of a light ray caused by the change of the refractive index by stress. We first consider the magnitude of the deviation of light. The direction of light $\vec{\tau}$ given by the theory of Eikonal (Klein 1970) is governed by

$$\text{grad } s = n\vec{\tau} \qquad (1)$$

where s is the optical path length and n is the refractive index.

Designating the optical path length in the specimen as s_1 under no loading and as s_2 under loading, the difference Δs is

$$\Delta s = s_2 - s_1 = (n' - n_0)t + (n_0 - 1)\Delta t \qquad (2)$$

In eq (2), n_0 and n' are the refractive indices of the specimen before and after loading, respectively, t is the thickness of the specimen and Δt is the increase in thickness due to loading. Using eq (2), eq (1) can also be written as

$$\text{grad } \Delta s = \frac{\vec{w}}{z_0}. \qquad (3)$$

$\vec{w}$ is the position vector of light ray at the screen and z_0 is the distance between the specimen and the screen. The magnitude of the deviation of the light ray $\vec{w}$ can be obtained by calculating the value of Δs using eq (3). Assuming that the light is transmitted under plane stress conditions in an optically isotropic material such as a perspex sheet, Δs is given by

$$\Delta s = c_0 t(\sigma_1 + \sigma_2) \qquad (4)$$

where $c_0 = \frac{1}{2}(c_1 + c_2) - \frac{\nu}{E}(n_0 - 1)$. $\qquad (5)$

In eq (5), c_1 and c_2 are the photoelastic constants, and E is Young's modulus, and ν is Poisson's ratio of the material. We can calculate the deviation of a light ray using eqs (3) - (5). Denoting the deviation by the angle ϕ and assuming that only the stress σ_y varies in the y direction, the deviation angle ϕ can be written as

$$\phi = \left| \text{grad } \Delta s \right| = c_0 t \frac{d\sigma_y}{dy}. \qquad (6)$$

The value of c_0 is usually negative and therefore ϕ has a negative value for the positive stress gradients. Figure 1 shows values of ϕ calculated using eq (6) for $c_0 = -1.045 \times 10^{-10}$ m^2/N, obtained for the perspex plate in transmitted caustics (Shimizu et al. 1978). The broken line in this figure shows the values when immersion fluid is used. In this case, Δs is obtained by putting $\Delta t = 0$ in eq (2), and the constant c_0 is calculated to be $\neg 0.42 \times 10^{-10}$ m^2/N using E = 2940 MPa, $\nu = 0.37$, and n = 1.5 for the perspex plate.

The deviation of a light ray given by ϕ influences the results of the conventional photoelastic experiment and gives rise to some error when a large stress gradient exists and the specimen is comparatively thick. Such a problem is considered in a paper on integrated photoelasticity (Aben et al. 1984).

Deviation of light is the fundamental principal behind the method of caustics. The caustic pattern on the screen can be calculated by using eqs (3) and (4). By measuring the maximum diameter D of this pattern, the value of the stress intensity factor K_I can be determined through the equation (Manogg 1966; Theocaris 1970)

$$K_I = \frac{1.671}{z_0 t(-c_0)} \times \frac{1}{\lambda^{3/2}} \times \left(\frac{D}{3.17}\right)^{5/2}. \qquad (7)$$

The initial curve r_0, defining the position of the light ray on the specimen forming the caustic pattern on the screen, is given by

$$r_0 = D/(3.17\lambda) \tag{8}$$

where λ is the magnification factor of light if divergent light is used. This factor is defined by the relation $\lambda = (z_0 + z_i)/z_i$, where z_i is the distance between the point light source and the specimen.

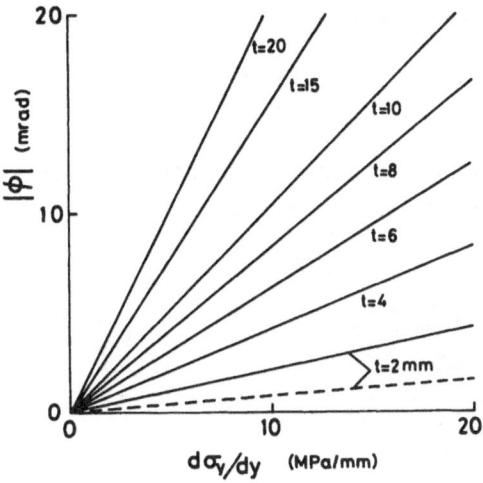

Fig. 1. Effects of the stress gradient and the thickness on the angle ϕ for light deviation (perspex plate) (———in the air,-----in the immersion fluid)

APPLICATION TO THE STRESS-FROZEN MODEL

The caustic pattern, under plane stress conditions, is formed by the sum of in-plane principal stresses and the thickness change. For a stress-frozen model suspended in immersion fluid, the caustic pattern formed on the screen is based on n´ alone as there is no thickness change Δt. In the following sections, we determine experimentally whether the method of caustics is as applicable to stress freezing as is photoelasticity.

Specimen and Experimental Procedure

We employed plates of epoxy, perspex, and polycarbonate. Each specimen was a rectangular shape having a single edge notch approximately 0.3 mm wide. Figure 2 shows the schematic diagram of the experimental apparatus. As shown in the figure, a collimated laser source was used as the light source. For viewing the caustic pattern during the stress-freezing process when the specimen is immersed in immersion fluid, a small box made of perspex plate is filled with immersion fluid and placed around the notch.

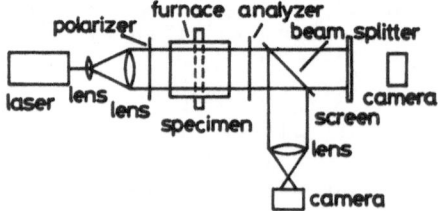

Fig. 2. Experimental setup for the observation of the caustic pattern during the stress-freezing process

Experimental Results

We describe the result of the epoxy resin Araldite D whose stress-freezing temperature was 52^0C. Figure 3 shows photographs of the

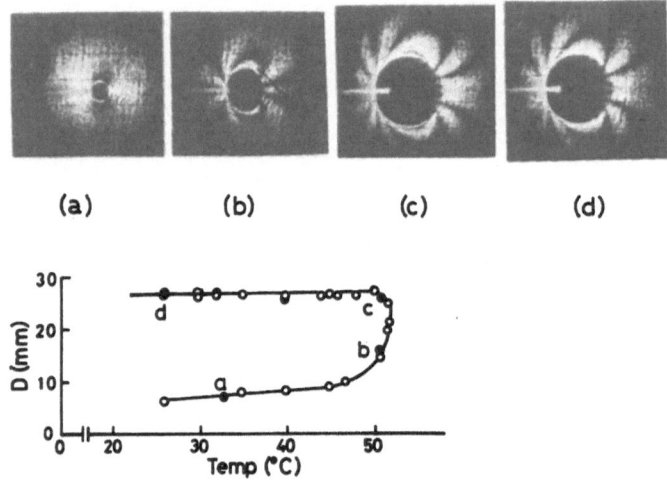

Fig. 3. Variation of the caustic pattern during the stress-freezing process (epoxy resin)

caustic patterns and the isochromatic fringe patterns obtained simultaneously without the use of the immersion fluid. The variation of the size D of the caustic pattern with temperature is also shown in this figure. Figure 3 (d) is the photograph after the stress-freezing process is finished and the load removed. It is seen that the size of the caustic pattern and the isochromatic fringe order is abruptly increased at the critical temperature, and there is little change in those patterns after that temperature. Although the caustic pattern for the optically anisotropic material such as epoxy resin has a double-circle shape at room temperature, no such pattern

existed on the specimen.   Next, we tried to observe the caustic
pattern obtained by putting the stress-frozen specimen into immersion
fluid; however, no caustic pattern was obtained, even though the
isochromatic fringe pattern remained.   An optical constant $c_0 = -82 \times 10^{-10}$ m^2/N was found for the stress-freezing condition.

As mentioned previously, when applying the caustics method to the
stress-frozen specimen, one observes the caustic pattern only in the
air atmosphere.   No pattern appears when the model is in immersion
fluid.   Consequently, the caustics method cannot be applied to slices
as in the usual stress-freezing technique of slicing.   It is felt,
however, that this method is useful in determining the value of the
stress intensity factor of a crack in a rotating disk.

The reason the caustic pattern cannot be observed using the immersion
fluid during the stress-freezing process (although the isochromatic
pattern appears in the epoxy and the polycarbonate plates) is that
the increments in the optical constants c_1 and c_2 during the
stress-freezing process are of opposite sign and therefore, a large
increase in the value c_0 cannot be expected.   This is easily seen
from eq (5).   The caustic pattern during the stress-freezing process
is formed mainly by the thickness changes and it takes the shape of
single circle, not double, even for those materials.

APPLICATION TO THE DETERMINATION OF K-VALUE IN ROTATING DISKS

Possible fracture of a rotating disk is a very critical problem in
numerous machine structures.   Ishida (1981) presented a theoretical
paper concerning the stress intensity factor of rotating disks, but
there is no corresponding theoretical study for disks with a rim.
There exists, however, some experimental work on this problem; viz,
the work of Blauel et al. (1977) which also includes results for
disks without a rim.   In this section, we evaluate the stress
intensity factor of the cracks in rotating disks with a rim using the
photoelastic stress-freezing technique.   Moreover, we apply the
method of caustics as presented in the preceding sections to this
problem.   If the caustics method as suggested in the preceding
section can be effectively used for these cases, then it can be used
to evaluate the corresponding K-value for such problems.

Specimen and Test Procedure

The material used was an epoxy resin prepared by mixing Araldite B
and a curing agent HT901 to a weight ratio of 100:30.   The
fundamental shape of the specimen is shown in Fig. 4.   Specimens
with various rim thickness, b, were used.   Four cracks were located
in one specimen.   A disk of diameter D = 200mm and some disks with
cracks oriented at 45^0 to the radial direction were prepared.   The
notch tip was made by hand using the thin blade of a high-speed
cutter.   The surface of some specimens was polished by using a water
proof abrasive paper and an abrasive agent.

A see-through furnace was employed to observe the development of the
photoelastic fringe pattern.   A variable speed motor was mounted
outside the furnace with its shaft extending inside the furnace.   A
keyed disk was attached to the end of this shaft.   The experimental

conditions were determined as 125^{0}C for the stress-freezing temperature, 1300rpm for the rotating speed and 3^{0}C/hr for the cooling rate.

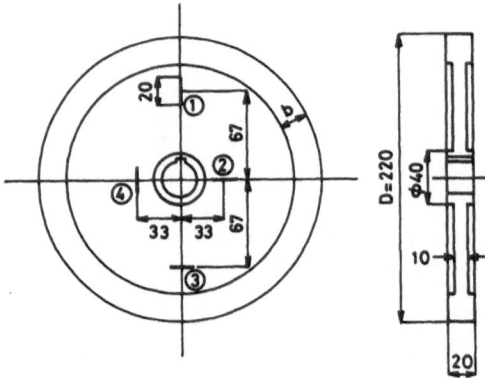

Fig. 4. Dimensions of the rotating disks

Experimental Results

Examples of the isochromatic fringe pattern and the caustic pattern are shown in Fig. 5. The caustic pattern could not be observed in the immersion fluid for reasons as described in the preceding section. The stress intensity factor was determined from the photoelastic pattern by use of the Taylor-series Correction Method (TSCM), (Schroedl et al. 1974) and a computer. The K-value was obtained from the caustic pattern using eq (7). For this case, the constant $c_0 = -176 \times 10^{-10}$ m^2/N. The effect of the width, b, of the rim on the stress intensity factor is shown in Fig. 6(a). The values for b = 0 were obtained by Ishida (1981). As seen from Fig. 6(a),

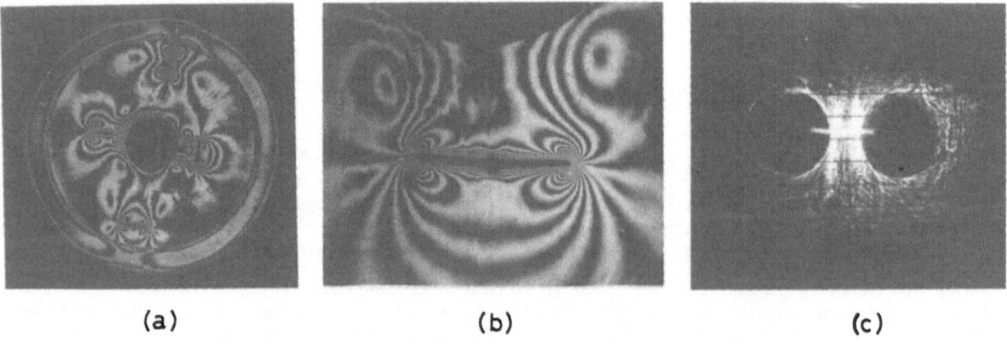

| (a) | (b) | (c) |

Fig. 5. Examples of the photoelastic fringe pattern and the caustic pattern(D=200 mm) (a) whole field pattern (b) crack No. 4 (c) caustic pattern of crack No. 4

cracks near the boss have larger K-values and cracks in the radial direction have larger K_I than circumferential ones. The value of K_{II} is approximately $0.1K_I$ for the crack at a 45^0 direction and is zero for radial and circumferential cracks. This shows that the effect of K_{II} is minimal in rotating disks. Examples of K_I obtained by using the caustics method are shown in Fig. 6(b). The figure shows the variation of K_I with the crack position along the radial direction.

As shown in Fig. 6(b), the results obtained using caustics are slightly lower than those obtained using photoelasticity. One reason may be that although the initial curve r_0 was to be more than half the thickness t of the specimen, it is possible that the r_0 value is not large enough and there is a "r_0 effect" (Shimizu et al. 1985).

Another reason is that the effect of the deviation of light will be larger in caustics than in photoelasticity for thick specimens, say for $t \geq 10$ mm. As shown in the preceding section, the caustics method uses the deviation of light directly. Figure 6(b) shows that the caustics method can be effectively used in determining the K-values for rotating disks when considering the points mentioned above.

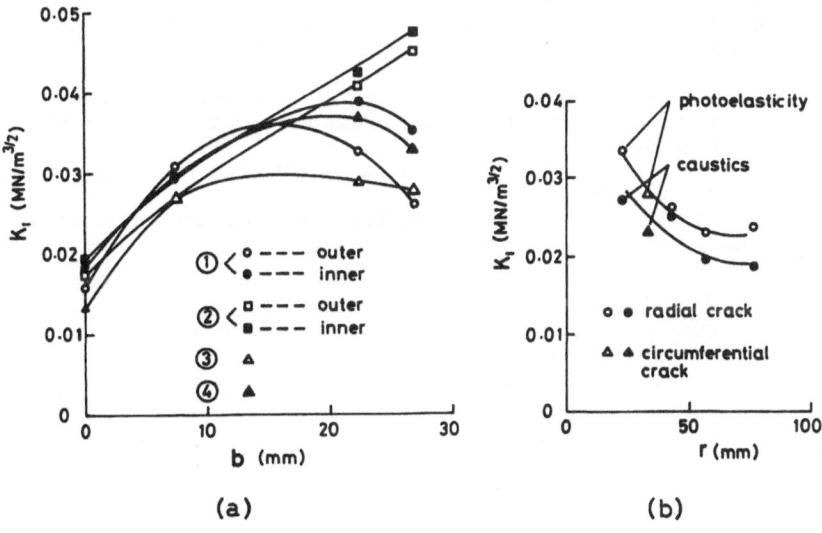

(a)   (b)

Fig. 6. (a) Effect of the width of the rim on the stress intensity factor (D=220 mm) (b) Comparison of the results obtained by caustics with those by photoelasticity (D=200 mm)

CONCLUSIONS

(1) The deviation of a light ray in the caustics method was displayed graphically.
(2) The caustics method can be applied to a stress-frozen model only in air atmosphere, and this method can be effectively utilized in determining the K-values for cracks in rotating disks.

(3) The Mode I stress intensity factor for rotating disks increases
monotonically with the width of the rim, particularly for radial
cracks near the boss.

REFERENCES

Aben HK, Krasnowski BR, Pindera JT (1984) Nonrectilinear light
  propagation in integrated photoelasticity of axisymmetric bodies.
  Trans CSME 8:195-200
Blauel JG, Beinert J, Wenk M (1977) Fracture-mechanics investigations
  of cracks in rotating disks. Exp Mech 17:106-112
Ishida M (1981) (in Japanese). Trans Jap Soc Mech Eng 47:229
Klein MV (1970) Optics. Joh Wiley & Sons, Inc, p 28
Manogg P (1966) Die Lichtablenkung durch eine elastisch beanspruchte
  Platte und die Schattenfiguren von Kreis- und Risskerbe.
  Glastechnische Berichte 39:323-329
Schroedl MA, McGowan JJ, Smith CW (1974) Determination of
  stress-intensity factors from photoelastic data with applications
  to surface-flaw problems. Exp Mech 14: 392-399
Shimizu K, Shimada H (1978) Determination of stress intensity factor
  by the method of caustics (in Japanese). J Jap Soc for
  Nondestructive Inspection 27:399-406
Shimizu K, Takahashi S, Shimada H (1985) Some propositions on
  caustics and an application to the biaxial-fracture problem. Exp
  Mech 25:154-160
Shockey DA, Kalthoff JF, Klemm W, Winkler S (1983) Simultaneous
  measurements of stress intensity and toughness for fast-running
  cracks in steel. Exp Mech 23: 140-145
Theocaris PS (1970) Local yielding around a crack tip in plexiglas.
  Trans ASME Ser E 37: 409-415
Theocaris PS (1982) Complex stress intensity factors in bent plates
  with cracks. Trans ASME Ser E 49:87-96

Application of the Caustic Method to an Environmental Crack-Craze Growth Problem

K. Takahashi, N. Takeda and A.E. Abo-El-Ezz

Research Institute for Applied Mechanics, Kyushu University, Kasuga-koen, Kasuga, Fukuoka, 816 Japan

INTRODUCTION

The stress intensity factor $K_I(c)$ is a controlling parameter for craze initiation and growth at crack tips of linear glassy polymers in environmental liquids (Marshall 1970). However, as the craze grows larger than the one which the Dugdale model (Dugdale 1960) assumes, linear fracture mechanics fails to describe the craze growth behavior. The caustic method is applied to a study of the environmental crack-craze stress field in poly(methyl methacrylate)(PMMA). The change of the caustic shape and size reflecting the nonuniform stress state along a craze is experimentally correlated with the craze growth behavior (Abo-El-Ezz). The caustic method was originally based on an elastic assumption (Mannog 1966; Theocaris 1970) and later was applied to materials displaying a large amount of plasticity and strain-hardening (Theocaris 1973, 1974). The two-step stress distribution model along a craze is introduced for a quantitative analysis by the elasto-plastic caustic theory (Takeda). The theoretical caustic shape and size based on this model is then compared with the experimental results.

EXPERIMENTAL

Single-edge-cracked PMMA specimens with dimensions of 200 x 40 x 2 or 5 mm were cut from a commercial sheet (Acrylite S). The middle part of the specimen containing a crack was immersed in methanol and uniaxially loaded by a constant tensile load at 20°C. Caustics were displayed on a screen plate behind the specimen in a transmitted optical system while the craze length was measured by a traveling microscope and the caustic patterns were photographed. A correction was made for the size of the caustic of a specimen in a liquid to obtain the real size in air.

The value of $K_I(c)$ was given by the linear fracture mechanics equation (Brown 1966). The caustic theory correlated the experimental value of the stress intensity factor $K_I(s)$ to the caustic diameter ϕ and the specimen thickness d as:

$$K_I(s) = A(\phi^{5/2}/d) \tag{1}$$

where A is an experimental constant. For a PMMA crack containing a small craze, the agreement between $K_I(c)$ and $K_I(s)$ is satisfactory (Sakurada 1981). In this case the Dugdale model predicts the average craze stress $\sigma_{c\ av}$ as:

$$\sigma_{c\ av} = \sqrt{(\pi/8)}\ K_I(s)/\sqrt{R} \qquad R/a \ll 1 \tag{2}$$

where R is the craze length and a, the crack length. As the craze be-comes larger, however, σ_C may no more be expressed by equation (2). In the present study, the craze stress σ_C was assumed to be a function of the caustic diameter ϕ as a first approximation as:

$$\sigma_C(x,t) = F(\phi), \quad \phi = \phi_{crack} \text{ at } x = 0 \text{ and } \phi = \phi_{craze} \text{ at } x = R \quad (3)$$

where x is the distance from the crack tip, t is the total elapsed time from the application of loading, and ϕ_{crack} is the caustic diame-ter at the crack tip and ϕ_{craze} is that at craze tip.

RESULTS AND DISCUSSION

Figure 1 shows an example of the measured craze length versus time under a comparatively low level of $K_I(c)$, while Fig. 2 shows caustic patterns at different craze lengths, each mark corresponding to the same one in Fig. 1. In Fig. 3, values of ϕ at the craze tip under dif-ferent levels of $K_I(c)$ were plotted against the craze length for 2 mm thick specimens. The change of ϕ may be divided into three stages. First, as the craze increases in length the value of ϕ decreases to a certain level designated ϕ_m. Second, the craze continues to grow with this ϕ_m until it reaches the equilibrium length depending on the level of $K_I(c)$. Third, the value of ϕ at the craze tip begins to increase without noticeable change of the craze length. In the first stage the decrease in ϕ may be caused by the craze growth, which reduces the craze-tip stress. The second stage may well be explained using the results of a thicker specimen (5 mm) shown in Fig. 4, where the change of ϕ at the crack tip is also given. While the value of ϕ at the craze tip remains almost constant (ϕ_m) that at the crack tip continues to decrease, probably due to the plasticization effect in the craze. The increase of ϕ in the final stage might be explained as follows: because the plasticization continues to take place particularly nearer the crack tip, the stress level at the craze tip should be raised without any craze growth. Another possible way to interpret this increasing stage is to assume the onset of the out-of-plane deforma-tion taking place without enhanced stress intensity at the craze tip. Figure 5 reveals that the change of ϕ at the craze tip versus the craze speed dR/dt is well fitted in one master curve even under dif-ferent levels of $K_I(c)$.

For specimens with the same thickness, the value of ϕ_m seems to be nearly equal, independent of the value of $K_I(c)$ and the craze length. Moreover, even for specimens with different thicknesses, the values of $\phi_m^{5/2}/d$ (see eq. (1)), can be shown to be almost equal. This means that one may define the *critical equivalent stress intensity factor* $K_I(s)_{c\ eq} = A(\phi_m^{5/2}/d)$ at the craze tip. The value of $K_I(s)_{c\ eq}$ for the present material is 0.15 $MN/m^{3/2}$. This value may well be compared with the minimum value of $K_I(c)$, presently 0.12 $MN/m^{3/2}$, below which no environmental craze growth would occur, and that given by Marshall et al (1970) as 0.06 $MN/m^{3/2}$.

On the basis of eq. (3) and sizes of the caustics at both crack and craze tips (Fig. 4), the stress distribution along a craze is assumed as shown in Fig. 6. It is suggested that: (1) the maximum stress ($\sigma_{C\ max}$) exists at the craze tip while the minimum ($\sigma_{C\ min}$) at the crack tip and that (2) during craze growth, both $\sigma_{C\ max}$ and $\sigma_{C\ min}$ decrease, with a larger rate for the latter.

THEORETICAL CONSIDERATIONS

The experimental method of caustics has proved to give valuable information on the stress distribution along a craze. An elasto-plastic theory of caustics, however, is necessary for quantitative analysis to correlate the caustic pattern to the stress distribution along a long environmental craze. The two-step stress distribution model(Williams 1984), or the modified Dugdale-Barenblatt (DB) model (Fig. 7) is used to predict the theoretical caustic pattern (Takeda). The Westergaard stress function for the modified two-stage DB model is

$$z = \frac{2\sigma_o}{\pi}[\lambda\cot^{-1}\{\frac{[z^2-(a+R)^2]^{1/2}}{z[(1+R/a)^2-1]^{1/2}}\}$$

$$+(1-\lambda)\cot^{-1}\{\frac{(1+\mu R/a)[z^2-(a+R)^2]^{1/2}}{z[(1+R/a)^2-(1+\mu R/a)^2]^{1/2}}\}] \qquad (z = x + iy)$$

(4)

with the required condition:

$$\frac{\sigma}{\sigma_o} = \frac{2}{\pi}[\lambda\cos^{-1}(\frac{1}{1+R/a}) + (1-\lambda)\cos^{-1}(\frac{1+\mu R/a}{1+R/a})]. \qquad (5)$$

The equation of the initial curve of the caustic is

$$|\frac{2C}{\lambda_m}Z''(z)| = 1 \qquad (6)$$

where C is the global optical constant and λ_m is the magnification factor given by $\lambda_m = (z_o + z_i)/z_i$, z_o and z_i being the distances between the specimen and either the image plane or the focus of the impinging laser light beam. The caustic is expressed by

$$W = \lambda_m(z + \frac{2C}{\lambda_m}\overline{Z'(z)}) \qquad (7)$$

where the complex variable z satisfies eq.(6) and W is the complex coordinate on the image plane. The experimental constants required for computation are $\lambda_m = 3.5$, $C = -1.127$ mm^4/N (d = 4.9 mm), a = 6.65 mm, $\sigma_o = 40$ N/mm^2, $\lambda\sigma_o = 3.4$ N/mm^2, and $\sigma = 2.184$ N/mm^2 (after the plate-width correction). The computed initial curves and caustics are shown in Fig. 8. The computed caustics agree well with the experimental ones shown in Fig. 2 both in shape and in size. Thus the modified two-stage DB model is promising in environmental craze stress analysis.

CONCLUSIONS

The caustic method was applied to a study of the environmental crack-craze growth problem of PMMA in methanol. The craze growth behavior was correlated with the change of the caustic size and shape reflecting the nonuniform stress along a craze with its maximum at the craze tip. The concept of a *critical equivalent stress intensity factor* below which no craze growth occurs in liquid was defined. The two-step stress distribution model along a craze was introduced for a quantitative analysis by the elasto-plastic caustic theory. The theoretical caustic size and shape agreed well with the experimental ones.

REFERENCES

Abo-El-Ezz AE, Takeda N, Takahashi K (to be published) Caustics obser-
vations for a study of environmental crack-craze stress fields.
J Mater Sci
Brown WF Jr, Srawley JE (1966) Plane-strain crack toughness testing
of high-strength metallic materials. ASTM STP 410
Dugdale DS (1960) Yielding of steel sheets containing slits. J Mech
Phys Solids 8: 100-108
Mannog P (1966) Die Lichtablenkung durch eine elastisch beanspruchte
Platte und die Schattenfiguren von Kreis-und Risskerbe.
Glastechnische Berichte 39: 323-329
Marshall GP, Culver LE, Williams JG (1970) Craze growth in polymethyl-
methacrylate: a fracture mechanics approach. Proc Roy Soc London
A319: 165-187
Sakurada Y, Takahashi K (1981) Measurement of the stress-intensity
factor for poly(methyl methacrylate) cracks by using the method of
caustics. Kobunshi Ronbunshu 38: 369-375 (in Japanese)
Takeda N, Abo-El-Ezz AE, Takahashi K (to be submitted) The modified
theory of caustics for evaluation of the environmental craze stress
distribution.
Theocaris PS (1970) Local yielding around a crack tip in plexiglas.
J Appl Mech Trans ASME Ser E 37: 409-415
Theocaris PS (1973) Stress intensity factors in yielding materials by
the method of caustics. Int J Frac 9: 185-196
Theocaris PS, Gdoutos E (1974) The modified Dugdale-Barenblatt model
adapted to various fracture configurations in metals. ibid 10:
549-564
Williams JG (1984) Fracture mechanics of polymers. Ellis Horwood Ltd,
Chichester, p 191

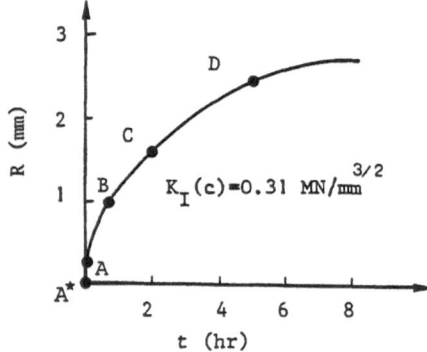

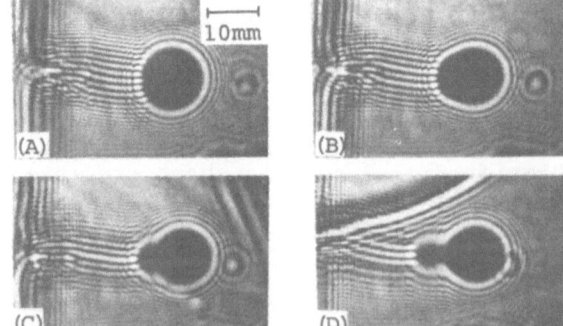

Fig. 1. An example of change
of the craze length (R) with
time (t) under a constant
load for the specimen thickness of 5 mm.

Fig. 2. Caustic patterns at different
craze lengths. Stages A to D correspond
to those in Fig. 1.

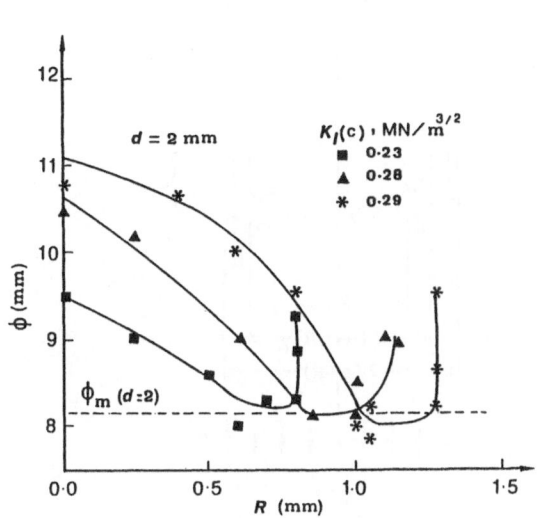

Fig. 3. Values of φ at the craze tip vs R for the specimen thickness of 2 mm under different levels of $K_I(c)$.

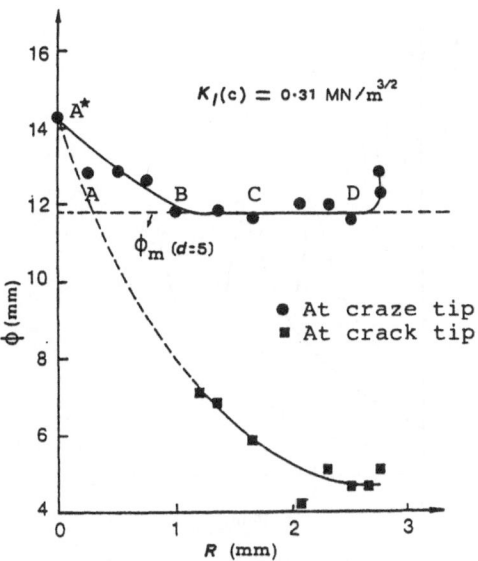

Fig. 4. Values of φ at both craze and crack tips vs R for the specimen thickness of 5 mm.

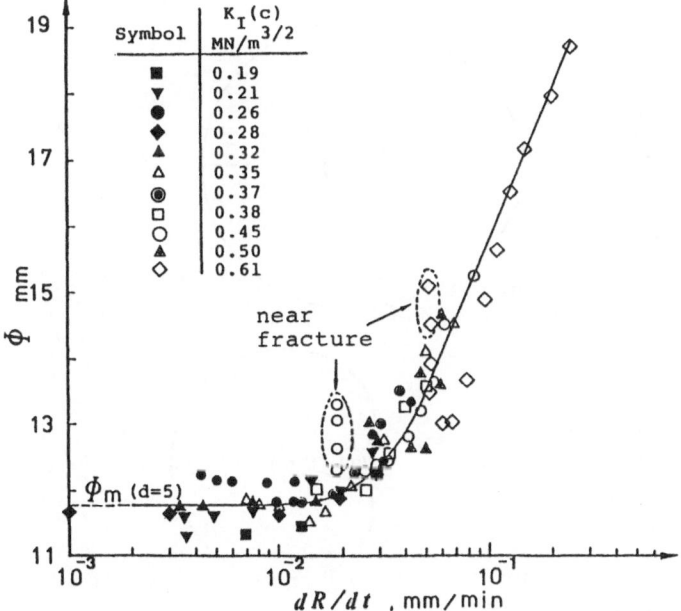

Fig. 5. A master curve for the change of φ at the craze tip vs the craze speed (dR/dt) under different levels of $K_I(c)$ (specimen thickness d = 5 mm).

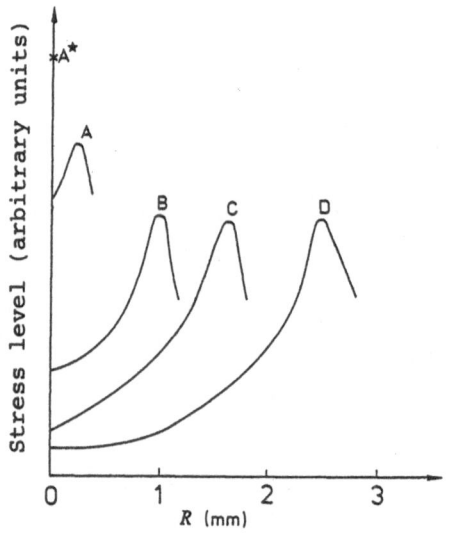

Fig. 6. Probable stress distribution along a growing craze in methanol predicted by the results in Fig. 4. Stages A to D correspond to those in Figs. 1, 2 and 4.

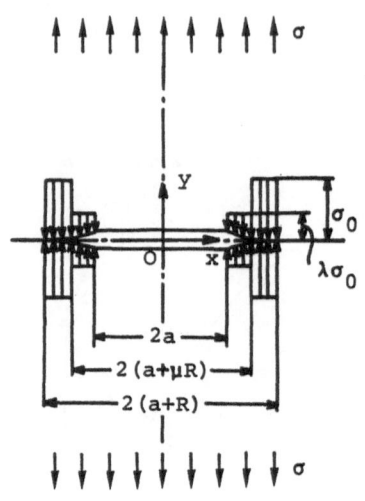

Fig. 7. The modified two-step DB model.

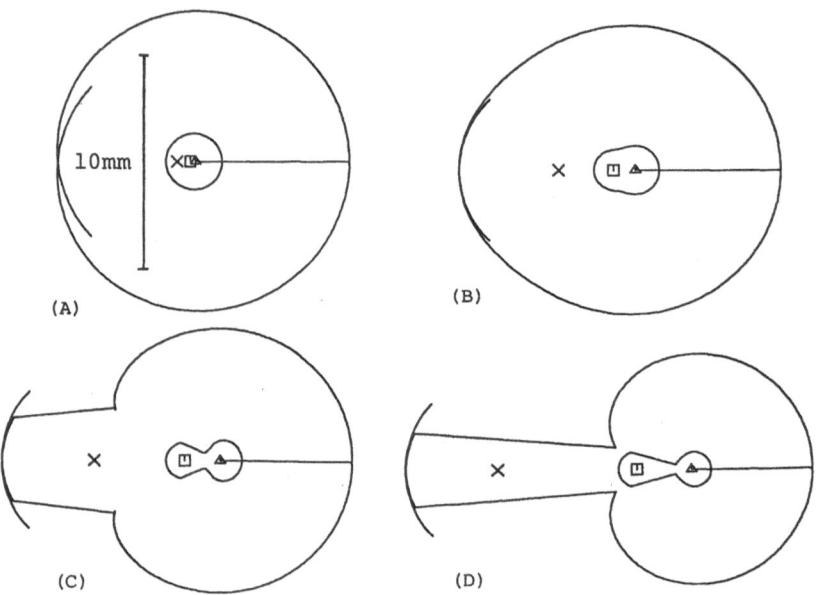

Fig. 8. Theoretical initial curves and caustics based on the modified two-step DB model. Stages A to D correspond to those in Fig. 2.

Initial curves □ crack tip, △ craze tip

Caustics × crack tip, △ craze tip

Topics in Photomechanics
Cracks, Waves and Contacts

H.P. Rossmanith

Institute of Mechanics, Technical University Vienna, Vienna, Austria

ABSTRACT

High-speed recording techniques in conjunction with methods of photomechanics serve
as a means for visualizing the highly complex interaction processes between elastic
stress waves and geometrical discontinuities such as boundaries, interfaces, cracks
and contact surfaces. Four topics of importance and general interest will be addres-
sed here: the *interaction between elastic stress waves and static or running cracks,
mixed-mode crack propagation, dynamic contact of impacting bodies* and *stress wave
focussing effects*. Experimentally recorded isochromatic fringe patterns and sequences of caustics assist and confirm numerical and analytical computations.

INTRODUCTION

With the development of advanced high-speed photographic recording methods and data
processing techniques dynamic photomechanics emerged from its state of infancy to
become a highly developed and due to its prosperity most attractive field of research with great potential for industrial applications. New solutions to complex problems hitherto believed unmanagable have been developed in fields as seemingly so
unrelated as removal of natural resources in mining engineering, instabilities in
the flow of grain in large hoppers and dental preventive care by ultrasound. The
body of scientific literature pertaining to photomechanics and its applications to
the engineering world is expanding at ever-increasing rate and it seems virtually
impossible for a single individual to keep track of the latest developments in all
fields where photomechanics' results are appreciated.

All of the problems dealt with in the 'applied' technical papers can be reduced to
very few basic problems of fundamental research. Extensive literature research performed by the author indicates a 90%-coverage of practical problems by the fundamental set: *interaction between elastic waves with inhomogeneities or discontinuities,
mixed-mode crack propagation, stress-wave focussing and dynamic contact problems*.
This of course refers to dynamic elasticity problems only.

CRACK-WAVE INTERACTION

When elastic waves are generated and propagated e.g. during blasting or an earth-
quake phenomenon, they interact with geometric discontinuities or acoustical imped-
ance mismatch zones and are reflected, refracted and diffracted and often give rise
to a high elevation of local stresses. These stress concentrations become extremely
severe when the discontinuity is a static or moving crack. The result of waves reflecting, diffracting and refracting about the crack tip may lead to unstable crack
behavior and fracture initiation (Rossmanith,1983). The incident wave can be either
a P-wave (polarized in the direction of the incoming wave) or an S-wave which may be
decomposed into an SV-wave (polarized parallel to the two-dimensional specimen plane
) and an SH-wave (polarized normal to the specimen plane). Plane dynamic photomechanics deals with the propagation of P-and SV-waves.

In connection with (obliquely) impinging P-and SV-waves normal opening and tangential sliding modes of fracture occur in various combinations in isotropic materials.
General wave diffraction at crack tips gives rise to a mixed-mode transient fracture
problem which for purposes of analysis is conveniently divided into a symmetric part
(mode I) and an antisymmetric part (mode II). Anti-plane mode of fracture is associated with SH-wave effects.

The incident wave will be diffracted and scattered about the crack tip and regardless of the type of incident wave, reflection and diffraction will give rise to a diffracted-reflected (scattered) stress field $\sigma_{ij,s}(x,y,t)$ which consists of both, P-and SV-waves.

The total stress field $\sigma_{ij,t}(x,y,t)$ will thus be composed of the stress field of the incident wave $\sigma_{ij,i}$ and ^{ij,t} the interaction field $\sigma_{ij,s}$:

$$\sigma_{ij,t}(x,y,t) = \sigma_{ij,s}(x,y,t) + \sigma_{ij,i}(x,y,t) \tag{1}$$

where the scattered field $\sigma_{ij,s}$ must satisfy the condition $\sigma_{ij,s} \rightarrow 0$ as $\sqrt{(x^2+y^2)} \rightarrow \infty$. The associated wave function must satisfy the Helmholtz equation (Graff,1975). In addition, boundary conditions appropriate for stress-free crack walls are to be prescribed (Sih,1977).

In the general crack-wave interaction process as shown in Fig.1 for the diffraction of a P-wave about an embedded crack of finite length the following elementary phases can be distinguished (Rossmanith-Shukla,1981):

1   *incident P-wave*   6   *S-wave reflection*
2   *P-wave reflection*   7   *primary S-wave diffraction at tip A*
3   *primary P-wave diffraction at tip A*   8   *secondary S-wave diffract. at tip B*
4   *secondary P-wave diffraction at tip B*   9   *Rayleigh-wave generation*
5   *incident S-wave*   10 *higher order diffraction*

A variety of elastic waves is generated and the total wave field for the diffraction problem may be constructed as the linear superposition of an incident wave field and various reflection and diffraction wave fields. Because the scattered wave systems will contain both, P-and SV-wave contributions both stress intensity factors K_1 and K_2 are present in the general case. The $1/\sqrt{r}$-stress singularity and the associated singular stress distribution about the crack tip is the same as under static loading for harmonic wave diffraction with the complex stress intensity factor $K^*=K_1-iK_2$ being time-dependent. Hence, the instantaneous isochromatic fringe patterns generated during the diffraction process can be directly compared with mixed-mode fringe patterns as obtained (Rossmanith,1979) for statically loaded cracks. A close-up of the experimentally recorded isochromatic pattern in the near crack-tip zone about B is compared in Figure 2 with an analytically generated mode-II fringe pattern, where the effect of a compressive stress field which acts parallel to the crack line has been taken into account (Rossmanith-Irwin,1979; Rossmanith,1979).

From a fracture mechanics viewpoint the stress-wave amplitude distribution of the wave pattern is of importance because it yields not only the locus but also the time of fracture. Even when the amplitude of the incident wave is too low to initiate fracture, the joint action of the various waves of the diffraction/reflection process may still cause crack extension at either one of the crack-tips or, when Rayleigh-wave effects are included, fracture initiation may take place somewhere along the crack walls (Brock,1975).

A dynamic photoelastic study (Rossmanith-Shukla,1981b) of the interaction of normal and obliquely incident elastic stress waves with running cracks shows a branching enhancement or suppression effect associated with shear wave interaction that depends on the geometrical configuration.

Formations encountered at quarry sites and oil shale deposits form stacks of layered rock with bedding planes and sets of joints present. Upon detonation of an explosive the wave pattern generated in layered media is extremely complicated, where incident and reflected wave systems not only interact with running cracks but also with existing cracks, flaws, inclusions or other inhomogeneities within the formation. Stress wave scattering about the tips of stationary interface cracks in a layered medium and the associated stress-wave induced fracture were investigated by means of dynamic photoelasticity and linear elastic fracture mechanics to analyse regions of high stress intensity as possible sources for crack initiation. The phenomenon of partial load transmission across closed crack walls and imperfect joints and its effect on fracture initiation was also studied (Rossmanith-Fourney,1982; Rossmanith-Knasmillner, 1983).

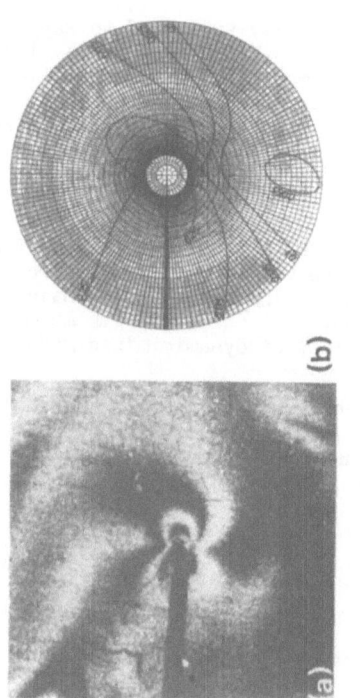

(b)

Fig.2 Blow-up of near crack-tip area B of Fig.1 with mode-2 isochromatic fringe pattern: a) experimentally recorded, b) analytically generated

Nomenclature for various types of waves:

P longitudinal wave R Rayleigh wave
S transverse wave V von Schmidt wave

X_d^{Zy}, X_r^{Zy} diffracted (d) and reflected (r) wave of type X due to an incident wave of type Y

R_+, R_- Rayleigh wave travelling along upper (+) or lower (−) crack wall

P incident longitudinal wave

S_d^A primary diffracted shear wave generated by incident P-wave at crack tip A

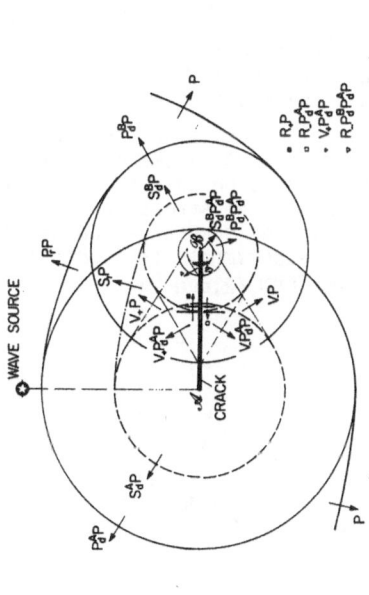

Fig.1 Diffraction of a cylindrical P-wave about a finite-length crack: a)dynamic isochromatic fringe pattern, b) analytical wave front pattern

The problem of a surface-breaking crack or a near-surface cavity subjected to Rayleigh-wave stress loading has been the subject of many contributions (e.g. Dally,1977; Rossmanith-Dally,1983; Cardenas,1983) and has derived its importance from the field of nondestructive testing. Results show that Rayleigh-waves may extend a surface-crack (Rossmanith,1985; Cardenas,1983) or initiate a crack at sharp reentrant corners (Rossmanith,1984).

MIXED-MODE CRACK PROPAGATION

Recently considerable attention has been devoted to the problem of the existence of a mode-2 component during dynamic crack propagation, i.e. the validity of a fracture criterion of the form: $K_{2,dyn}=0$. Despite all efforts the problem appears to remain unsolved to date. Because the problem is a fundamental physical one, there is much disagreement among scientists about the physical possibility of dynamic mixed-mode crack propagation.

Consider a mathematically sharp semi-infinite curved crack in an infinite plane sheet of isotropic, homogeneous and elastic material as shown in Fig.3. The crack contour is given by the parametric representation $\{\xi(t),\eta(t)\}$. Then, crack speed and the curvature of the crack path are given by

$$c(t) = \sqrt{\dot{\xi}^2 + \dot{\eta}^2} \quad , \quad R = \frac{1}{\ddot{\xi}\dot{\eta}}\{\dot{\xi}^2 + \dot{\eta}^2\}^{3/2} \tag{2}$$

The stress distribution in the vicinity of a moving crack tip subjected to combined loading conditions may be represented in the form

$$\sigma_{ij}(c,t) = \sum_{k=1,2} \frac{K_k}{\sqrt{2\pi r}} f_{ij,k}(c,\theta;t) + F(r^o,\sqrt{r};\sigma,\tau;a,R;c,\theta) + O(r) \tag{3}$$

where the K_k are the dynamic stress intensity factors which depend on the curvature of the crack path, and F is the lowest order regular stress function which is related to crack stability.

A second-order dynamic crack propagation theory is called for when dynamic crack path stability is to be investigated. It is mandatory that this theory focusses on the propagation of a running crack along a curved path subjected to time-dependent stress fields. No such theory has been developed sofar. Freund's (1972) approach to the running crack problem addresses the singular part of the crack tip stress field only.

Unlike in the static case where upon reaching critical conditions a crack initiates at an angle with respect to the direction of the tangent to the crack-tip segment, a running crack seems to be able to adjust its path simultaneously to cope with the changed new condition of stress and strain ahead of the moving crack-tip. For slowly varying stress fields such as encountered with a crack moving in a statically loaded specimen caustic experiments (Kalthoff,1978-85) support a dynamic crack propagation criterion of the form $K_{2,dyn}=0$.

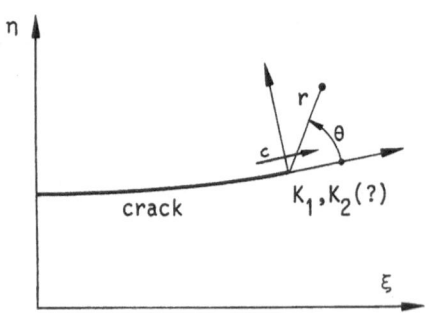

Fig.3 Coordinate systems associated with a high speed running curved crack

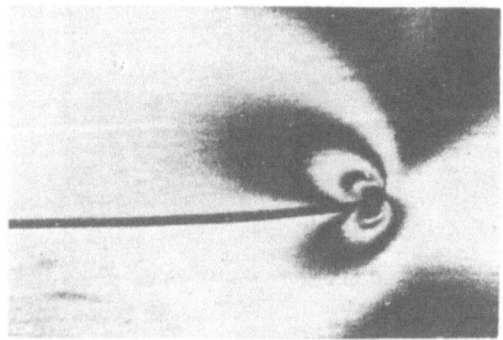

Fig.4 Dynamic isochromatic fringe pattern about mixed-mode loaded running crack

A continuous correction to maintain $K_{2,dyn}=0$ along the crack trajectory would imply that the character of the global and local stress fields be essentially different, with the global field exhibiting a mixed-mode stress pattern and the local field pertaining to mode-1. Dynamic photoelastic recordings such as shown in Fig.4 for a DCB-specimen exhibit an 'apparent' dynamic mixed-mode situation when referred to the global fringe pattern. A pure mode-1 situation is approached when the data region is shrunk to the near-tip region. For this and several other reasons associated with fringe clarity etc. in the crack tip vicinity dynamic photoelasticity may fail to 'prove' the existence of a dynamic mode-2 component in curved crack propagation.

Assuming crack propagation governed by $K_{2,dyn}=0$ as a dynamic fracture criterion the associated K_1-stress intensity factor may be determined by means of an 'apparent' mixed-mode index $\mu_m=K_{2m}/K_{1m}$ (Rossmanith,1983b):

$$K_1 = K_{1m} \sqrt{1 + \mu_m^2\, g_2(c)/g_1(c)} \qquad (4)$$

where K_{1m} and K_{2m} pertain to an arbitrary set of data points of the global 'apparent' mixed-mode isochromatic fringe pattern, and g_1 and g_2 are velocity correction functions (Freund,1972).

The situation becomes entirely different when the running crack interacts with stress waves that have been generated, e.g. by an explosive. Here, dynamic mixed-mode shadow spots, identified by their asymmetrical shape, may exist temporarily during the crack-wave interaction phase. Recent experimental observations in polymeric materials where stress waves impinge on fast running cracks (Rossmanith-Shukla,1981) indicate the existence of a strain-rate induced and/or stress-gradient dependent dynamic mixed-mode index as a function of the transient stress field.

The mixed-mode crack extension complexity extends directly into the fascinating but still vaguely understood field of high-speed crack branching. Investigations pertaining to crack-path directional stability and crack branching can be greatly simplified by making use of dynamic stability charts as developed for inclusion of the lowest higher order regular stress terms (Rossmanith-Irwin,1979; Rossmanith,1981).

DYNAMIC CONTACT PROBLEMS

Photomechanics as a means for visualization of stress distribution and transfer of load across the contact zone of elastic bodies in contact has been utilized successfully for wave propagation studies in connection with granular materials. Although the global aspects of wave propagation in granular media differ considerably from classical wave propagation in continuum mechanics, the elementary process of dynamic contact of two bodies forms the fundamental subject of research.

Experimental and theoretical investigations in granular media have revealed that load transfer is channeled along determined but randomly distibuted zig-zag-type load paths extending into the bulk of the material, which are confined to a cone-shaped region with the apex at the point of load application (see Fig.5). Nonsymmetrical contact isochromatic fringe patterns refer to obliquely contacting disks where normal (P) and shear (T) forces are transmitted (Rossmanith-Shukla,1982).

The magnitude and direction of the (static) contact forces can be evaluated by measuring the geometry of the isochromatics in the vicinity of the contact points and employing equations derived on the basis of elliptical distributions for normal and shearing contact stresses for static problems (Poritzky,1950; Durelli-Wu,1983). Thus the problem is reduced to isochromatic data selcetion and solution of a set of equations of the type

$$\{(N_i f_\sigma /h)^2 - g(a,T/P;x_i,y_i) \}_j = 0 \qquad (5)$$

for the unknown contact area, normal and shear loads (N=fringe order, f_σ=material fringe value, h=model thickness, 2a=contact area, T=shear load, P=normal load; x_i,y_i= cartesian coordinates of data point i in set j of isochromatics)(Rossmanith, 1985). The reader is reminded that sofar static equations have been applied to experimentally recorded dynamic isochromatic fringe patterns.

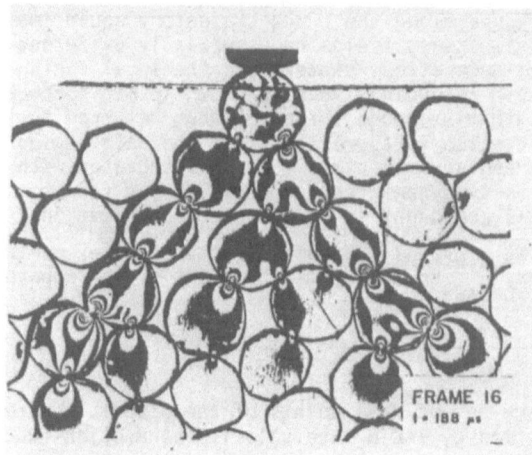 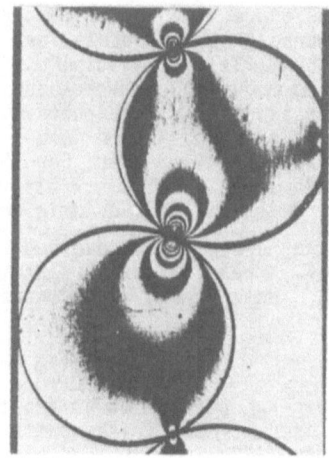

Fig.5a Dynamic isochromatic recording of wave propagation in granular material

Fig.5b Dynamic isochromatic recording of oblique dynamic contact of two elastic disks

General dynamic contact of two elastic bodies during collision or impact represents an extremely complex dynamic physical process the mathematical model of which is complicated by the time-dependent contact area. The extension of contact is not known apriori and therefore it becomes part of the solution of the problem. In general, numerical modelling techniques have to be employed for study of transmission, reflection, refraction and diffraction of elastic waves across the contact zones and about the time-dependent moving contact edge during increasing and receding contact.

A relatively simple dynamic situation is encountered in the well-controlled problem where a half-plane is initially in point contact with a disk which is dynamically loaded in its center (Shukla-Rossmanith,1984). Wave diffraction about the contact yields a system of bulk waves in the disk as well as in the half-plane and two pairs of Rayleigh-waves that originate at the contact site and propagate along the free surfaces. These Rayleigh-waves carry most of their energy concentrated within a thin layer just underneath the free surface and propagate with no dispersion along the straight surface (Cardenas,1983). During the passage of the surface wave this energy concentration causes appreciable thickness variations whithin the boundary layer at points close to the free surface.

The lateral deformation (and for transparent materials the change of refractive index of the material) gives rise to the formation of a time-dependent contact caustic (see Fig.6) given by the image equation

$$\vec{w}(x,y) = \int_{-\infty}^{\infty} P''(\tau) \frac{C_1\eta + C_2(\xi-\tau)}{(\xi-\tau)^2 + \eta^2} \, d\tau \qquad (6)$$

where $\vec{w}$ is the deflection vector for the light rays, $P(\tau)$ is the time-dependent loading function, and $x, y, \xi,$ and η are cartesian coordinates of the ray intersection points. For simplicity, a time-dependent line-force $P(\tau)$ has been selected, but equ. (6) can easily be extended to arbitrary contact pressure distributions. Figure 7 shows the excellent agreement between experimentally recorded and numerically generated Rayleigh-pulse caustics. The associated unknown contact pressure distribution is given in Figure 7c. The size of the caustic is directly proportional to the load magnitude (as shown in equ.(6)) and inversely proportional to the fourth power of the pulse width as measured in the central region. Employing and applying overdeterministic data reduction procedures as common in isochromatic work to the Rayleigh-pulse caustic curve the unknown coefficients of the series expansion of the pressure

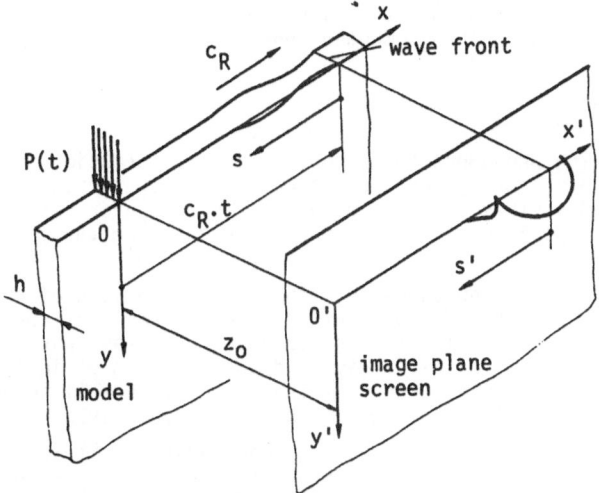

Fig.6 Optical arrangement for the Rayleigh-wave caustic experiment

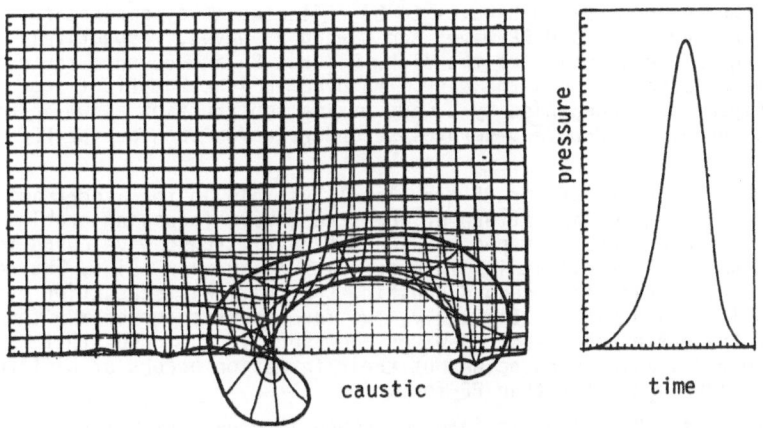

caustic

pressure

time

Fig.7 Numerically generated Rayleigh-wave caustic (a) and pressure-versus-time plot for load function (c)

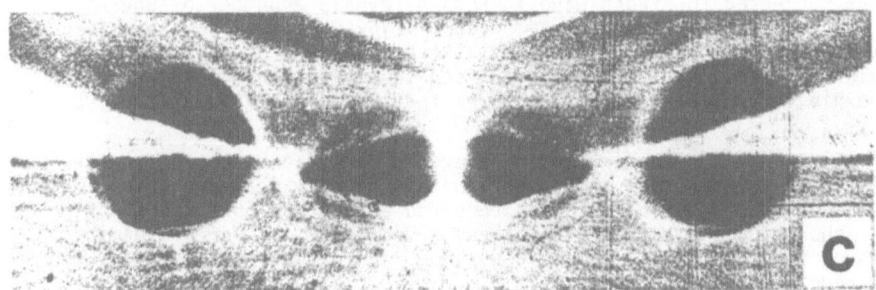

Fig.7b Experimentally recorded Rayleigh-wave caustic for dynamic contact of an explosively loaded disk with a half-plane

function P(t) can be determined. An iterative procedure is recommended, where the initial pulse is estimated and the computer-generated caustics are compared with the experimental recordings.

STRESS WAVE FOCUSSING

One of the most striking phenomena in ray optics is the formation of caustics, i.e. sharp bright light concentrating surfaces to which the light rays are tangential. Light refracted through a stress-loaded transparent specimen or reflected from the warped surface of the specimen yields caustics as known from the method of shadow patterns.

Superposition of stress waves produced by either the reflection of a single wave front or the interaction of two or more reflected wave fronts at convexly shaped boundaries or interfaces gives rise to the formation of stress wave caustics. This stress wave focussing effect in solids may lead to formation of internal fractures. A particular instructive example taken from the field of medical science is the crushing of kidney-stones under shock wave impact (Takayama,1985).

Using ray theory, commonly employed in optics and acoustics, it is possible to predict the positions and extent of these internal fractures (Al-Hassani-Silva Gomes, 1979; Silva Gomes-Al-Hassani,1977; Williams et al.,1983).

In two dimensional models such as a disk, upon detonation of a small explosive charge at the rim of the model, two initial stress waves, a longitudinal (P)-wave and a transversal or shear (S)-wave, emerge from the center of detonation. The curved surface causes the incident P-and S-waves to be reflected as a combination of longitudinal and shear waves, PP-and SP-waves and SS-and PS-waves, respectively, with the rays focussing in the interior of the disk. Figures 8a and 8b show the dynamic event prior to P-wave passage across the disk. The dynamic isochromatic fringe pattern recording in Fig.8a is accompanied by a wave maxima construction shown in Fig.8b. Multiple reflection of the P-wave generates secondary waves such as the PPP- and SPP-waves.

The wave cusps labeled PP in Fig.8b are clearly visible in the isochromatic pattern of Fig.8a. Each isochromatic pattern of the stress wave focussing problem yields two symmetrically located wave cusp points. The set of all these wave cusps forms the PP-caustic line which itself at its cusp exhibits a singular point. The reflected shear SP-wave gives rise to a second caustic, the critical focussation event of which is shown in Fig.8c and Fig.8d. Similarly, two more caustic curves are formed by focussation of the reflected SS-and PS-waves. However, it is found that the PP-and SS-caustic curves spacially coincide but their formation occurs at different times with SS-formation much later than PP-formation.

If the localized shock loading is intense enough internal fractures may originate and initiate along the caustic curves to produce significant cardioid crack configurations as has been reported (Williams et al.,1983) for explosively loaded hemispherically end-shaped cylinders. Post mortem examination of the explosively loaded specimens subjected to stress wave focussing reveals a number of unusual features in terms of fracture formation. The immediate vicinity of the loading site contains a crater, slanted craze formation regions, Hertzian wedge or cone cracks, cracking perpendicular and radial to the wave front and single curved cracks whereas the far region exhibits typical tensile fractures that have initiated at the cusp of the caustic.

Most of the mechanisms leading to these exotic multiple bifurcated crack formations are only vaguely understood at present. Detailed analytical, numerical and experimental work on stress wave focussing is required to improve our understanding of the singular phenomena associated with stress wave focussing. Here, dynamic photoelasticity proves a most powerful and suitable tool for experimental studies (Rossmanith-Knasmillner,1984).

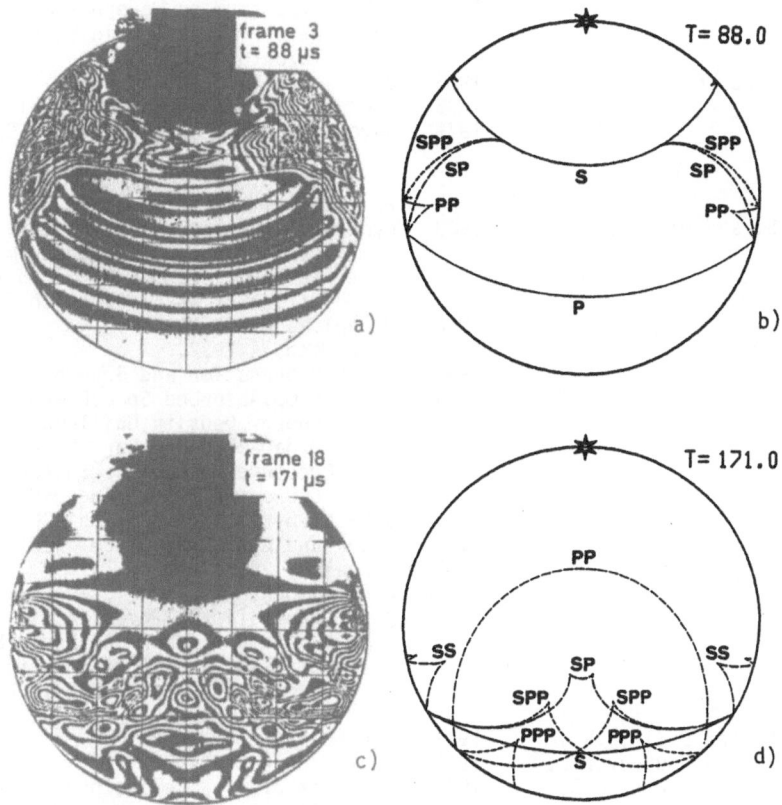

Fig.8 Stress wave focussing in an explosively loaded elastic disk
a) dynamic isochromatic fringe pattern at time t=88 μsec after detonation
b) associated wave maxima pattern
c) dynamic isochromatic fringe pattern at time t=171μsec after detonation
d) associated wave maxima pattern

ACKNOWLEDGEMENT

The author would like to express his sincere thanks to Mr. R.E.Knasmillner and
Dr. A.Shukla for stimulating discussions in the course of this research project.
The research program was financially supported by the Austrian Science Foundation
(Fonds zur Förderung der wissenschaftlichen Forschung in Österreich) under Con-
tract No#4532.

REFERENCES

Al-Hassani,S.T.S. and Silva Gomes,J.F.	1979	Inst.Phys.Conf.Ser.No.47,Chapt.2,187-196.
Brock, L.M.	1975	Int.J.Eng.Sci. 13,851.
Cardenas, J.F.	1983	On Rayleigh-Waves and Rayleigh-Wave extension of Surface Micro-Cracks. PhD Thesis, University of Maryland.

Dally, J.W.	1977	IUTAM-Symposium on Modern Trends in Wave Propagation in Solids, Toronto, Canada.
Durelli, A.J. and Wu,D.	1983	J.Appl.Mech.50,288-296.
Freund, L.B.	1972	J.Mech.Phys.Solids 20,129 and 141.
Graff,K.F.	1975	Wave Motion in Elastic Solids. Ohio State University Press; Clarendon Press, Oxford.
Kalthoff, J.	1982	Private Communication.
Kobayashi, A.S. and Ramulu,M.	1980	Dynamic Stress Intensity Factors for Un-symmetric Dynamic Isochromatics. Technical Report No.37 to the Office of Naval Research,January.
Poritzky, H.	1950	J.Appl.Mech.17,191-201.
Rossmanith, H.P.	1979	Acta Mech.34,1-32.
	1981	Crack Propagation and Branching. Proc. Symp.Absorbed Spec.Energy/Strain Energy Density,Guillemot-Symp., Budapest,September,283-294.
	1983a	Rock Fracture Mechanics (Editor) CISM-Course No.275,Springer-Verlag.
	1983b	J.Mech.Phys.Solids 31,3,251-260.
	1984	Theor.and Appl.Fract.Mech.1,257-269.
	1985a	Meccanica 20,127-135.
	1985b	Dynamic Contact of Solids- A Dynamic Photoelastic Investigation. Proc. IUTAM-Conf.Macr-and Micro-Mechanics of High Velocity Deformation and Fracture of Solids (Ed.K.Kawata), Tokyo, August.
Rossmanith,H.P. and Dally,J.W.	1983	Strain 19,7-13.
Rossmanith,H.P. and Fourney,W.L.	1982	Rock Mechanics 14,209-233.
Rossmanith,H.P. and Irwin, G.R.	1979	Analysis of Dynamic Isochromatic Crack-Tip Stress Patterns. University of Maryland, Department of Mechanical Engineering Report.
Rossmanith,H.P. and Knasmillner,R.E.	1983	Spallation, Break-Up and Separation of Layers by Oblique Stress-Wave Incidence. Proc.1st Int.Conf.Rock Fragmentation by Blasting, Lulea Sweden,August,149-168.
	1984	Stress Wave Focussing in Elastic Solids. Proc.1st DANUBIA-ADRIA Symp. Exp. Stress Analysis, Stubicke Toplice, Yugoslavia,October.
Rossmanith,H.P. and Shukla,A.	1981a	J.Mech.Phys.Solids 29,5/6,397-412.
	1981b	Exp.Mech. 21,2,415-422.
	1982	Acta Mech. 42,211-252.
Shukla,A. and Rossmanith, H.P.	1984	J.Mech.Phys.Solids (submitted)
Sih, G.C.	1977	Mechanics of Fracture (Ed.G.H.Sih),Vol.4, Chapts.1 and 3, Noordhoff, Leyden.
Silva Gomes,J.F. and Al-Hassani, S.T.S.	1977	Int.J.Solids Structures 13,1007-1017.
Takayama, H.	1985	Private communication.
Williams,D.J., Walters,B.J. and Johnson,W.	1983	Int.J.Fract. 23,271-279.

The Pseudocaustics for the Evaluation of the Order of Singularity in Stress Fields

P.S. Theocaris

Department of Engineering Science, Section of Mechanics, The National Technical University, 5 Heroes of Polytechnion Avenue, Athens GR 157-73, Greece

INTRODUCTION

The only analytic techniques which have been used up-to-now for the evaluation of stress singularities are the *eigenfunction-expansion theory* and the *application of Mellin transform*. The first attempt to study stress singularities was made by Williams (1952), who derived the characteristic equations for wedges, by making the assumption that the order of singularity is a real number. The same assumption has been accepted by Kalandiia (1969), for evaluating the orders of singularities at corners. Bogy (1968) and Dundurs (1969) have derived solutions for stress singularities in wedges, by using a straightforward application of the *Mellin transform*.

Rice and Sih (1965) have treated the plane-extension problem of two bonded dissimilar media with cracks existing along their common interface and found the stress intensity factors, by using the eigenfunction-expansion technique for complex-variable integration. Hein and Erdogan (1971) have used the Mellin transform and the theory of residues to determine the stress singularities in a two-material wedge.

On the other hand, England (1971) used the method of complex variables to examine a group of boundary-value problems, and has found that the wedge geometry affects the value of singularity, whereas the imposed boundary conditions determine its type. Theocaris (1975) evaluated the stress singularities at the corners of multiwedges for arbitrary complex singularities.

However, until now, there does not exist an experimental method for the evaluation of the order of singularities developed in stress fields besides the method of pseudocaustics developed by Theocaris (1984) and Theocaris and Makrakis (1986).

In the present paper the theory of *pseudoanalytic functions* and *quasi-conformal mappings* were successfully used, in combination with the method of pseudocaustics as in the previous references for the evaluation of the order of stress singularities in elastic and plastic stress fields. Infinitesimal circles were drawn at the vicinity of the singular points, which created elliptic pseudocaustics during deformation. A single measurement of the rotation of the major axis of this ellipse, together with data from the caustics gave the order of the stress singularity. Experimental evidence with different kinds of singularities in elastic or elastoplastic fields indicated the validity and the accuracy of the method.

CONTINUITY LAW BETWEEN CAUSTICS AND PSEUDOCAUSTICS

For defining the mode of continuity at the common point of a caustic and a pseudocaustic we shall determine their tangent vectors at the common point. For this reason we consider the mapping:

$$f : R^2 \to R^2$$
$$f : (x,y) \to (u,v) \tag{1}$$

where:

$$u = W_x = \lambda_m x + CA(x,y)$$
$$v = W_y = \lambda_m y - CB(x,y) \tag{2}$$

and;

$$W = W_x i + W_y j \tag{3}$$

with $A(x,y)=Re\Phi'(z)$ and $B(x,y)=Im\Phi'(z)$ and $\Phi(z)$ the complex potential function, expressing the elastic field of the plate. The coefficient C denotes a multiplication factor depending on the experimental set-up, the optical properties and the geometry of the plate and on the distance z_0 between the middle plane of the plate and the reference screen where the caustics are formed, (Theocaris 1970,1981).

Relation (3) expresses the position of the projection P' of a generic point P on the Oxy-system attached to the plate, on the reference screen, where the deviation of the reflected light vector, W, is given in the respective Cartesian system O'x'y'.

The singular points of this mapping satisfy the algebraic equation:

$$J(x,y) = u_{,x}v_{,y} - u_{,y}v_{,x} = 0 \tag{4}$$

where commas before indices mean differentiation with respect to the index. This equation defines another mapping, h_1, in the following manner:

$$h_1 : R \to R^2$$
$$h_1 : t \to (t,\gamma(t)) \tag{5}$$

under the condition that $J(t,\gamma(t))=0$. This condition constitutes a necessary and sufficient condition for the creation of a singular curve, the *caustic*, on the screen, and it is expressed by:

$$J = \frac{(W_x,W_y)}{(x,y)} = W_{x,x}W_{y,y} - W_{x,y}W_{y,x} = 0 \tag{6}$$

which is of course the same as relation (4). Then, it is easy to show that the compound function $f(h_1)$ defines the equations of the caustic curve.

We consider now a generic curve in the elastic field, which can be expressed by the compound function:

$$q(x,y) = 0 \qquad (7)$$

and assume that this curve can be represented by the mapping:

$$h_2 : R \rightarrow R^2$$
$$h_2 : t \rightarrow (t,g(t)) \qquad (8)$$

together with the condition that $q(t,g(t))=0$. It can be readily shown that the compound function $f(h_2)$ defines the equations of the *pseudocaustic curve*, which is generated from this boundary.

The tangent vectors of the above curves are expressed by:

$$\nabla_J f = J \, \nabla f = (-J_{,y},J_{,x})((u_{,x},v_{,x}),(u_{,y},v_{,y})) = (-J_{,y}u_{,x}+J_{,x}u_{,y},-J_{,y}v_{,x}+J_{,x}v_{,y}) \quad (9)$$

$$\nabla_t f = q \, \nabla f = (-q_{,y},q_{,x})((u_{,x},v_{,x}),(u_{,y},v_{,y})) = (-q_{,y}u_{,x}+q_{,x}u_{,y},-q_{,y}v_{,x}+q_{,x}v_{,y}) \quad (10)$$

The vectors J and q are the tangent vectors of the curves, which are expressed by relations (4) and (7) respectively.

We construct now their vector-product $\nabla_J f \times \nabla_q f$, which by taking into account relation (4) yields for the common points of the caustics and pseudocaustics that:

$$\nabla_J f \times \nabla_t f = 0 \qquad (11)$$

Relation (11) proves that the caustics and pseudocaustics have common tangential directions with the same or opposite sense, at their common points (Fig.1). However, it is worthwhile indicating that the form of contact shown in Fig.1a is a cusp-like one, but it is not a real cusp point, because the necessary condition for the existence of a cusp point is given by (Lu, 1976):

$$\nabla_J f = 0 \qquad (12)$$

Relation (12) in this case does not hold because these points are in reality fold points. On the contrary the common points in Fig.1b possess a common tangent, since at these points the caustic and the pseudocaustic have tangents which lie on the same straight line.

Fig.1c presents the shapes of caustics and pseudocaustics formed by applying a uniformly distributed load along a part of the straight boundary of a half-plane under conditions of plane-stress. It is evident from this figure that in the case when the equations of the caustic and the pseudocaustic satisfy the condition $\lambda>0$ the respective curves possess a common point but not a real cusp point (case 1a), whereas when the caustics and pseudocaustics satisfy the relation $\lambda<0$ they have further a common tangent and either curve lies on either side of this tangent.

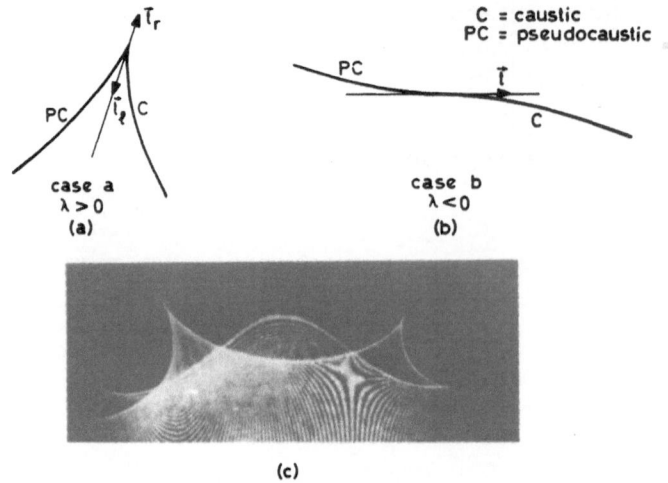

(a)

(b)

(c)

Fig. 1. Contact forms between caustics and pseudocaustics: (a) $\lambda>0$, (b) $\lambda<0$ and (c) experimental evidence in contact caustics.

In order to define the form of the contact at the common points of the caustics and pseudocaustics we construct the scalar product:

$$Q = \nabla_j f \cdot \nabla_t f \qquad (13)$$

which may be written in a matrix form as follows:

$$Q = x^T A x \qquad (14)$$

where $x^T = (u_{,x} v_{,y}\ u_{,y} v_{,y})$ and the matrix A is expressed by:

$$A = \begin{vmatrix} J_{,y}q_{,y} & \mu & 0 & 0 \\ \mu & J_{,x}q_{,x} & 0 & 0 \\ 0 & 0 & J_{,y}q_{,y} & \mu \\ 0 & 0 & \mu & J_{,x}q_{,x} \end{vmatrix} \qquad (15)$$

with;

$$\mu = -\frac{1}{2}(J_{,x}q_{,y} + J_{,y}q_{,x})$$

The matrix A is a symmetric and a non-singular, one, because of the validity of the relationship $|A| = 1/16\ (J_{,x}q_{,y} - J_{,y}q_{,x}) \neq 0$.

Taking into account that at the common points of caustics and pseudocaustics the inner product $J \cdot t = 0$ holds, it may be deduced at once that, whenever a singular point in the stress field lies on the curve considered in the stress field, as it is the case with singularities due to the externally applied concentrated loads or distributed loads presenting abrupt changes of their intensity or slope, it is valid that $q = 0$.

Considering now the eigenequation of the matrix A we have:

$$|A - \lambda I| = (\lambda^2 - \frac{1}{4}(J_{,x} q_{,y} - J_{,y} q_{,x})^2)^2 = 0 \tag{16}$$

which has as double roots the expressions:

$$\lambda_{1,2} = \pm \frac{1}{2}(J_{,x} q_{,y} - J_{,y} q_{,x}) \tag{17}$$

which generally are nonzero numbers.

Relation (17) indicates that the quadratic form **Q** changes sign. So, each of the contact forms, either the form of Fig.1a, or the form of Fig.1b, is possible for some value of λ_i. It may therefore be concluded that, *when the equations for the caustics and pseudocaustics satisfy the relationship $\lambda > 0$, they possess a common point, but not a real cusp point (case 1a). On the contrary, when the caustics and pseudocaustics satisfy the relationship $\lambda < 0$, they possess in addition a common tangent, and either curve lies on either side of this tangent (case 1b).*

The above analysis is valid also for every curve, which is parallel to the curve considered satisfying the condition $q = 0$, and therefore cuts the initial curve of the caustic.

Thus, any curve which satisfies the condition $q(x,y) = 0$ creates similar forms of contact at the common points of the caustic and the respective pseudocaustic created by this curve.

THE ORDER OF SINGULARITIES IN ELASTIC AND ELASTOPLASTIC STRESS FIELDS

We examine now the mode of formation of a caustic by considering the increment dW of the deviation vector **W** in the plane $O'x'y'$ as the polar angle θ from the singular point is increasing. This deviation dW may be expressed by the system:

$$MW_{,x,x} + NW_{,x,y} = W_{,y,y}$$
$$MW_{,x,y} - NW_{,x,x} = -W_{,y,x} \tag{18}$$

with values of the coefficients M and N given by:

$$M = \frac{W_{,x,x}W_{,y,y} - W_{,x,y}W_{,y,x}}{W_{,x,x}^2 + W_{,x,y}^2} \quad , \quad N = \frac{W_{,x,x}W_{,y,x} + W_{,x,y}W_{,y,y}}{W_{,x,x}^2 + W_{,x,y}^2} \tag{19}$$

It is evident from these relations that the sign of the function M depends on the sign of the function $J(x,y)=(W_{x,x}W_{y,y}-W_{x,y}W_{y,x})$. It has been shown, however, that in any elastic field with singular points the curve J=0 *represents the initial curve*. Then, the factor M remains positive when J>0, that is for all points which lie outside the initial curve.

Then the condition (19a) implies that the points of the stress-field satisfying this condition lie outside the initial curve of the caustic and at these points *the differential system (18) is of the elliptic type* and the complex-valued function $w(z)=(W_x+iW_y)$ *is a pseudoanalytic function of the second kind* (Courant and Hilbert, 1962).

Then, the following rule is valid: A complex-valued function is quasi-conformal, if and only if a real number δ can be found, with δ>1, for which the following inequality is valid:

$$(W_{x,x}^2+W_{x,y}^2+W_{y,x}^2+W_{y,y}^2) \leq 2\delta(W_{x,x}W_{y,y}-W_{x,y}W_{y,x}) \tag{20}$$

where again W_x and W_y are the real and imaginary parts of the deviation vector **W**.

Since every pseudocaustic generated by an arbitrary curve in the stress field, which lies outside the initial curve of the caustic surrounding the singularity, defines a pseudoanalytic function, the same curve defines also a quasi-conformal mapping.

However, the inequality (20) together with relations (2), can be written in the form:

$$\frac{\delta+1}{\delta-1}|\Phi''(z)|^2 \leq (\lambda_m/C)^2 \tag{21}$$

On the other hand, it has been shown, (Theocaris 1981), that the equation: $|\Phi''(z)|^2=(\lambda_m/C)^2$ describes the initial curve of the caustic. Then, all points outside the initial curve are given by the inequality:

$$|\Phi''(z)|^2 < (\lambda_m/C)^2 \ .$$

It is easy to define a real number $\delta(\lambda_m,C)$, δ>1, such that the inequality (21) be valid.

A fundamental property of the quasi-conformal mappings is that they map infinitesimal circles onto infinitesimal ellipses with uniformly bounded eccentricities, contrariwise to conformal mappings, which map circles onto circles.

Consider now an elastic field with a real singularity at the point z=0, of order p whose complex potential expressed by its dominant term, is given by:

$$\Phi(z) = Kz^{p+1} \tag{22}$$

An infinitesimal circle centered at the point z_κ of the singularity, with radius $\varepsilon, \varepsilon \to 0$, defined by $z=z_\kappa+\varepsilon e^{i\theta}$. We assume further that this circle lies outside the initial curve, so that its pseudocaustic obeys a quasi-conformal mapping. This circle is mapped on the screen $O'x'y'$ by:

$$W(z)-\lambda_m z_\kappa = r(z) = \lambda_m \varepsilon e^{i\theta}+CK(p+1)R^p e^{-ip\varphi} \tag{22}$$

where $R=|z_\kappa+\varepsilon e^{i\theta}|$ and $\varphi=\arg(z_\kappa+\varepsilon e^{i\theta})$.

Relation (22) yields for $\varepsilon \to 0$:

$$\tan\omega = \arg(r(z)) = \frac{r_y}{r_x} = -\tan(p\varphi_\kappa) \tag{23}$$

where:

$$r_x = \mathrm{Re}(r(z)) \;, \quad r_y = \mathrm{Im}(r(z))$$

and $\tag{24}$

$$\varphi_\kappa = \arg z_\kappa$$

Then an infinitesimal circle lying outside the initial curve of the caustic may be mapped onto an infinitesimal ellipse, whose major axis defines the angle φ_κ yielding the value of the order of singularity by means of relation (23).

EVALUATION OF THE ORDER OF SINGULARITY

Since in the praxis the infinitesimal circle must have some finite radius we express the polar radius $r(z)$ of the ellipse derived by the quasi-conformal mapping in terms of ε as follows (Fig.2):

$$r(\theta) = (\lambda_m \varepsilon)^2+S^2 R^{2p}+2\lambda_m \varepsilon S R^p \cos(\theta+p\varphi) \tag{25}$$

where:

$$R = ((x_\kappa+\varepsilon\cos\theta)^2+(y_\kappa+\varepsilon\sin\theta)^2)^{\frac{1}{2}} \tag{26}$$

and;

$$S = C(p+1)K \tag{27}$$

We define now the angle θ_{max} in the plane of the mapped ellipse $O'x'y'$, corresponding to the major-axis of the ellipse by satisfying the condition for an extremum that is $dr/d\theta=0$. Then, we have:

$$pS^2 R^{2(p-1)}+\lambda_m \varepsilon S R^{(p-2)}-\lambda_m S R^{(p-2)}\{R^2+p[(x_\kappa\cos\theta+y_\kappa\sin\theta)+\varepsilon^2]\}\,\frac{\sin(\theta+p\varphi)}{y_\kappa\cos\theta-x_\kappa\sin\theta} = 0 \tag{28}$$

with;

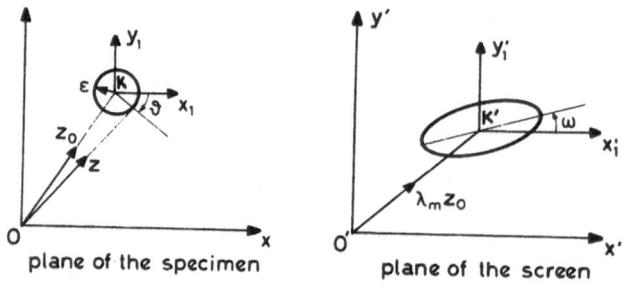

Fig. 2. Coordinate systems in the plane of specimen and the reference plane. The scribed circle in the first is transformed to an ellipse in the second system.

$$R^2 = (x_\kappa + \varepsilon\cos\theta)^2 + (y_\kappa + \varepsilon\sin\theta)^2 \quad \text{and} \quad \varphi = \arctan\left(\frac{y_\kappa + \varepsilon\sin\theta}{x_\kappa + \varepsilon\cos\theta}\right) \tag{29}$$

The slope of the generic polar radius of the ellipse is given by:

$$\tan\omega = \frac{\lambda_m \varepsilon\sin\theta - SR^p\sin(p\varphi)}{\lambda_m \varepsilon\cos\theta + SR^p\cos(p\varphi)} \tag{30}$$

where:

$$\varphi = \arctan\left(\frac{y_\kappa + \varepsilon\sin\theta}{x_\kappa + \varepsilon\cos\theta}\right) \tag{31}$$

Then, the following iteration procedure, for the evaluation of the order of the stress singularity may be applied. *In this procedure the only experimental data, which are required are the slope of the major axis of the ellipse and the characteristic diameter of the caustic.* The iteration procedure follows the steps:
i) Define an initial value of the p by means of Eq.(23).
ii) Evaluate the stress intensity K using the equation of the caustic:

$$D_{max} = \lambda_m \left(\frac{\lambda_m^2}{C^2 p(p+1)K}\right)^{\frac{1}{p-1}} \cos\theta_{max} + CK(p+1)\left(\frac{\lambda_m^2}{C^2 p(p+1)K}\right)^{\frac{p}{p-1}} \cos(p\theta_{max}) \tag{32}$$

where θ_{max} is the root of the transcendental equation:

$$C\sin\theta + \lambda_m\sin(p\theta) = 0, \quad \text{for} \quad -\pi \le \theta \le \pi \tag{33}$$

and D_{max} is the maximum diameter of the caustic.
iii) Define the quantity S from relation (27). Using now relation (28) the angle θ_{max} corresponding to the maximum polar radius, r_{max}, of the ellipse is evaluated.
iv) Solve relation (30) for a successive value of p. If this value is

in good agreement with its previous value, the iterative process may be stoped, otherwise it must be continued, using for the next step (i) the new value of p, until it is succeeded to obtain infinitesimal deviations between successive values of p.

EXPERIMENTAL EVIDENCE

In order to verify the effectiveness of the method a series of experiments was executed concerning the evaluation of the order of singularity. In previous papers (Theocaris 1984 and Theocaris and Makrakis 1986) the method has been applied successfully to two particular problems, where the elastic fields contained two different types of singularities. The first problem was the evaluation of the stress singularity at the position of a concentrated load applied on a half plane, whereas the second was referred to the evaluation of the singularity at the tip of an edge crack existing in an infinite plate under conditions of generalized plane stress. Either case used two different materials. The one was polymethyl-methacrylate (PMMA), which behaved as ideally elastic material almost up to fracture. The second was polycarbonate (PCBA) and this material was behaving as an elastic-perfectly plastic material.

Furthermore, two possibilities exist for preparing of the specimens for these experiments. Either a number circles of infinitesimal radius ($\varepsilon \to 0$) may be scribed at the vicinity of the crack tip in a region anticipated lying outside the respective initial curve of the caustic or a number of reference circles printed on a flat glass plate may be projected at the desired point of the stress field near the crack tip. Another alternative is to project a circular grating interposed in the light beam of the optical setup of a high density say 20 lines per millimeter. This last possibility presents the advantage to dispose practically an infinite number of circles and choose from them the most convenient.

Fig.3a presents the distorted image of a scribed circle of a radius $\varepsilon \simeq 0.001$m whose circumference was lying near the stress singularity created at the tip of an edge crack in a plate submitted to simple tension at infinity. The material of the plate was either plexiglas for the elastic case, or polycarbonate for the elastic plastic case. The amount of loading and the optical set-up was arranged to create an initial curve of the caustic tangent to the circumference of the scribed circle. The direction of the major axis of the elliptically deformed circle during loading gave the angle φ_{κ} for the evaluation of the order of singularity at the crack tip. Similar tests were executed at the vicinity of the stress singularity created by a concentrated load applied normally on the straight boundary of a half plane.

The results from these tests and by applying the iterative process previously described are concentrated in Table I for the cracked plate and the concentrated load in the case of an elastic stress field and an elastic-plastic one.

Similar tests were run with the method of projected infinitesimal circle from a screen interposed in the light beam of the optical set-up. Fig.3b shows a pattern of the deformed projection of the circle at the vicinity of the singularity. The results compare satisfactorily with the respective cases with scribed circles. But the projected

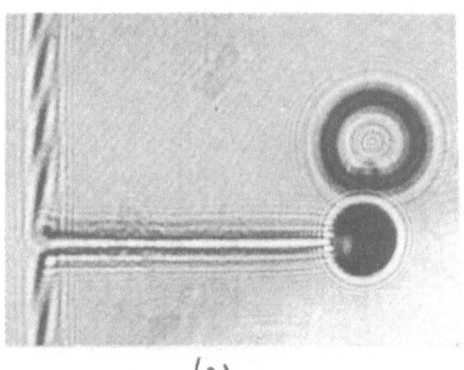
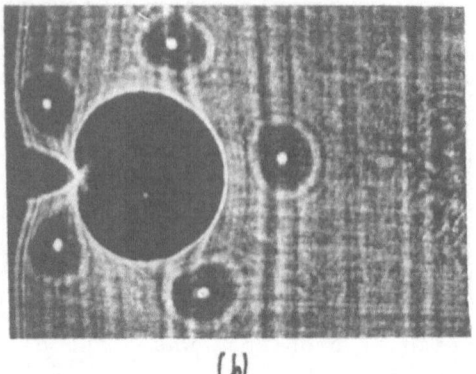

(a) (b)

Fig. 3. (a) The distorted images of circles scribed at vicinity of
a crack tip in a PCBA-specimen and (b) The distorted projected image of
a family of circles.

Table I. Experimental values of the orders of singularities in the case of a
concentrated normal load applied at a straight boundary of a half-plane and the
crack tip of an edge crack existing in an infinite plate. In both cases the
materials of the plates were either polymethyl-methacrylate (a brittle material), or
polycarbonate of bisphenol A (an elastic quasi-perfectly plastic material).
Experimental values were confronted with theoretical values where the real stress
situation was taken into account.

		ELASTIC FIELD					
polymethyl-methacrylate (PMMA)	Cracked plate	Theory	Experimental Values				
		order of sing.	-0.500	-0.495	-0.510	-0.490	-0.490	-0.500
		error (%)	-1	-1.0	0	-2.0	-2.0	0
	Concentrated load	Theory	-0.910	-0.870	-0.870	-0.870	-0.870
		Exp.	-0.920	-0.860	-0.860	-0.860	-0.860
		error %	1.0	1.0	1.0	1.0	1.0
		PLASTIC FIELD					
polycarbonate (PCBA)	Cracked plate	Theory	-0.435	-0.426	-0.414	-0.402	-0.392
		Exp.	-0.420	-0.420	-0.400	-0.390	-0.380
		error %	-3.40	-1.40	-3.30	-2.90	-3.00
	Concentrated load	Theory			-0.720		
		Experim.	-0.730	-0.730	-0.720	-0.720	-0.710
		error %	1.30	1.30	0	0	-1.30

circle technique presents a higher flexibility than the scribed circle
technique, since the experimenter is now free to move the projection
of the circle everywhere around the crack tip and along the crack-axis
and thus to optimize the evaluation of angle ω.

Furthermore, Fig.4 presents the +1 and -1 orders of diffraction of the
caustic, due to diffraction phenomena created by the interposition of
the dense circular grating. All three diffraction orders are enveloped
in a cone with apex the centre of the circular grating. Again, the
distortion of the higher order diffractions of the caustic yield all
the data needed for evaluating p.

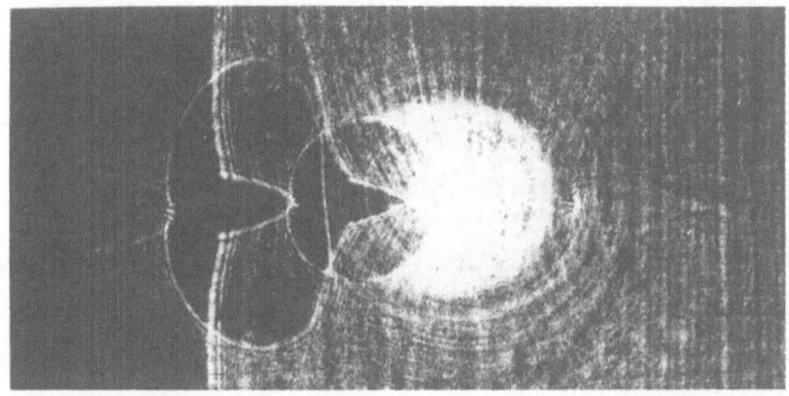

Fig. 4. The (±1) diffraction orders with a density of 20 lines/mm circular grating projected on the crack tip of a PCBA-plate. The shape and dimensions of the distorted images of the circles yield all the necessary information for evaluating accurately the actual order of singularity at the crack tip.

The experimental values of the orders of stress singularities for both types of singularities were always in good agreement with their respective theoretical values.

It may be derived from these results that, in the case of an edge crack, the experimental values for the PMMA-plates are in excellent agreement with their respective theoretical values, the errors not overpassing 2 percent. This was because the material behaved as an elastic material and the theory of elasticity yielded satisfactory results even for the singular zone. On the contrary, in the case of PCBA-plates a deviation between the experimental values and the theoretical ones appears. These deviations in reality indicate the influence of plasticity in the singular zone, where the linear theory of elasticity is inadequate to predict the actual behaviour.

This deviation between predictions of the ideal theory of elasticity and reality proves the necessity of disposing appropriate experimental tools to define the real state of stress around singularities when plasticity or viscoelasticity phenomena are influencing these singular zones. In these cases the experimental method of pseudocaustics combined with information taken from the caustics proves to be very efficient.

Concerning the order of elastic and plastic-stress singularities developed near concentrated loads applied to half-planes, or other forms of specimens, the theory of elasticity evaluates these singularities to be equal to p=1.0. The deviations found between theory and experiment for PMMA may be explained by the fact that the application of a concentrated-point load can only be achieved by a blunt indenter of some infinitesimal radius r=a. Then, the stress distribution around the indenter, which only at the early beginning of loading could be elastic, develops some kind of constraint, depending on the shape of the indenter, which changes drastically the state of stress near the singularity because of the constraining plastic zone. This plastic zone, which for PMMA may be very limited, becomes a large one in the case of PCBA, which presents an elastic-plastic behavior, and this influences more severely the state of stress at the vicinity of the singularity.

Taking into account that the order of stress singularity represents in reality the velocity of the variation of stresses in the neighbourhood of the singular point, and the theoretical results developed by Hutchinson (1976), the order of singularity may be expressed by:

$$p = -1/(N+1) \tag{34}$$

where N denotes the hardening power-coefficient, evaluated by a simple tension stress-strain diagram of the material of the plate. The values of the hardening exponent N for PCBA were evaluated from the respective stress-strain diagrams of the material, when a series of typical tension specimens was submitted to simple tension. If the instantaneous values of the hardening exponent N are evaluated from a stress-strain diagram of the material for each loading step of the cracked or indented plate and the respective plastic stress singularity is evaluated, and this value is compared with the value derived experimentally, a good coincidence of results may be established, the deviations not overpassing in all cases 1 to 4 percent (Theocaris and Makrakis, 1986).

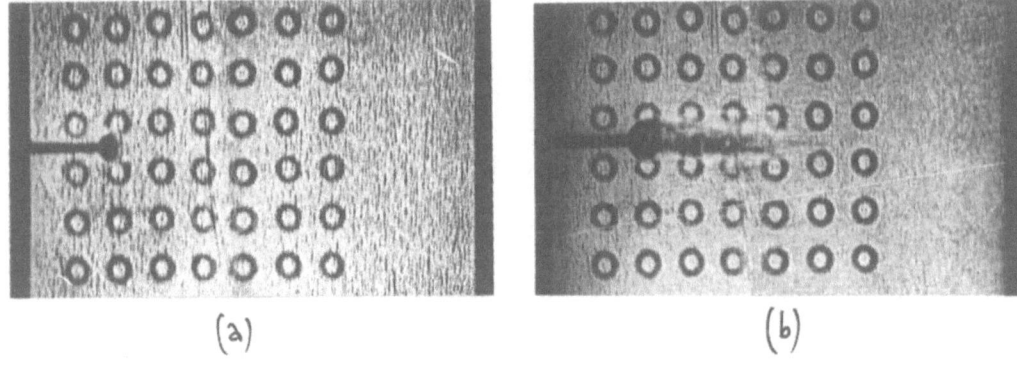

(a)   (b)

Fig. 5. The projected array of circles for the experimental evaluation of the order of singularity in a cracked aluminium plate (a) before the formation of plastic zones, and (b) after the formation of plastic zones.

Fig.5 presents an application of the technique of the projected circles on an aluminium cracked plate submitted to a tensile load up to fracture. An array of equidistant circles of a radius ε=0.001m and an intercenter distance d=0.006m was projected along lines parallel to the crack axis. The distances of neighbouring rows from the crack axis were equal to 0.003m and the angles of the ellipses formed at the vicinity of the instantaneous positions of the crack tips were measured. The order of singularity (p+1) was derived by using the relation.

$$\tan\omega = -\tan(p\varphi)$$

where φ is the angular position of the center of the neighbouring circle to the instantaneous position of the crack tip relatively to the crack axis.

From the evaluation of the variation of the order of singularity with loading presented in Fig.6 it may be derived that there are four distinct zones of values of the stress-singularity:

i) The zone of constant values with $\lambda=0.50$ extending up to 60 percent of the yield limit for the cracked plate.
ii) The zone where λ varies from $\lambda=0.50$ to $\lambda=0.40$ and corresponding to a transition region in the elastoplastic domain of loading of the plate.
iii) The zone of constant value of $\lambda=0.40$ during which large plastic zones develop in front of the crack-tip, and iv) The zone of an abrupt reduction of the value of the stress singularity due to excessive plasticity and impending fracture. However, the values of λ in this zone are doubtful since in this zone the elastic theory ceases to be valid.

In the same figure appear the values for the order of singularity given by Swedlow, Williams and Yang (1965). These values follow a similar variation as the experimental one but they are smaller. These differences are expected since the assumptions used in each case are different. The experimental values are based on an elastic analysis while the Swedlow et al. analysis is based on an elastic-plastic numerical solution with finite-elements.

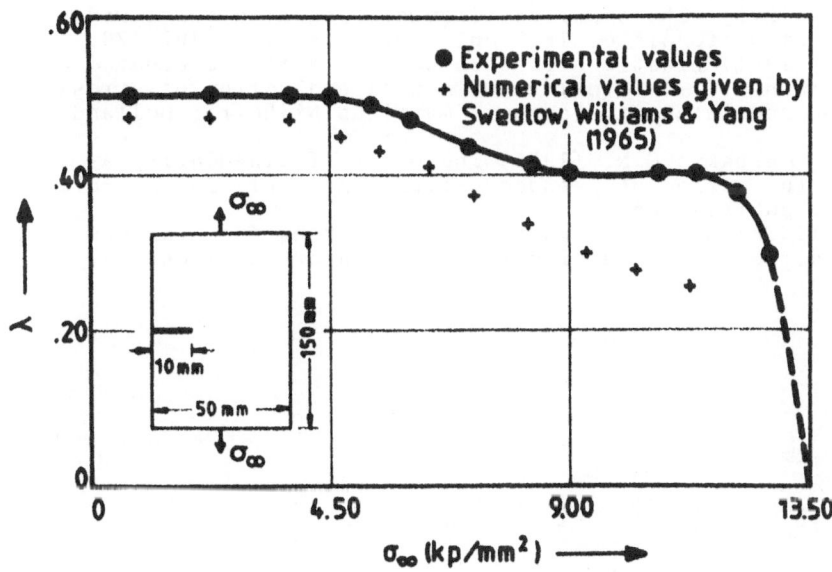

Fig. 6. Experimental (PSEUDOCAUSTICS) and numerical (FEM) values of the order of singularity.

REFERENCES

Bers L., (1956) On the theorem of Mori and the definition of quasi-conformality. Trans. Amer. Math. Soc. 84:78-93.

Bogy D.B., (1968) Edge-bonded dissimilar orthogonal elastic wedges under normal and shear loading. J. Appl. Mech. 35:460-466.

Courant C., Hilbert G. (1962) Methods of mathematical physics. (Interscience) II:375-389.

Dundurs J., (1969) Discussion of "Edge-bonded dissimilar orthogonal elastic wedges under normal and shear loading". J. Appl. Mech. 36:650-652.

England A.H., (1971) On stress singularities in linear elasticity. Int. Jnl. Engrg. Sci. 9:571-585.

Hein V.L., Erdogan F. (1971) Stress singularities in a two-material wedge. Int. Jnl. Fract. Mech. 7:317-330.

Hutchinson J.W., (1976) Plastic stress and strain field at a crack tip. J. Mech. and Phys. Solids. 16:337-347.

Kalandiia A.I., (1969) Remarks on the singularity of elastic solutions near corners. J. Math. Mech. 33:132-135.

Lu Y.C., (1976) Singularity theory and an introduction to catastrophe theory. Springer-Verlag, Berlin Heidelberg New York.

Rice J.R., Sih G.C. (1965) Plane problems of cracks in dissimilar media. J. Appl. Mech. 32:418-423.

Swedlow J.L. Williams M.L., Yang W.H. (1965) Elasto-plastic stresses and strains in cracked plates. In:Yokobori T., Kawasaki T., Swedlow G.L. (eds), Proceedings of the first international conference on fracture held in Sendai (Japan), p.259.

Theocaris P.S. (1970) Local yielding around a crack tip in plexiglas. J. Appl. Mech. 37:409-413.

Theocaris P.S. (1975) The order of singularities at a multi-wedge corner of a composite plate. Int. Jnl. Engrg. Sci. 12:107-120.

Theocaris P.S. (1981) Elastic stress intensity factors evaluated by caustics. In:Sih G.C. (editor) Experimental evaluation of stress concentration and intensity factors. Martinus Nijhoff., Holland, p.212.

Theocaris P.S., Makrakis G.N. (1985) The order of singularity as evaluated by the method of pseudocaustics. Int. Jnl. Engrg. Sci. Submitted for publication.

Williams M.L. (1952) Stress singularities resulting from various boundary conditions in angular corners of plates in extension. J. Appl. Mech. 19:526-528.

Some New Developments in Photoelasticity

L.S. Srinath

Indian Institute of Technology, Madras-600036, India

In this report we discuss the results of two investigations that are under progress. The first investigation deals with the possibilities of utilising Raman scattering to measure the stress fields in the area of fracture mechanics. The results reported are of a very tentative nature. The second investigation deals with the development of a new semi-destructive method to determine the complete state of stress in a three dimensional photoelastic model using the shear difference method. In this case also, the results reported are based on the extent of the investigations carried out so far.

PRELIMINARY STUDIES ON STRESS ANALYSIS
USING RAMAN SCATTERING

INTRODUCTION

When a wave of light passes through a transparent substance (solid, liquid or gas), light is scattered by individual or group of atoms or molecules that are much smaller in linear dimension than the wavelength of the incident light. This scattering is known as secondary radiation and Rayleigh observed that if the incident light is monochromatic, the scattered light is also unchanged in frequency. This scattering is due to the fact that the incident light causes the electrons in the particle to oscillate periodically in response to the vibrating electric vector of the incident light. The intensity of this Rayleigh scattering is proportional to λ^{-4} where λ is the wavelength of the incident light. From classical electromagnetic theory, it was shown that if the electrons that scatter light are in motion, then the light that is scattered should contain other frequencies in addition to that of the incident light. This secondary radiation discovered by Raman is known as Raman Scattering (Raman 1928).

The usual scattered light techniques used in stress analysis make use of Rayleigh scattering, whose intensity is much higher than the Raman scattering. But, Raman scattering has some properties not possessed by Rayleigh scattering and possibilities exist in utilising these properties advantageously for stress analysis. Two possibilities of utilising Raman scattering for stress analysis are:

(i) Changes in the frequencies of Raman scattering (or lines) due to stresses.

(ii) Changes in the intensity of Raman lines due to stresses.

In this paper, preliminary studies utilising the frequency variations of Raman lines as a measure of stresses are discussed.

In Raman spectroscopy, the spectrum is analysed in terms of its wave number. If λ is the wavelength in cm, $1/\lambda$ is the wavenumber (ν) which is used as a measure of frequency. Further, the difference between the incident frequency and the frequency of the scattered light is known as the Raman frequency. For example, if the exciting beam has a frequency 19436 cm^{-1} (corresponding to 5145 angstrom units) and if the frequency of the Raman line is 19336 cm^{-1}, one calls this line as a line with a Raman frequency of 100 cm^{-1}

EFFECT OF HYDROSTATIC PRESSURE

Extensive investigations on the effect of hydrostatic pressure on Raman frequencies have been carried out (Whalley 1974, Vu 1974). In the case of diamond, for example, the line with a Raman frequency of 1334 cm^{-1} changes to 1341 cm^{-1} with the application of a pressure of 22.8 kilobar. Thus, there is a shift of 7 wave numbers with a pressure of 22.8 kbar. In other words, the sensitivity is roughly 1 wave number per 3 kilobar. In the case of solid nitrogen at 4.2° Kelvin, a shift of 1.5 to 2 cm^{-4} per kilobar is observed.

Our interest in this experiment was to investigate whether a state of stress other than hydrostatic will cause considerable shift in the Raman frequencies.

LASER RAMAN SPECTROMETER

In order to observe shifts in Raman frequencies, one requires a highly sensitive spectrometer (Tobin 1971). The modern Laser Raman Spectrometer manufactured under the trade name "Spex Ramalog" was used for these investigations. The basic units of this spectrometer are:

(1) Laser source; (2) Sampling system, (3) Monochromator, (4) Photomultiplier, (5) Signal processing and recording system

Generally speaking, no laser is universally useful for all samples. For example, when He-Ne (6328 angstroms) line is used to excite a smaple, the spectrum that needs to be observed due to Raman scattering lies between 6000 and 9000 angstroms, and no photo cathode material presently available is very efficient in this region. But, from stability and long-life points of view, He-Ne laser is desirable and its maximum power is limited to 80 mw. Argon laser is useful because of its high output. In Spex Ramalog, both He-Ne and Argon lasers (4880 angstroms) are used.

Figures 1 and 2 show respectively, the block diagram of the various units and the elevation view of sample holding system with pre-slit optics. While photographic recording is still used in the high resolution spectroscopy of gases, the vast majority of work is done to-day with a scanning grating monochromator and photoelectric detection. The monochromator is based on the Czerney-Turner mount, Fig. 3. Light coming through the entrance slit is reflected on to the grating. The dispersed light is reflected and focussed by the second mirror to the exit slit. In Spex Ramalog, there are three monochromators in succession.

MATERIAL SELECTION

Since the material generally used in scattered light photoelasticity is araldite, experiments were conducted on this sample material. The sample was excited by He-Ne laser. In the entire range of possible Raman lines (50 cm^{-1} to 3000 cm^{-1}), no appreciable results could be seen. The D.M.S. Atlas (1974) gives a list of organic polymers that have well defined Raman spectra. Two of them are polycarbonate (carbonic acid polyester) and polymethyl methacrylate (trade name perspex). For our investigation, PMM material was chosen. A small piece (5 x 5 x 10 mm) was kept in the sample holder and Raman spectrum was taken under the following conditions:

Laser	Argon Laser (4880 A)
Slit opening	250μ
Chart speed	5mm/sec
Scan rate	0.2 cm^{-1}/sec
Time constant	0.8 sec.

Full scale of the chart was 1000 counts of photons.

The following Raman lines were observed:

Raman frequency (cm^{-1})	Intensity	
810	1150 counts	(sharp line)
969	20 "	
1450	30 "	
2963	1453 "	(broad base)

Our initial investigations were, therefore, confined to the line having a Raman frequency of 810 cm^{-1}.

EFFECT OF STRESS ON RAMAN FREQUENCIES

To study the effect of stress, beams of rectangular cross section (1.38 x 0.8 x 12.15 cm) were machined from commercially available perspex sheets. The beams were subjected to bending and portions in the tensile region were examined. To induce high stresses, a V-notch was made on the tensile side of the beam and a point very near the notch tip was exposed to the measuring instrument. The model was placed in such a manner that the σ_1 axis ($\sigma_1 > \sigma_2$) was perpendicular to the direction of observation. The incident light, which was perpendicular to the direction of observation was linearly polarized with the direction of polarization parallel to σ_1 axis. It should be noted that Raman scattering observed at 90° to the direction of incident light is polarized like in the case of Rayleigh scattering.

Attention was focussed on the line having a Raman frequency of 810 cm^{-1}, with the beam in an unloaded condition. With the model under load, the spectrum of the 810 line was again observed to see if there was any frequency shift. Initial observations have given poor results. We were able to observe Raman frequencies of 810.4 cm^{-1} for the beam under stress. The investigations are continuing and possibilities exist in improving the quality of the sample material and also the shift due to stresses.

REFERENCES

DMS Raman/IR Atlas of Organic Compounds, Ed. Schrader B and Maier W, Vol 2 Verlag Chemie GmbH

Raman CV, (1928) A New Radiation. Indian Jr. of Physics Vol 2: 387-398

Tobin MC (1974) Laser Raman Spectroscopy, Wiley Interscience.

Vu H, Jean-Louis M et al (1974) IR and R Spectra of Simple Molecular Solids upto 15 kbar at 4.2 K, Proc. 4th Int. Conf. on High Pressure, Kyoto.

Whalley E, (1974) High Pressure Raman Spectroscopy, Proc. 4th Int. Conf. on High Pressure, Kyoto.

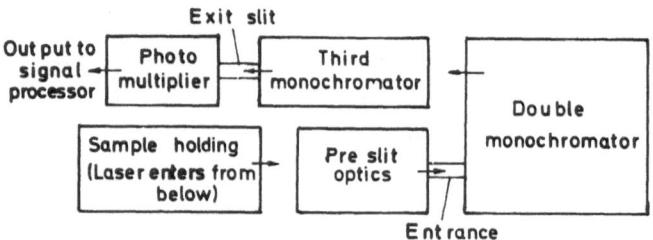

FIG.1 BLOCK DIAGRAM (Plan view) SHOWING THE VARIOUS UNITS OF SPEX RAMALOG

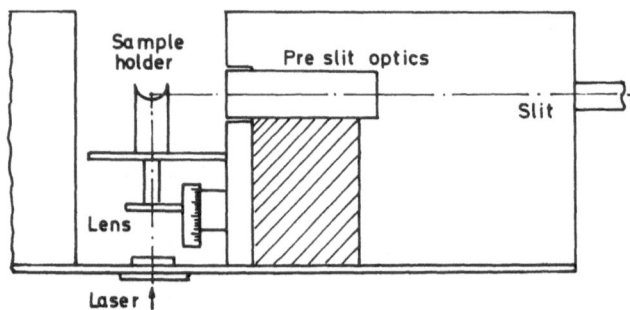

FIG.2 SAMPLE HOLDING SYSTEM WITH PRE SLIT OPTICS

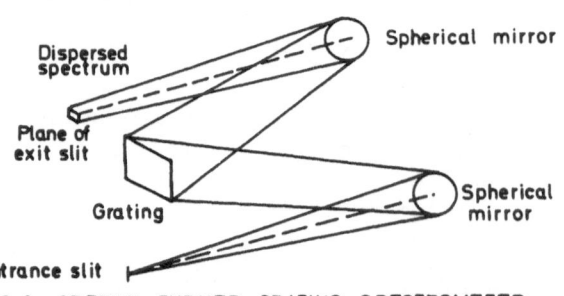

FIG.3 CZERNY TURNER GRATING SPECTROMETER

A NEW SEMI-DESTRUCTIVE METHOD
FOR THREE DIMENSIONAL PHOTOELASTICITY

INTRODUCTION

In order to determine the complete state of stress at a point in a general three dimensional photoelastic model, one adopts the shear difference method (Frocht and Guernsey, 1952). In this method the model with stresses locked in is first cut into a thin slice containing the point of interest, Fig. 1. One normal and two oblique photoelastic observations made on this slice and through the point of interest yield the values of

$$(\sigma_x - \sigma_y), \quad (\sigma_y - \sigma_z), \quad \tau_{xy}, \quad \tau_{yz} \text{ and } \tau_{zx} \qquad \ldots (1)$$

In order to separate the normal stresses, the slice is again cut in the form of a rectangular parallelepiped containing the point of interest. Photoelastic observations made in the y-direction on the front and rear faces of the subslice yield the values of τ_{zx} on these faces, thus giving the variation of τ_{zx} across the thickness of the model. Using the following equation of equilibrium

$$\frac{\partial \sigma_x}{\partial x} + \frac{\partial \tau_{xy}}{\partial y} + \frac{\partial \tau_{zx}}{\partial z} = 0 \qquad \ldots (2)$$

and with the known value of σ_x at the boundary point, the equation is integrated step-by-step along the x-axis, thus giving the value of σ_x at the point of interest. One can observe two points in this regard.

(a) When a slice is taken from the model and photoelastic observations are made on it, one assumes that the slice is thin enough so that the photoelastic observations yield the values of $(\sigma_x - \sigma_y)$, $(\sigma_y - \sigma_z)$, τ_{xy}, τ_{yz} and τ_{zx} at the mid-plane of the slice.

(b) When the slice is cut into subslice and observations are made on the front and rear faces, one assumes that the slice is thick enough to yield different values of τ_{zx} on these two faces so that one can evaluate $\partial \tau_{zx}/\partial z$.

These two points are in some sense contradictory and it is not always easy to estimate how thin or thick a slice should be cut from the model. This paper presents a new approach to the problem. Without cutting the slice into a subslice, the complete investigation can be conducted on a fairly thick slice, thus making the method semi-destructive in nature.

THE METHOD

A slice (which can be slightly thicker than the normal slice taken in the usual subslice method) is cut from the stress-frozen model and photoelastic observations (at normal and oblique incidences) are made on it.   One assumes that the secondary rectangular stress components causing photoelastic effect vary linearly along the light paths.   For example, for normal incidence parallel to z-axis (for a given x), one can assume

$$\sigma_x - \sigma_y = a_o + az \quad \text{and} \quad 2\tau_{xy} = b_o + bz \qquad \ldots (3)$$

The slice with stresses locked in is subjected to known live loads at room temperature, Fig. 2.   The additional stress field introduced by these live loads can be determined by taking an auxiliary stress-free slice which is geometrically identical to the stress frozen slice and subjecting it to the same known live loads at room temperature. Assuming the auxiliary slice to be a plane stress model, the auxiliary stresses σ_x^*, σ_y^*, τ_{xy}^* can be determined by the conventional photo-elastic techniques.   When these are superposed on the existing stresses in the stress frozen model, one gets a new stress distribution along the light path. For normal incidence, these would be

$$\sigma_x - \sigma_y = a_o + (\sigma_x^* - \sigma_y^*) + az = a_o + \alpha L + az \qquad \ldots (4a)$$

$$2\tau_{xy} = b_o + 2\tau_{xy}^* + bz = b_o + 2\beta L + bz \qquad \ldots (4b)$$

where α is the value of $\sigma_x^* - \sigma_y^*$ and β the value of τ_{xy}^* for a unit live load ($L = 1$) on the auxiliary slice. By changing the value of L, different states of stress along the light path can be realised and the corresponding characteristic parameters can be determined.   Since the number of unknowns along a given light path still remains four, one can obtain as many equations as one needs, involving the four unknowns, to solve for them.   One possible method of solving these equations is the following:

Using the calculus of Jones (1941), the optically equivalent model can be represented by the following matrices:

$$M \equiv \begin{bmatrix} \cos\gamma & -\sin\gamma \\ \sin\gamma & \cos\gamma \end{bmatrix} \begin{bmatrix} \cos\theta & -\sin\theta \\ \sin\theta & \cos\theta \end{bmatrix} \begin{bmatrix} e^{i\psi} & 0 \\ 0 & e^{-i\psi} \end{bmatrix} \begin{bmatrix} \cos\theta & \sin\theta \\ -\sin\theta & \cos\theta \end{bmatrix}$$

$$= \begin{bmatrix} J + iK & -L + iN \\ L + iN & J - iK \end{bmatrix} \qquad \ldots (5)$$

where $J = \cos\psi \cos\gamma$ $\qquad\qquad K = \sin\psi \cos(2\theta + \gamma)$

$\qquad\qquad L = \cos\psi \sin\gamma \qquad\qquad N = \sin\psi \sin(2\theta + \gamma) \qquad \ldots (6)$

Since $J^2 + K^2 + L^2 + N^2 = 1$, only three of these are independent. It can be shown that (Srinath and Bhave 1974)

$$\frac{d}{dz} \begin{bmatrix} J + iK \\ -L + iN \\ L + iN \\ J + iK \end{bmatrix} = \frac{ic}{2} \begin{bmatrix} \sigma & o & \tau & o \\ o & \sigma & o & \tau \\ \tau & o & -\sigma & o \\ o & \tau & o & -\sigma \end{bmatrix} \begin{bmatrix} J + iK \\ -L + iN \\ L + iN \\ J + iK \end{bmatrix} \quad \ldots (7)$$

where we have put

$$\sigma = \sigma_x - \sigma_y = a_o + \alpha L + az$$

$$\tau = 2\tau_{xy} = b_o + 2\beta L = bz \qquad \qquad \ldots (8)$$

In the above equations for σ and τ we have used the expressions for normal incidence. For any oblique incidence, one has to use the appropriate secondary rectangular stress components along the light path. The problem now is to determine the values of σ and τ from the known values of J, K, L and N at the entrance and at exit. This can be obtained by using the Peano-Baker method (Ince 1956), and the solution is

$$\begin{bmatrix} J + iK \\ -L + iN \\ L + iN \\ J + iK \end{bmatrix}_z = (G) \begin{bmatrix} J + iK \\ -L + iN \\ L + iN \\ J + iK \end{bmatrix}_{zo} \qquad \ldots (9)$$

(G) is called the matrizant and is given by

$$(G) = \text{matrizant} = (1) + (mu) + (mumu) + (mumumu) + - - - - \qquad \ldots (10)$$

$$u = \frac{ic}{2} \begin{bmatrix} \sigma & o & \tau & o \\ o & \sigma & o & \tau \\ \tau & o & -\sigma & o \\ o & \tau & o & -\sigma \end{bmatrix} \qquad \ldots (11)$$

(mu) represents the integration of u taken between the limits z_o and z (i.e. the entry point and the exit point). To get (mumu), one multiplies u and mu in the order umu and integrates the result between the limits z_o and z. The values of J, K, L and N at the entry point are respectively 1, 0, 0 and 1. At the exit point, these are determined experimentally for each value of the live load. By changing the value of the live load L on the slice, as many equations as required can be obtained.

An alternative set of equations similar to Eqs. (7) are the equations of Aben (1966). These are

$$\begin{bmatrix} E_x \\ E_y \end{bmatrix} = -\frac{ic}{2} \begin{bmatrix} \sigma & \tau \\ \tau & -\sigma \end{bmatrix} \begin{bmatrix} E_x \\ E_y \end{bmatrix} \qquad \ldots (12)$$

where E_x and E_y are the light vectors along the x and y axes. These equations can be obtained in a manner similar to the one used in getting Eqs. (7). The problem once again is to determine σ and τ along the light path from known values of E_x and E_y at entry and at exit points. Here also, one can use Peano-Baker method to obtain the solution.

REFERENCES

Aben HK, (1966) Optical Phenomena in Photoelastic Models by the Rotation of Principal Axes. Exptl. Mech. 6 (1), 13-22.

Frocht MM, Guernsey R, (1952) NACA Technical Note 2822

Ince EL, (1956) Ordinary Differential Equations, Dover Publications, Inc.

Srinath LS, Bhave SK, (1974) A New Nondestructive Method for Three-dimensional Photoelasticity, Exptl. Mech. 14 (9), 367-372.

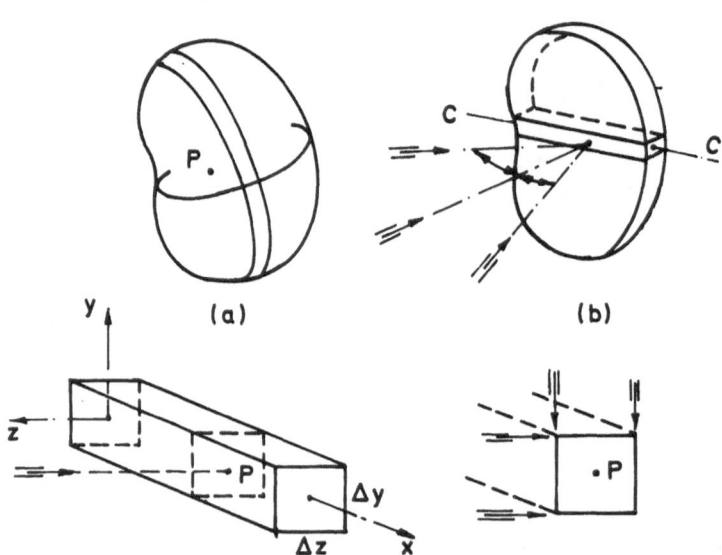

FIG. 1 FROZEN STRESS MODEL, SLICE AND SUB-SLICE

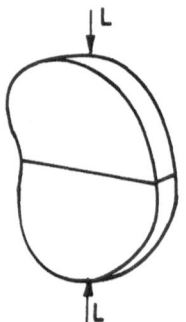

FIG. 2 AUXILIARY SLICE UNDER LIVE LOAD L

Micro-Photoelasticity and its Picture Processing

Ryuichi Shintani, Toshiharu Yoshikawa and Toshimitsu Funayoshi

Department of Physics, Faculty of Science, Kwansei Gakuin University, Uegahara, Nishinomiya, 662 Japan

SYNOPSIS

Two micro-photoelastic equipment models, reflective and transparent model, are made by improving of a metallurgical microscope. The sensitivity of the former model is twice that of the latter one. Photoelastic fringes can be observed clearly in a monofilament with 200 to 400 μm in diameter. Its fringes were analyzed by numerical integration. Some preliminary tests were also conducted with polymer monofilaments. Moreover, the equipment has been developed to continuously and automatically measure the coefficient of birefringence in monofilaments. Fringe patterns are recorded taken by a TV camera, then its video signals are fed to a microcomputer after being converted into digital signals by an A/D converter. The results are consistent with those from photographs.

1. INTRODUCTION

Photoelastic experiment have been used to find stress in transparent matter but only small test pieces with a microscope. For example, there is fishing line, optical fiber and so on. Recently, these are being mass-produced and their qualities are also being improved. To assist in their quality control and to study optical properties, two types of micro-photoelastic equipment are made for trial by improving a metallurgical microscope(1984).
One study of polymer chains in a monofilament(1953)(1954) used a polarizing microscope in a linearly polarized field but could not observe fine curved fringes in a bent monofilament. We can clearly observe complicated photoelastic fringes through the circularly polarized field. The material used in the present study was copolymer with polyamide and nylon-6 which were supplied by UNITIKA LTD.

2. EQUIPMENT

Two trial equipment models for micro-photoelasticity are reflective and transparent. Both have a light source with interference filters (432, 542, and 634 nm). Photographs are taken so that direct magnifications are comparatively small (25 to 50 times), to make distance between an object lens (x5 or x10) and a test piece longer. For precise observation photoelastic fringes were magnified 100 to 400 times by enlarging the negatives.

2.1 Reflective model

The optical alignment of the reflective model is the same as the metallurgical microscope shown in Fig. 1(a). A test piece is soaked in a liquid bath which has the same refractive index to prevent reflection at its surface. The bottom of the liquid bath is a mirror. The sensibility is twice that of the transparent model, because the

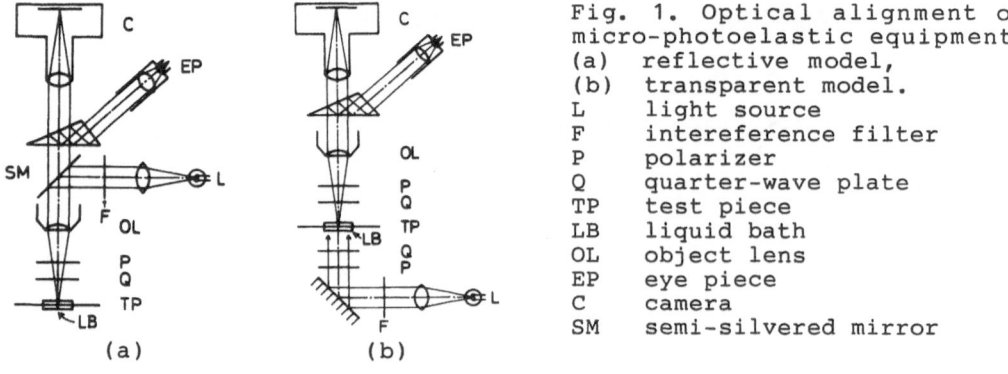

Fig. 1. Optical alignment of
micro-photoelastic equipment,
(a)   reflective model,
(b)   transparent model.
L   light source
F   intereference filter
P   polarizer
Q   quarter-wave plate
TP   test piece
LB   liquid bath
OL   object lens
EP   eye piece
C   camera
SM   semi-silvered mirror

(a)   (b)

light has double the path through the test piece, but the field of
vision is very dark, because incident and reflected beams become half
due to the semi-silvered mirror. We can only observe dark field,
since the phase of the incident beam is inverted by the bottom mirror
of the bath.

2.2 Transparent Model

A test piece is illuminated from the lower part of the microscope.
The bottom of the liquid bath is a transparent glass; its alignment is
shown in Fig. 1(b). We can more clearly observe photoelastic fringes
through the field which is set bright or dark by turning the analyzer.

3. TYPICAL MICRO-PHOTOELASTIC PHOTOGRAPHS

Many micro-photoelastic photographs are taken by the micro-
photoelastic equipment. Figure 2 shows photoelastic fringes in a bent
monofilament with a diameter of 410 µm. Figure 2 (a), (b) and (c) are
photographs by the transparent model using the interference filters
with wavelength of 432, 542, and 634 nm. Figure 2 (d), which has twice
the fringes, is a picture by the reflective model with a 542 nm
wavelength filter. When a few fringes are observed, it is convenient
to take color photographs. To analyze their patterns, the monochromatic picture is obtained by observing them through the desired interference filter. Figure 3 shows photoelastic fringes of a stretched
monofilament under dark field using the transparent model.   The test
piece with a diameter of 410 µm is stretched nearly 20% at a time at

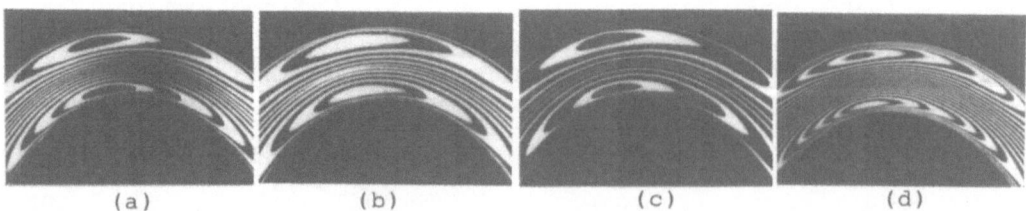

(a)   (b)   (c)   (d)

Fig. 2. Photoelastic fringes in a monofilament with the diameter of
410µm by bending. Photos (a), (b), and (c) are taken by the transparent model with the interference filter 432, 542, and 634 nm in
wavelength. Photo (d) is a picture by the reflective model with a
filter of 543 nm.

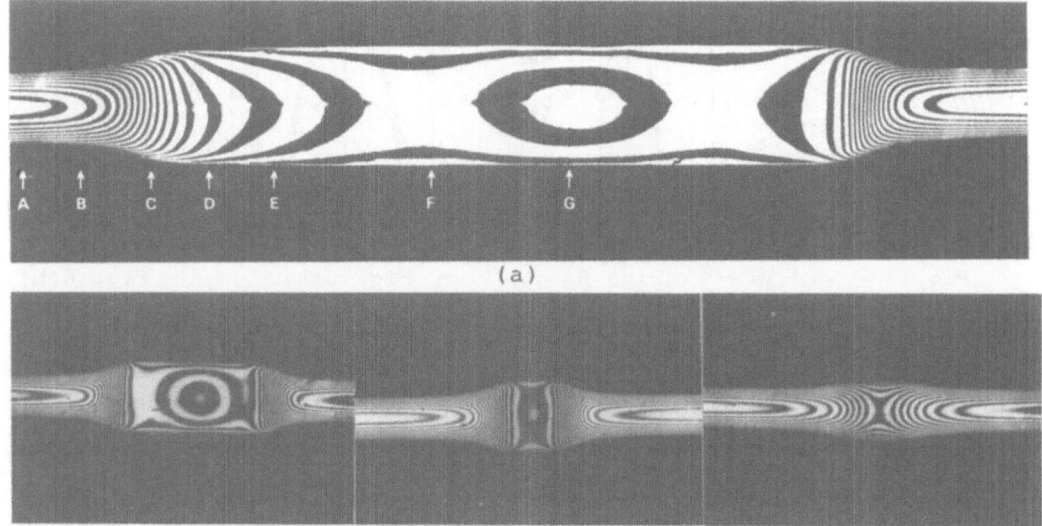

Fig. 3. Photoelastic fringes in an elongated monofilament. The initial diameter is 410 µm. Photo (a) is about 20% stretching. Photos (b), (c), and (d) show the fringes for stretching in increments of an additional 20%.

about 70 °C in moist air then is observed at room temperature. Its strains under loading are not uniform in places, and elongations happen suddenly at certain positions. The fringe patterns are very complex at the boundary of elongation and suggest to us that the radial distribution of the coefficient of birefringence B(ρ) is not uniform. The circular fringes shown in Figs. 3(a) and (b) may be seen in early stage of elongation. As the test piece is stretched more, it is elongated from both sides, as shown in Figs. 3(c) and (d). Finally, fringes become parallel to the direction of the monofilament and its diameter is 170 µm.

4. ANALYSIS OF FRINGES IN MONOFILAMENT

As mentioned above, the pictures in Fig. 3 suggest to us that the coefficient of birefringence B(ρ) estimated from the fringe pattern is not uniform in th necking part of the monofilament in the early stage of stretching. The fringe order N and B(ρ) at several positions A through G in Fig. 3(a) were therefor obtained by numerical integration. The positions of B, C, D, E, F and A are shown in table below.

Point	B	C	D	E	F	A
Position (mm)	0.21	0.43	0.65	0.87	1.41	1.85

Figure 4 indicates the cross section of a monofilament with radius R. The strain at any point on the radius vector may be given by the coefficient of birefringence B(ρ). The fringe order N through the element of vertical distance at the fixed point y is given by

$$dN = B(\rho)dz. \tag{1}$$

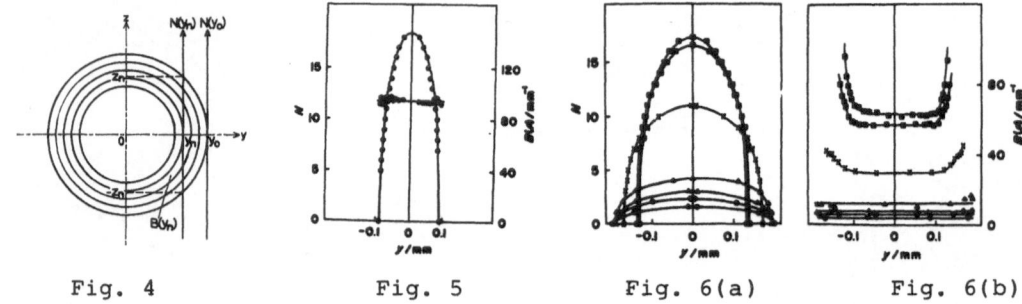

Fig. 4   Fig. 5   Fig. 6(a)   Fig. 6(b)

Fig. 4. Cross section of monofilament. Concentric rings in which B(ρ) may be uniform show division for numerical integration.

Fig. 5. Fringe order and coefficient of birefringence in fully stretched monofilament.
y radius vector, N ● fringe order, B(ρ) o coefficient of birefringence.

Fig. 6. Fringe order and coefficient of birefringence in the early stage of stretching at A(■), B(□), C(✗), D(▲), E(△), F(○), and G(●) in Fig. 3 (a). (a) fringe orders, (b) coefficients of birefringence.

The fringe order N through the vertical path at the point y_n is thus

$$N(y_n) = \int_{-z_n}^{+z_n} B(\rho)dz. \tag{2}$$

If $y_0 = R$ and the fixed point y_n is on the n-th ring from outer to inner side of a monofilament, Eq. (2) may be replaced with following equation

$$N(y_n) = 2 \sum_{i-1}^{n} B(y_i)\{\sqrt{y_{i-1}^2 - y_n^2} - \sqrt{y_i^2 - y_n^2}\} . \tag{3}$$

Then,

$$B(y_n) = \frac{N(y_n) - 2 \sum_{i-1}^{n} B(y_i)\{\sqrt{y_{i-1}^2 - y_n^2} - \sqrt{y_i^2 - y_n^2}\}}{2\sqrt{y_{n-1}^2 - y_n^2}}. \tag{4}$$

Using this equation, B(ρ) of fully stretched monofilament can be obtained nearly constant and is shown with the fringe order in Fig. 5. For stretching, the fringe orders at points A through G in Fig. 3(a) are shown in Fig. 6(a), and the coefficients of birefringence are shown in Fig. 6(b). In the early stretching first several percent, the value of B(ρ) is small and nearly constant, but it increases the middle stage and becomes considerably greater near the surface. Finally, it is nearly constant as shown in Fig. 5.

6. PICTURE PROCESSING

Equipment has been developed to automatically measure fringes and calculate coefficients of birefringence with a TV camera and a microcomputer. Figure 7 shows a block diagram of this equipment. The optical system is the transparent model of micro-photoelasticity. Both sides of the test piece are connected to a stainless steel wire with adhesives. The wires are attached to the chuck of the tension tester. Tensile stress is applied to the test piece as the cross-head moves upward. Fringe patterns are recorded by a TV camera, and the video signals are separated into vertical and horizontal synchronous pulses. The former acts as a control signal for a peripheral interface adaptor (PIA) to feed digital video data to the microcomputer; the latter generates the two pulses with widths of 5 μs and 0.2 μs by monostable multivibrators. The 5 μs pulse acts as starting signal for an A/D convertor. The other is added to the video signal, and a cursor appears on the monitor TV screen. The pulse position in a raster can be shifted, and we can determine the sampling point by the cursor on the screen. The flow chart for feeding video signals to the microcomputer is shown in Fig. 8. When the PIA detects the vertical synchronous pulse, one point in a raster is sampled, and the microcomputer reads 256 bits in 1/2 frame, in only 1/60 sec. The flow chart for processing digital data with the microcomputer is shown in Fig. 9. To detect the monofilament outline,

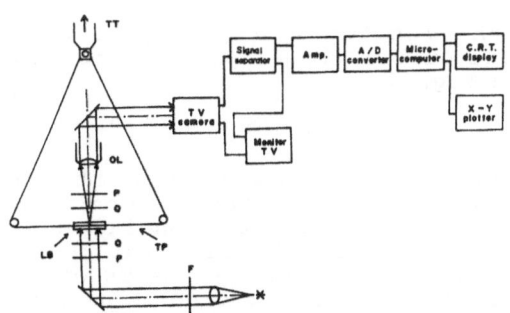

Fig. 7. Block diagram of micro-photoelastic equipment with the picture processing unit.
TT   tension tester
TP   test piece
LB   liquid bath
P   polarizer
Q   quarter-wave plate
OL   object lens

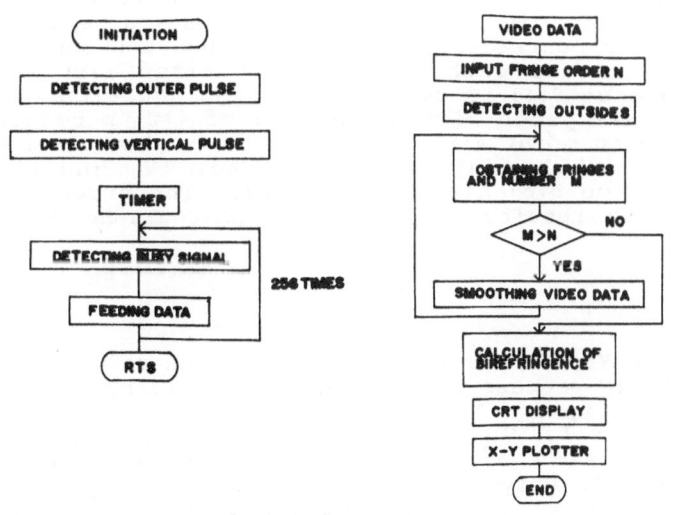

Fig. 8. Flow chart for feeding video signals to the picture processing unit.

Fig. 9. Flow chart for processing digital video data in the unit.

Fig. 8   Fig. 9

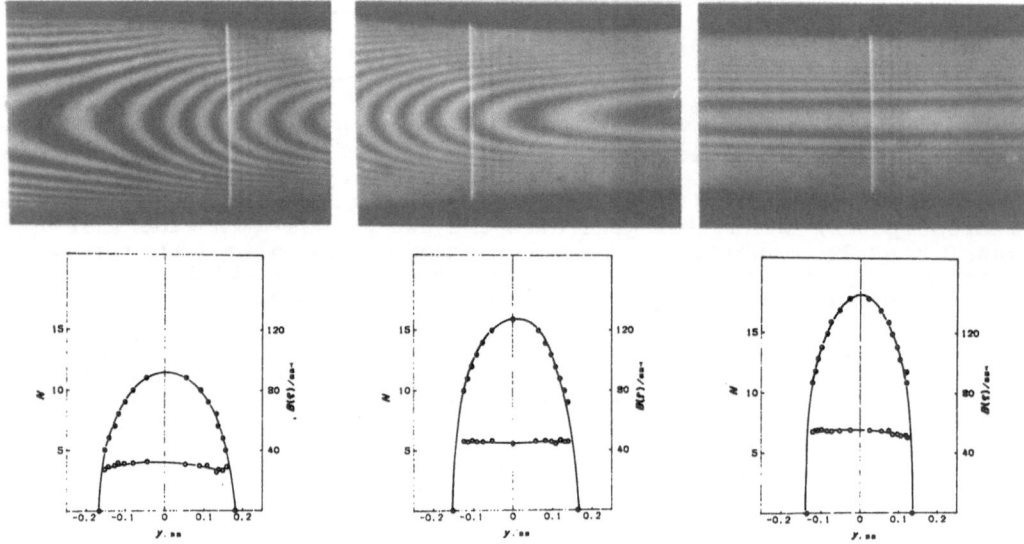

Fig. 10. Photoelastic fringes on the monitor TV (upper part), and coefficients of birefringence at the cursor (lower part). Axis name and symbols are the same as those in Fig. 5.

differences from neighboring data are calculated with the microcomputer, and the first position along the cursor at which the difference is positive is assigned to the outline of monofilament. Fringe order and its position are obtained in the same way. When fringes are close at the periphery of the monofilament, the TV camera cannot resolve the fringe pattern. The maximum fringe order N near the center of the monofilament must then be fed manually to the micro-computer. The number of fringes obtained M is compared with N, and if the former is greater than the latter due to noise, the digital data is smoothed by averaging three adjacent data. The process is repeated many times until there is no contradiction. The distance between the center of the monofilament and each fringe is obtained, and the coefficient of birefringence $B(\rho)$ is calculated using equation (4). In this experiment, the test piece with a diameter of $410\,\mu m$ before being stretched is treated in moist air. Figure 10 shows photoelastic fringes of stretched monofilament on the monitor TV, and the coefficient of birefringence $B(\rho)$ along the cursor. The processed results are consistent with results analyzed from photographs under the same conditions.

REFERENCE

Shintani R., Kubota T. and Yoshikawa T. (1984) Analyses of mono-filaments in stretching process by micro-photoelasticity. Proc. Jpn. Soc. Photoelasticity, 5 1-8.
Stein R. S. and Tobolsky A. V. (1953) Determination of the statistical segment size of polymer chains from stress-birefringence studies. J. Polym. Sci., 11 285-288.
Gurnee E. F. (1954) Theory of orientation and double refraction in polymers. J. Appl. Phys., 25 1232-1240.

A Stress Freezing Procedure for Less Photoelastic Model Distortion

M. Nisida and Y. Sawa

Faculty of Science and Technology, Science University of Tokyo, Noda, Chiba, 278 Japan

INTRODUCTION

Although the stress-frozen photoelastic technique provides an excellent and ingenious method for three-dimensional stress analysis in photoelasticity, there are some cases in which the distortion of the stress-frozen models is remarkable due to the considerably low modulus at the freezing temperature, leading to an unignorable error in the stress distribution.
This problem of large distortion of the stress-frozen model manifests itself especially when the model is comparatively slender such as beams under bending and twisting bars.
Regarding this characteristics of model materials, as is well known, "Figure of Merit" $Q = E/f_\sigma$ [1] (E, Young's modulus, f, stress fringe value) is usually used as an index. In Japan, however, $Q = \alpha$ E Fringe Order/mm replacing f_σ by α ; photoelastic sensitivity of the material = fringe order/stress kgw/mm^2/thickness mm, is used for this purpose.
In this investigation, a systematic experiment has been carried out to determine the relation between the figure of merit Q and the freezing temperature T of photoelastic material epoxy (Araldite B) commonly used in Japan.
The results show that, contrary to a commonly held belief, the most favorable freezing temperature Tm, from the view point of model deformation, exists at 30 to 35 °C below the phase transition point from glassy to rubber elastic state T_0 , and it has been found that 3.5 to 6.0 times higher value of Q compared to the Q at T_0 is attained at this temperature.
To demonstrate this fact, two stress frozen patterns are compared in Fig. 1. In the figure, A is the photoelastic pattern of the longitudinal symmetrical slice from a stress frozen cylinder model at phase transition point T_0 = 130°C under three-point bending, and B is the pattern frozen at Tm = 90°C for the same model material as A. The ratio Qm/Q_0 = 4.3 is attained.

SPECIMENS AND MEASUREMENT OF PHOTOELASTIC SENSITIVITY α AND YOUNG'S MODULUS E

Epoxy (Araldite B) is used as the photoelastic material for the investigation since it is the most common material for three-dimensional photoelasticity in Japan and available on the Japanese market.
To obtain all-comprehensive results for epoxy, about 30 specimens made of differently polymerized epoxy were examined, but it is found that the following one indicates the most favorable properties in view of the figure of merit Q.
    Araldite B polymerized with 23.1% in weight of hardner HT901 (phthalic acid anhydride) and heated at 130 °C for 170 hrs.

Two-dimensional beam type calibration specimens of this material were cut from a plate of large area which had been cast in a glass plate-mold from a kit and were heat treated taking special precautions to avoid chemical and physical inhomogeneity.

The dimensions and shape of the calibration specimens and load applied are shown in Fig. 2. The specimen is loaded in four-point-bending at elevated temperature in a furnace.

The deflection of the specimen is measured through grass windows of the furnace with a scale telescope, and the photoelastic pattern is photographed to determine α every minute.

After stress freezing is completed, E_f and α_f related to the frozen stress are obtained.

In Fig.3, two stress frozen photoelastic patterns of the calibration specimen are shown for comparision.

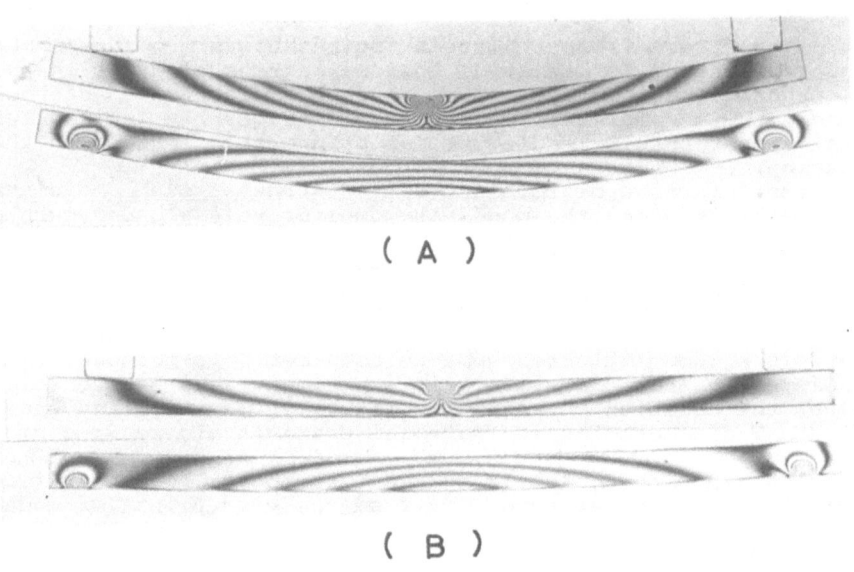

(A)

(B)

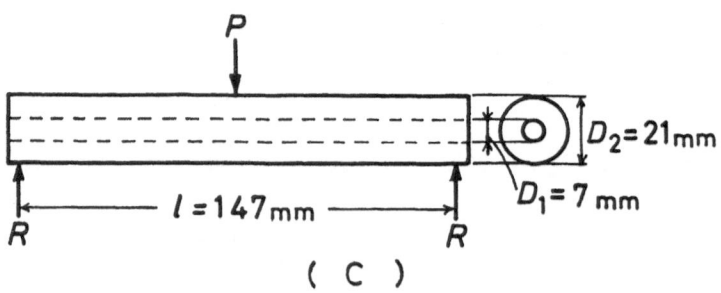

(C)

Fig. 1 Comparision of two stress frozen patterns of a cylinder model.
A: freezing temperature T = 135°C, P = 1.2 kgw,
$Q_0 = \alpha_0 E_0 = 30.4$ F.O/mm
B: freezing temperature Tm = 90°C, P = 47.0 kgw,
$Q_m = \alpha_m E_m = 130$ F.O/mm

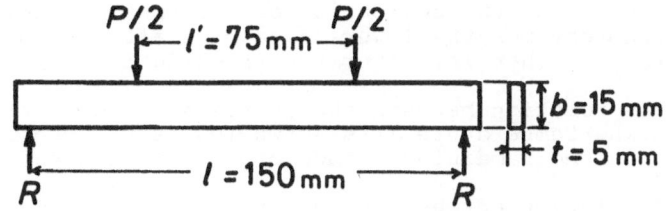

Fig. 2 Dimensions of and loading applied to the specimens.

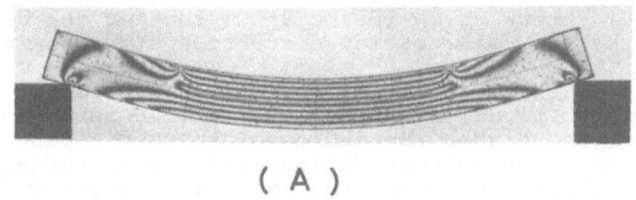

(A)

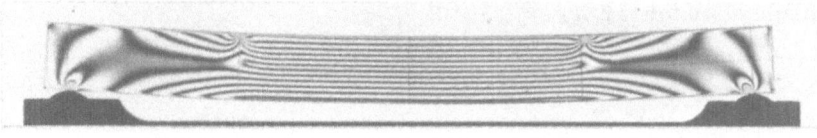

(B)

Fig. 3 Stress-frozen patterns for two different freezing
temperatures.
A: freezing temperature T = 135°C(T_0),
photoelastic sensitivity α_f = 43.0 F.O/kgw/mm,
Young's modulus E_f = 0.95 kgw/mm^2,
Q = α_f E_f = 40.9 F.O/mm
B: T = 95°C(Tm), α_fm= 0.33 F.O/kgw/mm,
E_fm= 530 kgw/mm^2, Qm = α_fm E_fm = 175 F.O./mm
In both cases, suffix f is related to the remaining distortion
and fringe after the freezing is completed at room temperature.

RESULTS

The results obtained by the experiment are arranged as shown in Figs. 4,
5 and 6.
In Fig. 4, the relations between the apparent and stress-frozen Young's
modulus E, E_f and the freezing temperature T are shown. E_f is obtained
from the deflection of the specimen after freezing is completed.

As is easily noticed, the deflection of the specimen beams shows a remarkable recovery for the lower range of T after removal of load at room, temperature, that is, after the freezing is completed.

In Fig. 5, the relation between the photoelastic sensitivity α , α_f and the freezing temperature T is shown (α_f is the photoelastic sensitivity based on the remaining fringe order after unloading at room temperature).
The required relation between the figure of merit $Q = \alpha_f E_f$ and the freezing temperature T is obtained as shown in Fig. 6.
It is interesting to note that neither α_f nor E_f indicate a sharply defined phase transition at point T_0 as is expected but show a gradual variation with temperature and that the figure of merit $Q = \alpha_f E_f$ exhibits a maximum value Qm at a temperature Tm below the phase transition point T_0. This means that it is far more desirable to freeze the model stress at Tm (85 to 95°C) than at the usual transition point T_0 from the viewpoint of low distortion of frozen-stress model.

The rate of increase of Q (Qm/Q$_0$) ranges from 4.0 to 6.0 depending on the degree of polymerization of the material. Results show that, generally speaking, the higher the degree of polymerization or the temperature and the hours for polymerization, the higher the absolute value of Qm = $\alpha_f E_f$ is while the lower the rate of increase Qm/Q$_0$ is.
For example, the highest value of Q can be expected at freezing temperature Tm = 95°C to be 15.5 F.O/mm provided that the material has been polymerized by a sufficient heat treatment. This value is about 4.5 times as high as that at T_0 = 135°C.

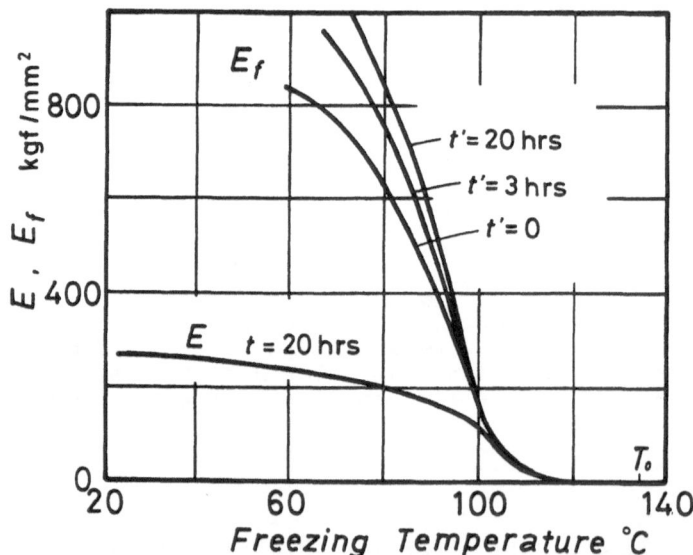

Fig. 4   Relation between Young's modulus and freezing temperature.
          E is the apparent Young's modulus of stress not yet frozen,
          E_f is the Young's modulus after the freezing is completed,
          t' being the time after unloading at room temperature.

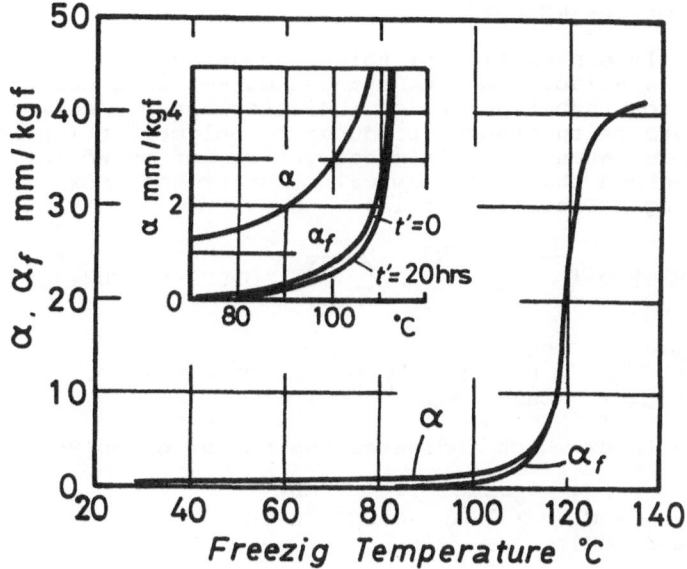

Fig. 5 Relation between α , α_f and freezing temperature T.
α is the photoelastic sensitivity based on the live fringe
order under loading and α_f is the one based on the fixed
fringe order after unloading at room temperature.

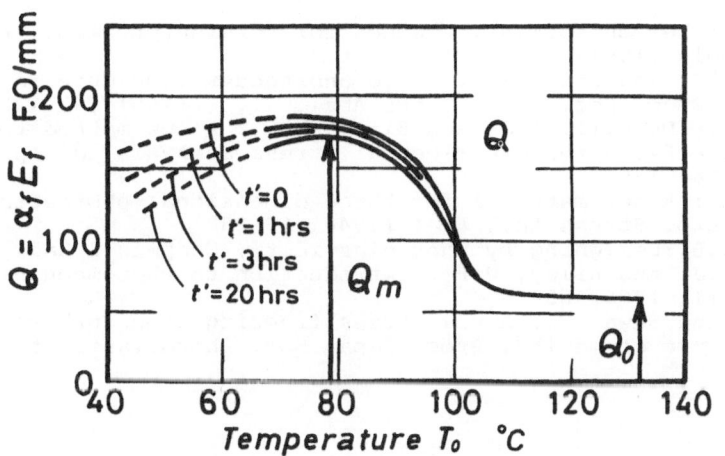

Fig. 6 Relation between the figure of merit $Q = \alpha_f E_f$ and the
freezing temperature T. The maximum value Qm is obtained
at Tm about 35°C below the transition point T_0.

CONSIDERATION OF THE RESULTS

Probably the only explanation of this unexpected fact is that the epoxy material available in the Japanese market does not consist of chemically uniform substance which indicates sharply defined transition but is a mixture of substances produced by polymerization in different degrees, that is, substances of great variety of molecular weight.
It has been learned that the chemical structure of epoxy is expressed by the following molecular formula:

where n is the index which indicates the number of polyglycidyl-ether hase

in other words the degree of polymerization.
Therefore, we are convinced that, for the actual material available in Japan, the value of n is different from molecule to molecule depending on the degree of polymerization during the manufacturing process and casting heat treatment.
By making use of this fact, it is possible to freeze stresses in three-dimensional photoelastic models with far higher value of figure of merit Q than that at the usual (apparent) phase transition point.

REFERENCE

1) Nisida, M.: New photoelastic method for torsion problems, Proc. Int. Symp. Illinois (1961), 109-121
2) Durelli, A.J. and Lake, R.L.: Some unorthodex procedure in photo-elasticity, Proc. Soc. Exp. Stress Anal., I, 1 (1951),97-122
3) Dally, J.W., Durelli, A.J. and Riley, W.F.: A new method to "lock-in" elastic effects for experimental stress analysis, J. App. Mech., 25 (1985), 189-195
4) Leven, M.M.: A new material for three-dimensional photoelasticity, Proc. Soc. Exp. Stress An., 6, 1 (1948) 19-28
5) Heywood, R.B.:Designing by photoelasticity, Chapman & Hall, 1952, 63
6) Durelli, A.J. and Riley, W.F.: Introduction to Photomechanics, Prentice Hall, 1965, 65-66
7) Nisida M. and Sawa, Y.: A new stress freezing procedure for im-proving "figure of merit". Proc. Japan Soc. Photoelasticity (in Japanese)

On Wooden Column under Torsion and Deflection of Membrane

Gengo Matsui[1], Teruaki Tanaka[2] and Hidetoshi Yokomise[2]

[1] Waseda University, Okubo, Shinjuku-ku, Tokyo, 160 Japan
[2] Kokushikan University, Setagaya-ku, Tokyo, 154 Japan

§ 1 Introduction

In general, a wooden column can be assumed as an anisotropic member that has different modulas of rigidity in the direction perpendicular to the annual ring[1].
By taking the stress function, an anisotropic wooden column loaded by torque can be analyzed and compared with the isotropic solution[2].
The isotropic solution of torsional problems has been related to the membrane analogy.
Hence, the corresponding torque acting on the column can be obtained by experimentation using the membrane analogy.
For anisotropic problems, the stress acting on the column can be investigated by comparing the experimental method to theory.

§ 2 Stress function and deflection of membrane

In discussing the problem, we take the centroidal axis of annual rings as the torsional axis and use polar coodinates r and θ to define the position of an element in the plane of a cross section.
The notation for shearing stress, modulus of rigidity, and shearing strain in the radical and tangential directions are denoted by τ_r, G_r, γ_r, and τ_θ, G_θ, γ_θ, respectively (Fig.1).
The relations between the components of stress and strain are

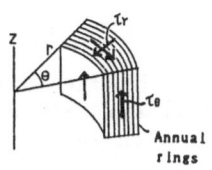

Fig.1

$$\gamma_\theta = \frac{1}{G_\theta} \cdot \tau_\theta \qquad \text{and} \qquad \gamma_r = \frac{1}{G_r} \cdot \tau_r$$

where u, v, and w are the displacements in the r, θ, and z directions, and ω is the angle of rotation per unit length of the cross section. Defining the angle of rotation as a function ϕ, we find[3]

$$u=0 \quad , \quad v=\omega \cdot z \cdot r \quad , \quad w=\omega \phi(r \cdot \theta)$$

The corresponding components of stress are

$$\sigma_r = \sigma_\theta = \sigma_z = \tau_{r\theta} = 0$$

$$\tau_\theta = G_\theta \cdot \omega (r - \frac{1}{r} \frac{\partial \phi}{\partial \theta}) \qquad \text{and} \qquad \tau_r = G_r \cdot \omega \frac{\partial \phi}{\partial r} \quad .$$

The differential equation can be expressed by taking $k = G_\theta / G_r$

$$\frac{\partial^2 \phi}{\partial r^2} + \frac{1}{r} \frac{\partial \phi}{\partial r} + k \frac{1}{r^2} \frac{\partial^2 \phi}{\partial \theta^2} = 0 \quad --------- (1)$$

where $k = 1$ is called an isotropic condition.
Substituting $r = r'$ and $\theta = \sqrt{k} \theta'$ into Eq. (1), we find

$$\frac{\partial^2 \phi}{\partial r^2} + \frac{1}{r} \frac{\partial \phi}{\partial r} + \frac{1}{r^2} \frac{\partial^2 \phi}{\partial \theta'^2} = 0 \qquad --------(2)$$

which is called a two-dimensional Laplace equation.
Considering the harmonic function ϕ instead of ϕ , the following equations are

$$\frac{\partial \phi}{\partial r'} = \frac{1}{r'} \frac{\partial \psi}{\partial \theta'} \qquad \text{and} \qquad \frac{\partial \psi}{\partial r'} = -\frac{1}{r'} \frac{\partial \phi}{\partial \theta'} \quad .$$

Assuming the stress function in the form,

$$\Psi = \sqrt{G_r G_\theta} \cdot \omega (\psi - \frac{1}{2} \sqrt{K} \, r^2)$$

the equilibrium equation is denoted by Ψ .

$$\frac{\partial^2 \Psi}{\partial r^2} + \frac{1}{r} \frac{\partial \Psi}{\partial r} + K \frac{1}{r^2} \frac{\partial^2 \Psi}{\partial \theta^2} = -2 G_\theta \omega \qquad --------(3)$$

We next obtain the stress components τ_θ and τ_r as shown in Eq.(4) and Eq.(5).

$$\tau_\theta = G_\theta \omega (r + \frac{1}{r} \frac{\partial \phi}{\partial \theta}) = -\frac{\partial \Psi}{\partial r} \qquad --------(4)$$

$$\tau_r = G_r \omega (\frac{\partial \phi}{\partial r}) = \frac{1}{r} \frac{\partial \Psi}{\partial \theta} \qquad --------(5)$$

The magnitude of torque M_t can be obtained by Eq.(6).

$$M_t = 2 \int \Psi \, dA \qquad --------(6)$$

Substituting $r = r'$ and $\theta = \sqrt{K} \, \theta'$ into Eq.(3), we arrived at

$$\frac{\partial^2 \Psi}{\partial r'^2} + \frac{1}{r'} \frac{\partial \Psi}{\partial r'} + \frac{1}{r'^2} \frac{\partial^2 \Psi}{\partial \theta'^2} = -2 G_\theta \omega \quad . \qquad --------(7)$$

If T is the uniform tension per unit length of its boundary and p is load per unit area of the membrane, the vertical deflection w can be written in the form[4]

$$\frac{\partial^2 w}{\partial r^2} + \frac{1}{r} \frac{\partial w}{\partial r} + \frac{1}{r} \frac{\partial^2 w}{\partial \theta^2} = -\frac{p}{T} \quad . \qquad --------(8)$$

From the deflection of the membrane,we can obtain values of Ψ by replacing the quantity $-p/T$ with $-2 G_\theta \omega$.

Hence, several types of problems in an anisotropic condition can be solved by membrane analogy.
The instruments for this membrane experimentation are as shown in Fig.2. The model, which is made of thin rubber of thickness 0.2mm is supported at the boundary where the atmospheric pressure is uniform.
Thus standard deflection can be determined by measuring the contour lines due to deflection of membrane in a circular cross section.

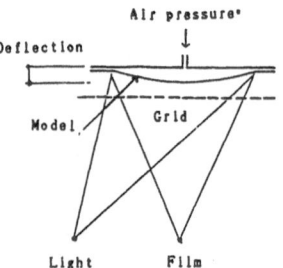

Fig.2 Instruments for measuring the deflection of membrane

§3 A circular cross section

The stress function is assumed in the form

$$\Psi = \frac{1}{2} G_\theta \omega (a^2 - r^2) \qquad --------(9)$$

where 'a' is a radical.
Shearing stress components τ_θ and τ_r are

$$\tau_\theta = -\frac{\partial \Psi}{\partial r} = \frac{2M_t}{\pi a^4} r \qquad \text{---------(10)}$$

and

$$\tau_r = \frac{1}{r}\frac{\partial \Psi}{\partial \theta} = 0 \qquad \text{---------(11)}$$

and the magnitude of torque M_t is

$$M_t = 2\int \Psi dA = G_\theta \omega a^2 \frac{\pi}{2} . \qquad \text{---------(12)}$$

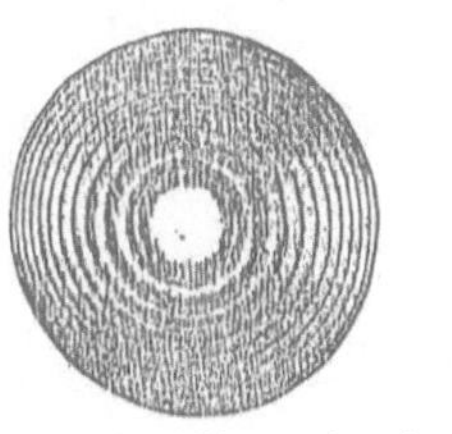

These stress components and the magnitude of torque correspond to the value of isotropic cross section with modulus of rigidity G_θ. The values for torsional stress function in a circular cross section are represented in Fig.3, and contour lines due to deflection of the membrane are shown in Photo.1, and these results can be used as standard model.

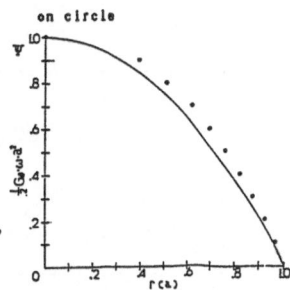

Photo.1 Contour lines of membrane on circle

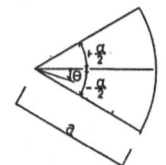

Fig.3 Torsional stress function and contour lines of membrane on circle

§4 A sector of a circle

In this case, the boundaries are given by radical $r = a$ and $\theta = \pm \alpha/2$ as shown in Fig.4. Equation (3) denotes the isotropic solution when $r = r'$, $\theta = \sqrt{k}\theta'$ and $\Psi = \sqrt{k}\Psi'$. That is, the function Ψ' in an isotropic condition when $r' = 0 \sim a$ and $\theta' = +\alpha/2\sqrt{k} \sim -\alpha/2\sqrt{k}$ can be reduced to the anisotropic solution for $r = 0 \sim a$ and $\theta = +\alpha/2 \sim -\alpha/2$.
Hence, by using the torsional stress function Eq. (13), shearing stress components τ_r and τ_θ, in the anisotropic condition can be obtained by Eq. (14) and Eq. (15)[5].

Fig.4 Sector of a circle

$$\Psi' = \frac{1}{2}\sqrt{G_rG_\theta}\cdot\omega\left\{-\left(1-\frac{\cos\frac{2}{\sqrt{k}}\theta}{\cos\frac{\alpha}{\sqrt{k}}}\right)+\frac{16a^2\alpha^2}{\pi^3 K}\sum_{n=1,3,5}^{\infty}(-1)^{\frac{n+1}{2}}(\frac{r}{a})^{\frac{\sqrt{K}\pi}{\alpha}}\cdot\frac{\cos\frac{n\pi}{\alpha}\theta}{n(n^2-\frac{4\alpha^2}{K\pi^2})}\right\} \quad\text{------(13)}$$

$$\tau_r = \frac{1}{2}\sqrt{K}\sqrt{G_rG_\theta}\cdot\omega\left\{r\left(-\frac{2\sin\frac{2}{\sqrt{k}}\theta}{\sqrt{k}\cos\frac{\alpha}{\sqrt{k}}}\right)+\frac{16a\alpha^2}{\pi^3 K}\sum_{n=1,3,5}^{\infty}(\frac{r}{a})^{\frac{n\sqrt{k}}{\alpha}\pi-1}\cdot\frac{\frac{n\pi}{\alpha}\sin\frac{n\pi}{\alpha}\theta}{n(n^2-\frac{4\alpha^2}{K\pi^2})}\right\} \quad\text{------(14)}$$

$$\tau_\theta = \frac{1}{2}\sqrt{K}\sqrt{G_rG_\theta}\cdot\omega\left\{2r\left(1-\frac{\cos\frac{2}{\sqrt{k}}\theta}{\cos\frac{\alpha}{\sqrt{k}}}\right)-\frac{16a\alpha^2}{\pi^3 K}\sum_{n=1,3,5}^{\infty}(-1)^{\frac{n+1}{2}}\frac{\sqrt{k}}{\alpha}\cdot\pi(\frac{r}{a})^{\frac{n\sqrt{k}}{\alpha}-1}\frac{\cos\frac{n\pi}{\alpha}\theta}{(n^2-\frac{4\alpha^2}{K\pi^2})}\right\} \quad\text{------(15)}$$

The magnitude of the torque M_t is given in Eq. (16).

$$M_t = 2\sqrt{K}\sqrt{G_rG_\theta}\cdot\omega\left\{-\frac{a^4}{4}(\frac{\alpha}{2}-\frac{\sqrt{K}}{2}\tan\frac{\alpha}{\sqrt{k}})-\frac{16a^4\alpha}{\pi^3 K}\sum_{n=1,3,5}^{\infty}\frac{1}{\frac{n\sqrt{k}\pi}{\alpha}+2}\cdot\frac{\alpha}{n\pi}\cdot\frac{1}{n(n^2\frac{4\alpha^2}{k\pi^2})}\right\} \quad\text{----(16)}$$

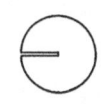

Fig.5 Slotted circle section

The torsion of the slotted cross section as shown in Fig.5 is equal to the solution of sector cross section for $\alpha = 2\pi$. From Eq. (14) and (15), each stress component can be obtained as shown in Fig.6.
Torsional stress component τ_r at the boundary (a = 0) for $\alpha = 2\pi$ are:

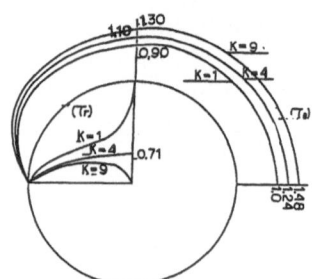

$$k < 4 \qquad \tau_r = \infty$$
$$k = 4 \qquad \tau_r = 0.71(M_t/a^3)$$
$$k > 4 \qquad \tau_r = 0$$

Torsional stress component τ_r uses the constant k = 4 to find the contour lines due to deflection of membrane as shown in the Photos. 2 to 4.

Fig. 6 Torsional stress on the slotted
circle section

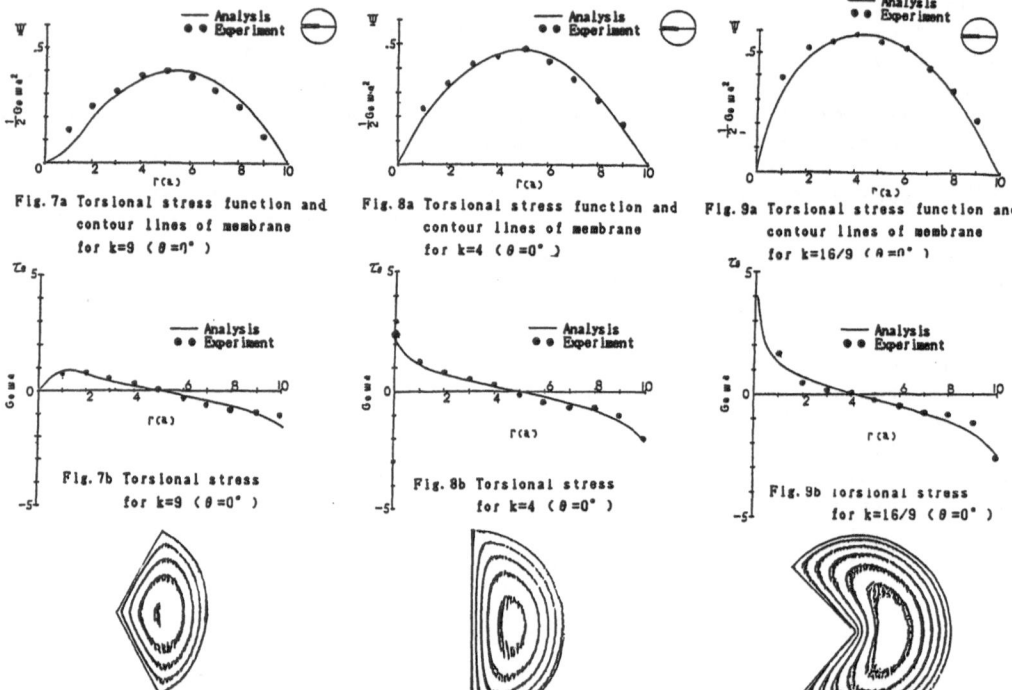

Fig. 7a Torsional stress function and
contour lines of membrane
for k=9 (θ =0°)

Fig. 8a Torsional stress function and
contour lines of membrane
for k=4 (θ =0°)

Fig. 9a Torsional stress function and
contour lines of membrane
for k=16/9 (θ =0°)

Fig. 7b Torsional stress
for k=9 (θ =0°)

Fig. 8b Torsional stress
for k=4 (θ =0°)

Fig. 9b Torsional stress
for k=16/9 (θ =0°)

Photo. 2 Contour lines of membrane
on slotted circle section (k=9)

Photo. 3 Contour lines of membrane
on slotted circle section (k=4)

Photo. 4 Contour lines of membrane
on slotted circle section (k=16/9)

Photos. 2 to 4 represent the anisotropic condition for k = 9, 4, and 16/9, and correspond to the torsional stress component for $\theta = ^1/_3 \pi$, $^1/_2 \pi$, and $^3/_4 \pi$ in an isotropic condition.
Torsional stress function $\Psi (\theta = 0)$, torsional stress component $\tau_\theta (\theta = 0)$ for k = 9, 4, and 16/9 in an anisotropic condition are as shown in Figs. 7 to 9.

§5 Membrane analogy for several types of cross section

It can be seen that the torsional stresses of the complicated anisotropic cross section are easily determined by using membrane analogy, which has proved simple to use and of sufficient accuracy. Several examples are shown as follows:

a) Square cross section

In a square cross section, by replacing $\theta = \sqrt{k}\,\theta'$ with $\theta' = {}^{1}/{}_{2}\,\theta$ for
$k = 4$, the boundary of the cross section correspond to octagen in an
isotropic condition as shown in Fig. 10 and also the boundary for $k = 9$
can be represented by dodecagon as shown in Fig. 11.
Photo. 5 shows the contour lines due to deflection of membrane for
actagonal boundary, and Photo. 6 shows it for dodecagonal boundary.
Torsional stress function Ψ, and torsional stress component τ_θ are
given as shown in Fig. 12 and 13.
The maximum torsional stress component for τ_θ occurred at a center
of the boundary as well as in an isotropic condition. And the values
do not change with k.
And as the ratio of k increases, the stress which obtained by contour
lines due to the deflection at the salient angle of the cross section,
decreases.

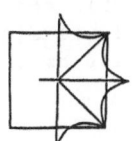

Fig. 10 Octagonal boundary

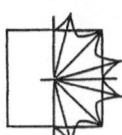

Fig. 11 Dodecagonal boundary

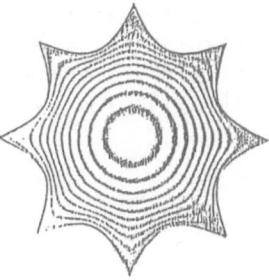

Photo. 5 Contour lines of membrane on
square section (k=4)

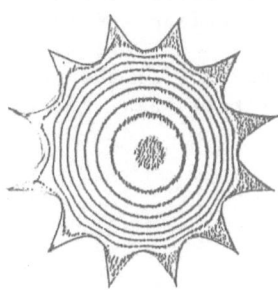

Photo. 6 Contour lines of membrane on
square section (k=9)

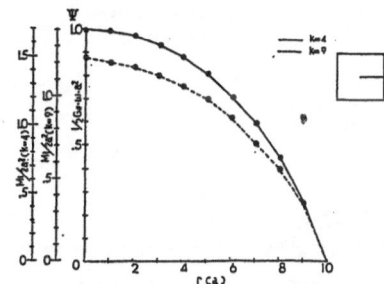

Fig. 12a Torsional stress function and contour lines
of membrane on square section ($\theta = 0°$)

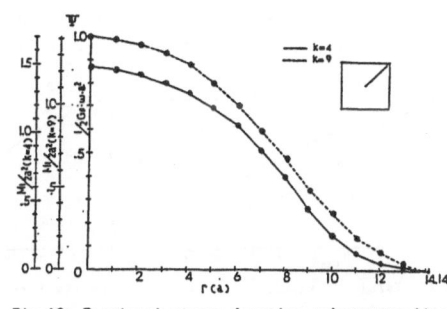

Fig. 13a Torsional stress function and contour lines
of membrane on square section ($\theta = \pi/4$)

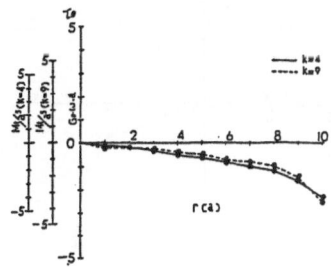

Fig. 12b Torsional stress on square section ($\theta = 0°$)

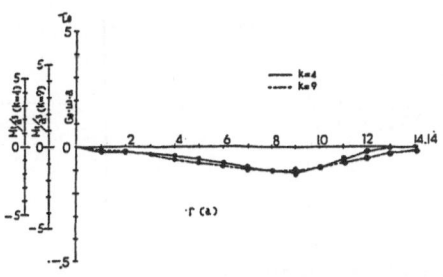

Fig. 13b Torsional stress on square section ($\theta = \pi/4$)

Hence, it can been seen that the form of the boundary approaches the circle condition.
From the volume which is bounded by the deflected membrane and plane, the relation between ω and M_t can be calculated.

$$k = 4 \qquad \omega \fallingdotseq 0.59 M_t / G_e a^4$$
$$k = 9 \qquad \omega \fallingdotseq 0.71 M_t / G_e a^4$$

b) Rectangular cross section

Taking the length $a \times 2a$ as sides of the square, contour lines due to deflection of membrane analogy for $k = 9$ and $k = 4$ are as shown in Photo 7 and 8.
As k increases, the stress obtained by contour lines due to deflection at the coner of the side 2a becomes less.
Torsional stress function Ψ ($\theta = 0$), Ψ ($\theta = \pi/2$), and Ψ ($\theta = \pi/4$) of the section and the stresses τ_e ($\theta = 0$), τ_e ($\theta = \pi/2$), and τ_e ($\theta = \pi/4$) are as shown in Fig. 14, 15, and 16, and torsional stress component τ_e at the center of the side 2a can be represented as three times larger than the center of the side a

Photo. 7 Contour lines of membrane on rectangular section(k=9)   Photo. 8 Contour lines of membrane on rectangular section(k=4)

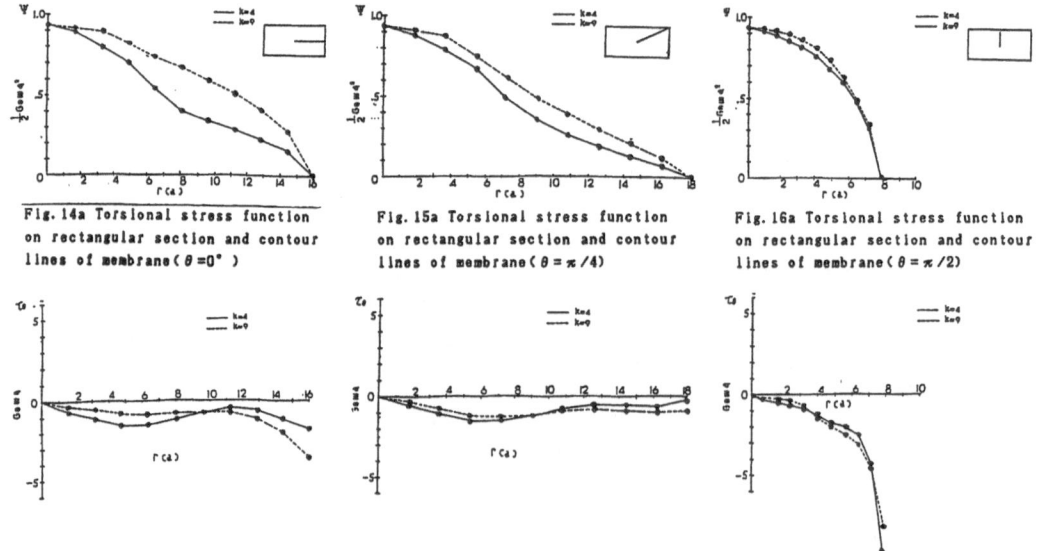

Fig. 14a Torsional stress function on rectangular section and contour lines of membrane($\theta = 0°$)

Fig. 15a Torsional stress function on rectangular section and contour lines of membrane($\theta = \pi/4$)

Fig. 16a Torsional stress function on rectangular section and contour lines of membrane($\theta = \pi/2$)

Fig. 14b Torsional stress on rectangular section ($\theta = 0°$)

Fig. 15b Torsional stress on rectangular section ($\theta = \pi/4$)

Fig. 16b Torsional stress on rectangular section ($\theta = \pi/2$)

§6 Conclusion

Since the torsional stress function is similar to the deflection of the membrane, the function Ψ can be obtained by experimentation for the deflection of the membrane.

Considering the ratio $k = G_\theta/G_r$ for modulas of rigidity in an anisotropic condition, the function Ψ correspond to the deflection of the membrane which has the boundary $\theta' = 1/\sqrt{k}\,\theta$. And by measuring the slope, the torsional stress component in a complexed cross section can be easily obtained. Generally, for the cross section which has the boundary $\theta' = 1/\sqrt{k}\,\theta$, the stress component at the re-entrant angle reaches an infinite value. And at the asalient angle, it becomes zero. And, on a wooden column, we usually find the slit to the annual rings or it to the back.

Thus, in the slotted cross section, if the ratio k has sufficient values each slit does not proceed by torque.

Reference

(1) 'Handbook of the Lumber Industry', Forestry Experiment Station ed, Maruzen. 131.
(2) to (5) Timoshenko.S.P. and Goodier.J.N., 'Theory of Elasticity', 3rd ed., McGraw-Hill Book Company, New York,1970. 293~318.

A Method for Detecting Isoclinics by Modified Isochromatics

S. Sakai and H. Okamura

Faculty of Engineering, University of Tokyo, Hongo, Bunkyo-ku, Tokyo, 113 Japan

INTRODUCTION

In the present paper, the authors propose a new method for detecting isoclinics by using modified isochromatics. The ordinary method of detecting isoclinics by a crossed plane polariscope has some difficulties in determining the precise location of isoclinic line because the fringe width of isoclinics is rather broad and isochromatics prevent observation of isoclinics, especially in high-stress regions. When we use circularily polarized light as incident light on the photoelastic model and observe the fringe pattern through an analyzer, we will get a fringe pattern of the isoclinics modified by isochromatics. If the isoclinics intersect the isochromatics at a large angle, the isoclinics can be detected more accurately than by the ordinary method and can be observed clearly in high-stress regions because this method uses modified isochromatics.

PRINCIPLE OF THE METHOD FOR DETERMINING ISOCLINICS

The optical system used to observe isoclinics by the present method is shown in Fig. 1. The isoclinics have been measured only by a plane polariscope. The system in Fig. 1, however, uses circularly polarized light as incident light on the photoelastic model. When we remove the second quarter wave plate Q_2 from the crossed circular polariscope, we can get the arrangement in Fig. 1. Assuming that the angle between the direction of a principal stress and the principal axis of P_1 is ψ and the angle between the direction of P_2 and the horizontal axis is φ, we can derive the light intensity after passing through P_2 as

$$I = I_0/2 \ (1-\sin 2\psi \ \sin \delta) \qquad\qquad [1]$$

where, I_0 is the maximum brightness and δ is the phase retardation.

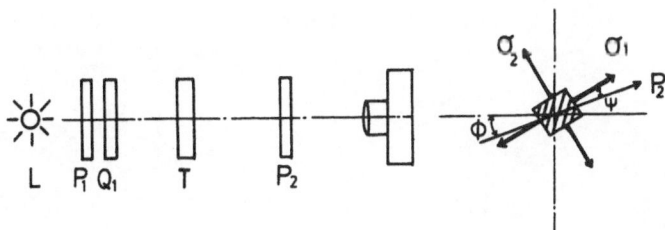

Fig. 1 Arrangement of photoelastic apparatus in the proposed method

Figure 2 is a schematic drawing of the distribution of light intensity "I" around the line for $\psi=0$, which is calculated by Eq. [1]. From this figure, it can be seen that the light intensity at the position of $\psi=0$ is constant ($1/2I_0$), independent of δ. The line for constant ψ is one of the isoclinics, because ψ means the angle between the direction of principal stresses and the direction P_2. Therefore, the line for $\psi=0$ in Fig. 2 is also one of the isoclinics. If we can distingush the line for $\psi=0$ from the photoelastic picture, the isoclinic line can be detected.

Figure 3(a) is an example of photoelastic picture around $\psi=0$ observed by the optical system shown in Fig. 1. This figure shows that the fringes around the isoclinic line can not be determined easily because the light intensity of isochromatics around the isoclinics is not necessarily zero. We can, however, distinguish the isoclinic line accurately by considering the distribution of light intensity around isoclinics.

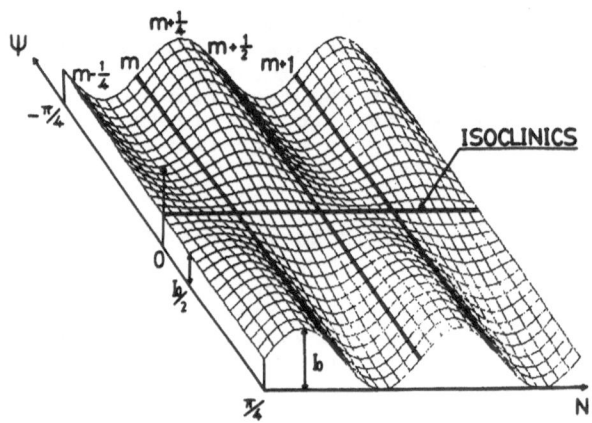

Fig. 2 Distribution of light intensity around isoclinics

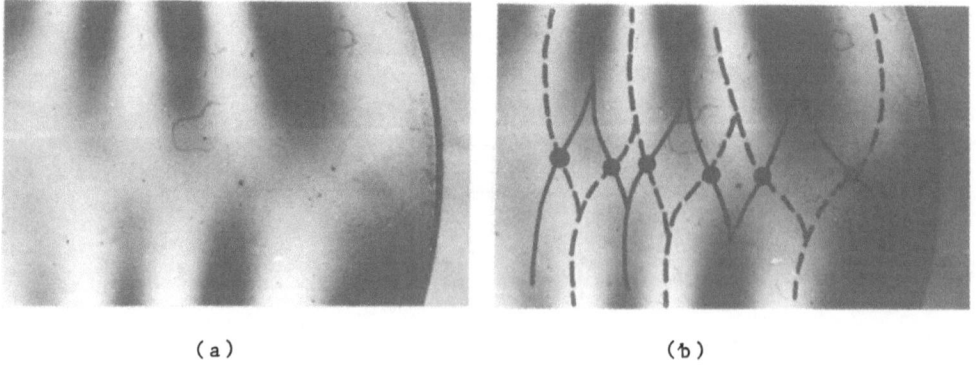

(a)   (b)

Fig. 3 Examples of the present method

If we draw solid lines through the points where the light intensity is
extreme maximum and draw dotted lines through the points where the
light intensity is extreme minimum, the photoelastic picture will be
shown as Fig. 4(a). These lines are of course one of the isochromatics, so we call these modified isochromatics. From Fig. 4(a), we can
recognize isoclinics drawn by the dotted and dashed line easily and
accurately. As is clear from Fig. 3(a), the actual picture near the
isoclinics cannot be observed as distinctly as in Fig. 4(a). If we
draw and connect lines as shown in Fig. 4(b), however, we can distinguish the points on isoclinics as the intersections of solid lines and
dotted lines which are shown as black circles in the figure. The
isoclinics detected by this method are drawn in Fig. 3(b).

EXAMPLES

Figure 5(a) is an example of the photoelastic picture observed by the
proposed method. The dots in the figure are plotted by the method
shown above. These points are shown to be on the theoretical lines
which are drawn by the solid lines in the figure. Corresponding isoclinics observed by the crossed plane polariscope are shown in Fig.
5(b). It is apparent that we can detect isoclinics more accurately by
Fig. 5(a) than in Fig. 5(b).

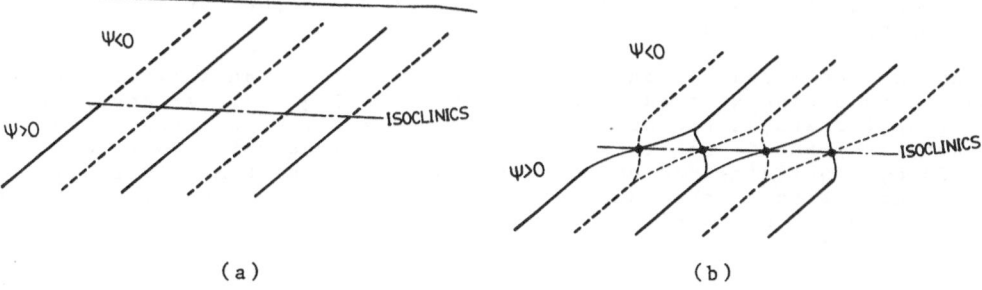

(a)   (b)

Fig. 4 Method to determine the precise location of isoclinics

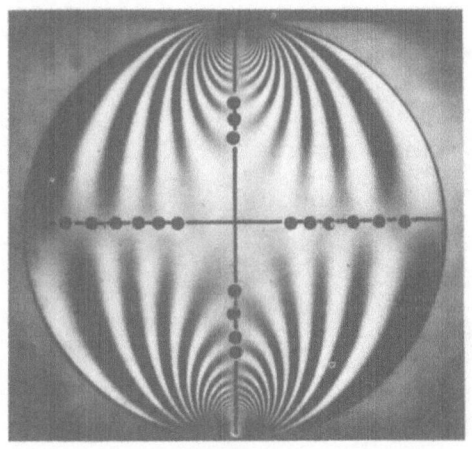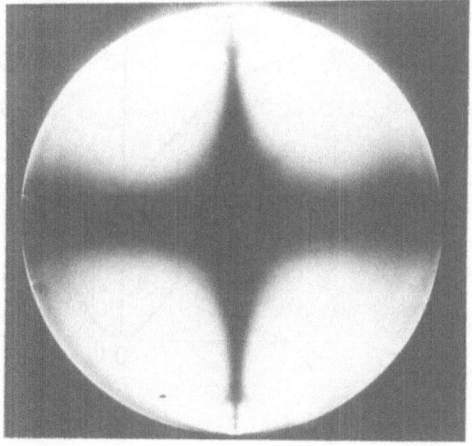

(a) Proposed method   (b) Crossed plane polariscope

Fig. 5 Comparison of isoclinics observed by two methods

CHARACTERISTICS OF THE METHOD

Comparison of the present method with that of the plane polariscope

The existance of isochromatics prevent to observation of the isoclinics by the plane polariscope. The present method, however, becomes more effective as isochromatics increase because the present method uses the modified isochromatics.
The width of isoclinics observed by both methods is considered next. Fig. 6 shows the distribution of brightness near the isoclinic line for $\psi=0$ observed by both methods, where I_{th} is the threshold level of brightness below which fringes can be observed. Isoclinics are observed at $I_{th} = I_0/2$ by the present method, but are observed at $I_{th}=0$ by the plane polariscope. When we use the plane polariscope, the distribution of brightness is tangent to the horizontal line at the position of isoclinics, and it is very difficult to establish I_{th} at the proper level. Contrary to this, by the present method the isoclinics can be observed distinctly and most definitely when $I_{th}=I_0/2$.

Applicable Range of the Method

The measuring accuracy of the present method is greatly affected by the crossing angle of isoclinics and isochromatics. Therefore, the relationship of these fringes is investigated and the applicable range of the proposed method is considered.

Around the concentrated load: For a typical case, an infinite plate presssed by a concentrated load, as shown in Fig. 7, is considered. Isochromatics around the origin O are circles which are tangent to the boundary at point O; isoclinics are straight lines which pass point O. Therefore, isoclinics can be discerned by the present method easily at the inner region of the specimen. Near the surface, however, it is hard to observe the isoclinics clearly because the isoclinics crosses the isochromatics at a small angle.

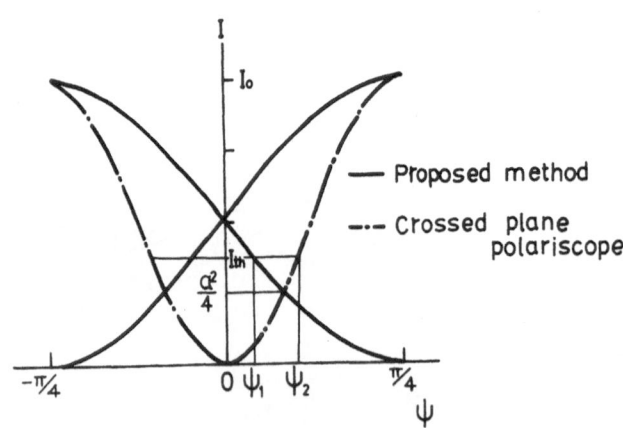

Fig. 6 Comparison of distribution of brightness of two methods

Around the zero point: The shape of the isochromatics near the zero point is considered. Assuming that the stress distribution near the zero point is proportional to the position (x,y) and the position of the zero point is set at the origin, we can describe the stress distribution near the zero point by considereing the condition of zero point, $\sigma_x = \sigma_y = \sigma_0, \tau_{xy} = 0$, as follows

$$\sigma_x = \sigma_0 + 6b_1 y + 2b_3 x$$
$$\sigma_y = \sigma_0 + 2b_0 y + 6b_2 x \qquad\qquad [2]$$
$$\tau_{xy} = -2b_0 x - 2b_3 y$$

where, b_0, b_1, b_2, and b_3 are constant. If Eq. [2] is substituted into the equation of the condition of isochromatics $\sigma_1 - \sigma_2 =$ constant, then the following relation can be derived.

$$ax^2 + 2hxy + by^2 = c^2 \qquad\qquad [3]$$

where,

$$b = (3b_1 - b_0)^2 + 4b_3^2$$
$$h = (3b_1 - b_0)(b_3 - 3b_2) + 4b_0 b_3$$
$$a = (b_3 - 3b_2)^2 + 4b_0^2$$

Equation [3] satisfies the condition of an elliptic curve (Sakai 1984). That is, the isochromatics around zero points become elliptical. For example, Fig. 8(a) shows isochromatics of circle plate with a center hole, pressed on both side. Shapes of isochromatics around zero points designated by a, b and c are elliptic. On the other hand, all isolinics pass through the zero point. Therefore, the crossing angle between isochromatic and isoclinic near the zero points is rather large and isoclinics can be observed clearly. Fig. 8(b) shows the isoclinics observed by the present method and it is confirmed that isoclinics are observed definitely around zero points.

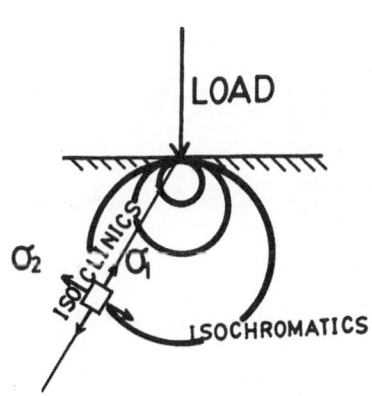

Fig. 7 Isochromatics and isoclinics around concentrated load

EVALUATING THE EFFECT OF Q ERROR ON THE MEASURING ACCURACY

Assuming that phase retardation of Q_1 is different from the ideal
value of $\pi/2$ and the error is $\Delta\delta$, light intensity after passing
through plate P_2 becomes

$$I = I_0/2 \ [1 - \sin2\psi \ \sin(\ \delta - \Delta\delta\ )].\tag{4}$$

From this equation, it is shown that $\Delta\delta$ is not included in the factor
"$\sin2\psi$"   and $\Delta\delta$ does not affect the position of isoclinics at all
but does affect one of the isochromatics. $\Delta\delta$ affects only the position
where the light intensity is extreme minimum. Therefore, the error of
Q_1 does not affect the accuracy of the locating isoclinics at all.

CONCLUSION

A new method of detecting the isoclinics by modified isochromatics is
proposed, and a few examples are presented. The method is effective in
high-stress regions where many isochromatic fringes cross isoclinics.
Singular points such as the concentrated load point and zero point are
also the cases.

REFERENCES

Sakai S. (1984) Improvement of Measuring Accuracy of Isoclinics in
   photoelastic stress analysis. Transactions of JSME 452 A:700-708

(a) Isochromatics   (b) Isoclinics

Fig. 8 Modified isochromatics around zero points

New Research Perspectives Opened by Isodyne and Strain Gradient Photoelasticity

Jerzy T. Pindera

Department of Civil Engineering and Institute for Experimental Mechanics,
Faculty of Engineering, University of Waterloo, 200 University Avenue, W., Waterloo,
Ontario N2L 3G1, Canada

INTRODUCTION: PHOTOELASTICITY AND ADVANCED EXPERIMENTAL MECHANICS

Photoelasticity, being one of the major fields of experimental research
in mechanics, is subjected to the same processes and changes which are
developing rapidly in other fields of experimental mechanics and in
experimental research in general.

These changes, which are basic and extensive, are rooted in the extraordinarily rapid progress in theories and techniques of measurements
and information collecting and processing systems. The growing influence of physical and mathematical sciences and the recent development
in their fields, together with demands of modern technology, have been
crucial for this development.

Of particular significance is the revolutionary progress in reliability, accuracy and resolution of measurements, and in the related progress of data collecting and processing. This progress can be exempli-
fied by the development of so-called intelligent instruments, automation of measurements, data reduction, factor analysis or principal component analysis which allows the unscrambling of signals produced by
overlapping sources, and the development of integrated measurement systems called hyphenated instruments. One of the most striking features
of this process is the wide introduction of spectral methods and procedures.

The progress in theories, techniques and procedures of measurements
exposed theoretical weaknesses in some procedures of experimental mechanics. They are rooted in dichotomy between the speculative-
phenomenological and the physical modelling of real events in engineering mechanics. In this paper the physical approach is chosen in
accordance with modern trends commonly denoted as advanced experimental
mechanics. In this approach all analytical relations and experimental
results represent responses of simplified models of reality. The
theory of modelling criteria provides the link between the prediction
of models and the actual physical processes. Thus all analytical,
numerical, and experimental results are always obtained within the
framework of chosen physical and mathematical models; their reliability
with respect to real physical events must be evaluated and assessed in
accordance with the known statement: "science is basically empirical."

It must be noted that two major factors influence the design or selection of physical and mathematical models of phenomena under investigation, and determine the reliability of model predictions:
- the correspondence principle formulated by Niels Bohr, which requires
  that a particular theory (model) must be, at most, a limiting case of
  a more general physical theory; and
- the concept of paradigm developed by Thomas Kuhn, which acknowledges
  psychological and social factors in the development of science.

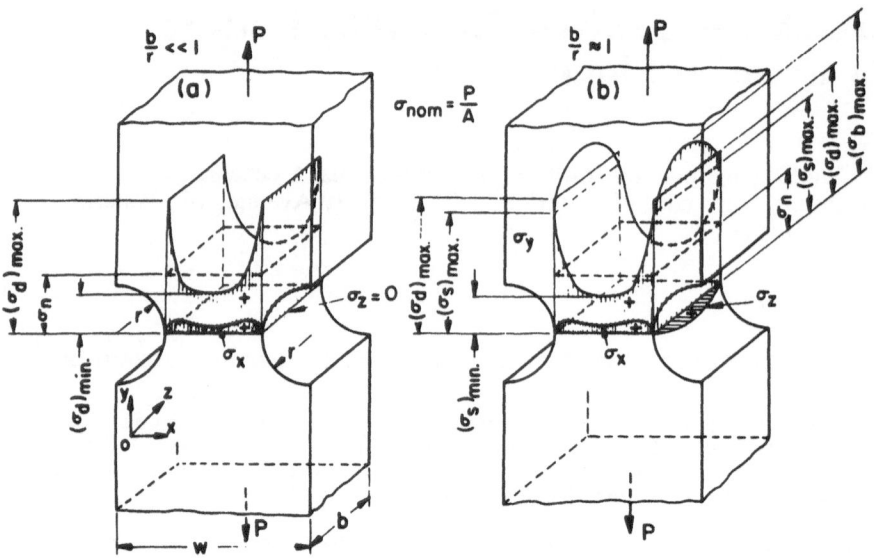

Fig. 1. Two incompatible models of stress states in plates with notches in tension: (a) a commonly used model; and (b) an advanced model

Requirements of advanced experimental mechanics expose the following modelling issues, the satisfactory solution of which will determine the usefulness of photoelastic methods for advanced requirements of modern technology and modern engineering sciences:
- reliability of models of stress/strain states in linear elastic and viscoelastic bodies
- ranges of practical applicability of linear models of stress/strain states with regard to geometric and material linearity, taking as a measure the reliability of results demanded by modern industry
- reliable and sufficiently comprehensive models of interaction between the information-collecting signals and stress/strain states, such as models describing the actual path of radiant energy in stressed and deformed bodies

Figure 1 illustrates the first issue. The actual value of a major stress component at the bottom of the notch in a plate can be up to 30% higher in the middle plane than at the plate surface. The fact that the predictions of the Airy stress function for circular disks loaded diametrically are not applicable in some regions of actual disks (Pindera 1981) illustrates the second issue.

This paper presents the actual and potential application of two new experimental methods which are based on more comprehensive models of the involved phenomena with regard to actual stress states in plates and symmetric bodies: isodyne and strain gradient photoelasticity.

Because of space limitations, in general no references are made in the text to the original sources. However, the pertinent concepts, models, information, and data are presented in reference books and research papers listed at the end of this paper.

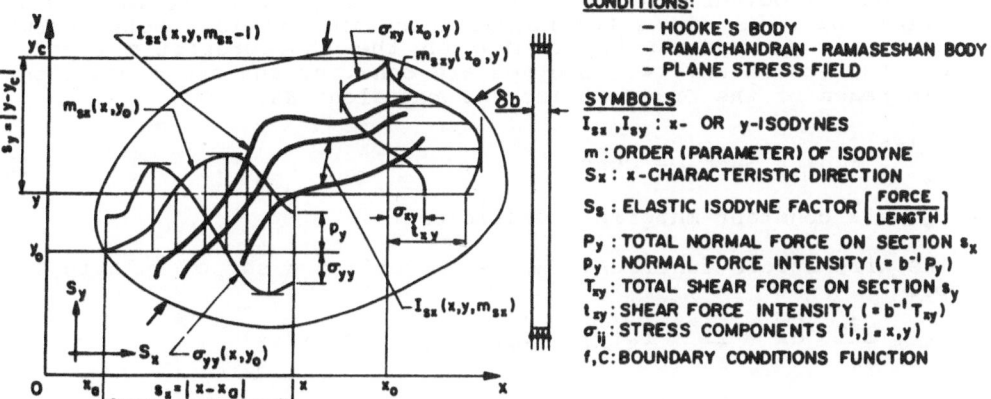

Fig. 2. Elastic isodynes. Concepts, definitions, symbols.

CONCEPTS, THEORIES AND TECHNIQUES OF ISODYNES

The concept of isodynes is related to plane stress fields in plates. It is generalized to actual three-dimensional stress fields in plates which occur in regions where the gradients of stress components are not negligible--in particular, in regions of stress concentrations. The concept of isodynes is applicable in some cases to general three-dimensional stress states.

Plane Elastic Isodynes

The derived relations pertain rigorously to a hypothetical homogeneous and elastic plate of an "infinitesimal" (vanishingly small, but finite) thickness δb, which is loaded by boundary forces coplanar with the plate. The (x,y) plane of the coordinate system is coplanar with the plate surface. For such a plate the general static equilibrium conditions reduce to two known relations between the first derivatives of σ_{xx}, σ_{yy} and σ_{xy} with respect to x and y. The coordinate z is meaningless in this model. By definition, only a two-dimensional stress state can exist in such a plate, so that all the stress components and their derivatives related to the direction z are equal to zero. The strain components related to the direction z cannot exist when both δb is infinitesimal and the plate satisfies plane equilibirum conditions. The above-stated restrictions can be considered satisfied for plates of a finite thickness b only when the stress state is homogeneous. Plane stress states in such hypothetical plates can be described by the Airy stress function $\Phi(x,y)$ which yields known expressions for the components σ_{xx}, σ_{yy} and $\sigma_{yx} = \sigma_{xy}$ of the stress tensor.

Isodynes are related to particular parameters of a plane stress field. The arbitrary direction x (or y) is called x- (or y-) characteristic direction, and all collinear lines are x- (or y-) characteristic lines. A plane containing the x-characteristic line is called the x-characteristic plane. The length of the x-characteristic line between any two arbitrary points within the stress field is the x-characteristic section.

Plane elastic isodynes are geometric loci of points within a plane stress field at which the total normal force intensities (total normal force per unit thickness) Δp_n acting on the characteristic sections of length Δs between any two isodynes are constant and are proportional to the increase of the order of isodyne Δm_s along Δs:

$$\Delta p_n = \int_s^{s+\Delta s} \sigma_n ds = S_s \Delta m_s = const \tag{1}$$

where S_s is a constant material coefficient.

Using the Airy stress function, one may establish a set of relations of the type:

$$\int \sigma_{yy} dx = \int \frac{\partial^2 \phi}{\partial x^2} dx = \frac{\partial \phi}{\partial x} + f_x(y) = p_y(x,y) = S_s m_{sx}(x,y) \tag{2}$$

where the function $f_x(y)$ depends on boundary conditions.

The function $p_y(x,y)$ and the analogous function $p_x(x,y)$--related to the x- and y-characteristic directions along which the stress function is differentiated--represent the x- and y-plane elastic isodyne surfaces spanned over the plate surface. The lines $m_{sx}(x,y) = const$ and $m_{sy}(x,y) = const$ on the isodyne surfaces represent the x- and y-plane elastic isodynes. Projections of these lines on the plate surface give fields of plane elastic x- and y-isodynes, Fig. 2. The values of m_{sx} and m_{sy} are the orders of isodynes.

The intersections of isodyne surfaces with planes normal to the plate surface and collinear with the x- or y-direction define isodyne functions, m_{sxx}, m_{sxy}, m_{syy} and m_{syx}. The slopes of the xx- and yy-isodyne functions are proportional to the normal stress components σ_{xx} and σ_{yy}. The slopes of the xy- and yx-isodyne functions are proportional to the shear stress components σ_{xy} and σ_{yx}. As a result, four independent pieces of information are obtained on the three components of stress tensor.

Differential Elastic Isodynes

It has been reported more than half a century ago that the stress states in real plates of finite thickness are pronouncedly three-dimensional in regions of notches and cracks when the radius of curvature of the boundary is of the order of magnitude of the plate thickness or less, Fig. 1. The regions of contact exhibit similar behavior. Such stress states can be described and subsequently analysed using the concept of differential elastic isodynes. Differential elastic isodynes are defined as geometric loci of points in the characteristic planes ($z = z_i$) of real plates of finite thickness along which the difference of the total normal force intensities acting on characteristic sections between two differential elastic isodynes in directions within and normal to the characteristic plane are constant. For the x-characteristic direction within the (x,y)-characteristic plane, the differential isodynes can be given by:

$$\Delta p_{y,z}(x,y,z=z_i) = \Delta p_y - \Delta p_z = \int_s^{s+\Delta s} (\sigma_y - \sigma_z) ds = S_s \Delta m_{s(y,z)} = const \tag{3}$$

197

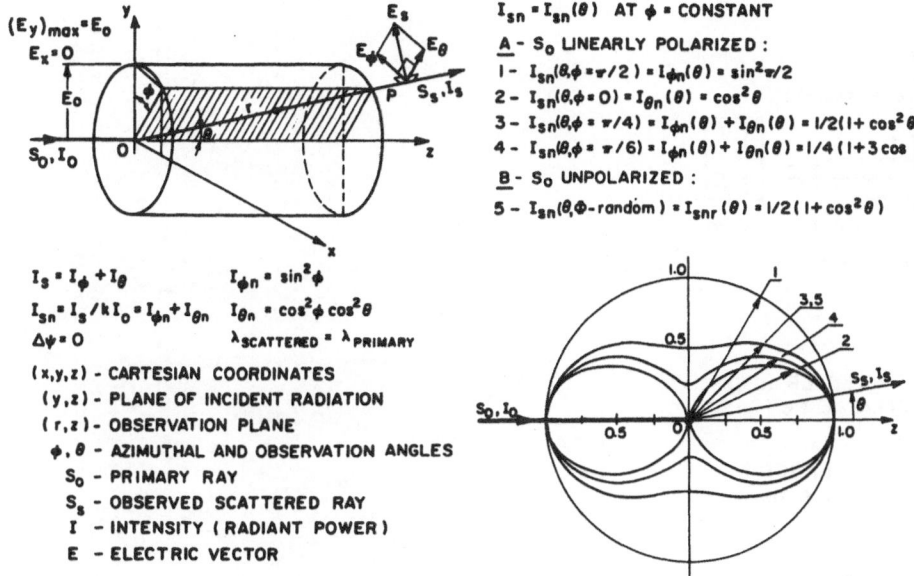

Fig. 3. Patterns of Rayleigh scattering at a point: mathematical model, definitions, intensity distributions

Photoelastic Isodynes

There are various ways for experimentally obtaining fields of plane elastic or of differential isodynes. Using a scattered light technique developed by the author, it is possible to obtain fields of isodynes--called photoelastic isodynes--when a particular set of conditions is satisfactorily approximated. The basic conditions are that the actual light scattering may be described by the Rayleigh model, Fig. 3, and that the stress/strain state modulates the flow of radiant energy in the manner close to that postulated in the phenomenological model of Ramachandran-Ramaseshan. The concept of secondary principal stresses is assumed to be applicable. These models readily yield conditions for the existence of a simple relation between the intensity of light I^x scattered along the path of the primary light beam I_0 propagating in, for example, the x-direction, and the normalized relative linear and angular (phase) retardations, $R/\lambda = m_{sx}$ and ϕ_{sx} respectively, such as:

$$(I^x)_n = I^x/I_O = \sin^2 \pi m_{sx} = \sin^2(0.5\phi_{sx}) = 0.5(1 - \cos \phi_{sx}) \qquad [4]$$

where the relative retardations are simply related to the secondary principal stresses σ_1^x and σ_3^x, and to the total normal force intensities p_1^x and p_3^x:

$$\phi_{sx} = 2\pi m_{sx} = 2\pi S_s^{-1}\int(\sigma_1^x - \sigma_3^x)dx = 2\pi S_s^{-1}(p_1^x - p_3^x) \qquad [5]$$

Scanning the primary light beam I_0 within the characteristic plane, and using techniques presented in the author's papers listed in the references, one can obtain fields of lines of constant scattered light intensities. For the x-characteristic direction and the (x,y)-characteristic plane, the pertinent relations are of the form:

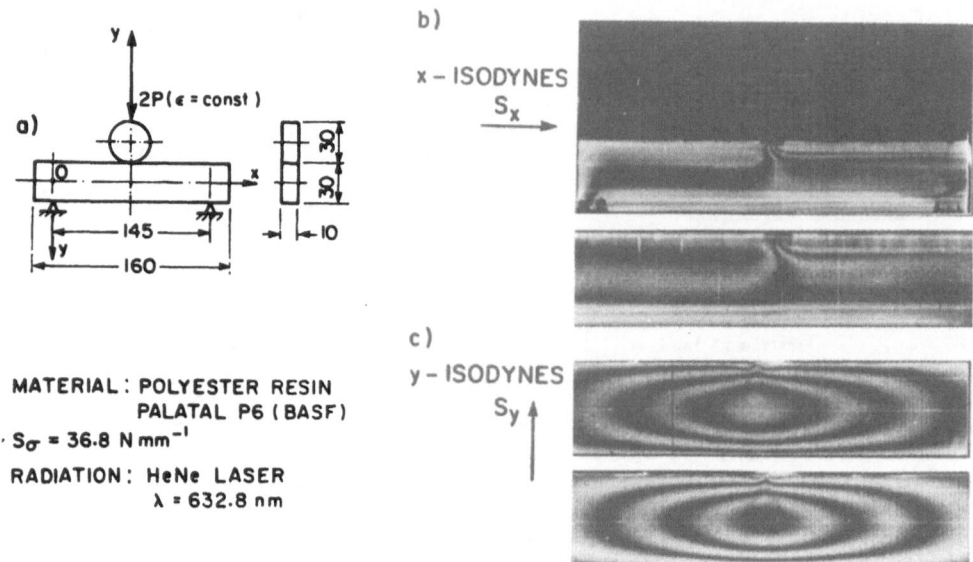

Fig. 4. Rectangular prismatic beam under three point bending. x- and y-fields of photoelastic isodynes in the middle beam plane, which can be used to test various analytical models

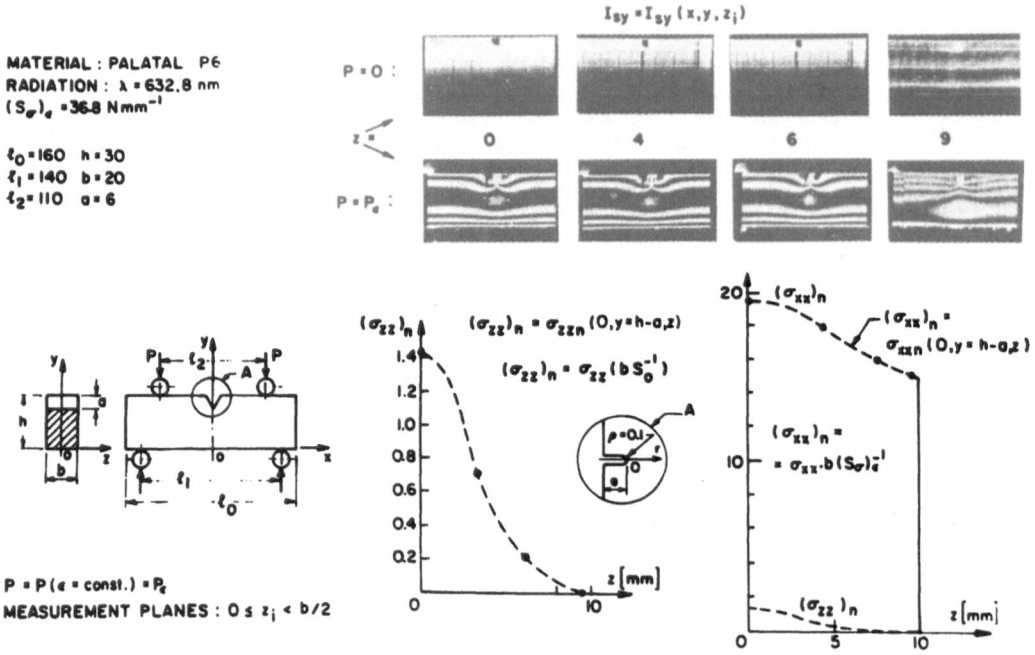

Fig. 5. Beam with a crack in pure bending. Determination of three-dimensional stress state along crack tip using photoelastic isodynes

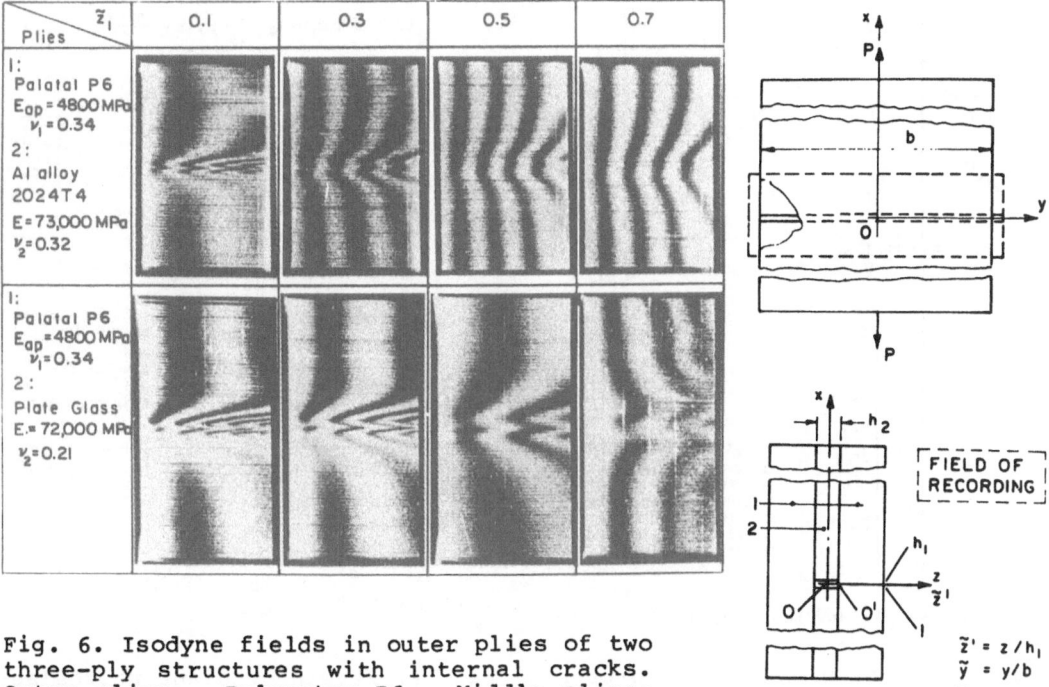

Fig. 6. Isodyne fields in outer plies of two three-ply structures with internal cracks. Outer plies: Polyester P6. Middle plies: Al alloy or plate glass. Influence of mismatch of Poisson ratios on isodyne fields

$(I^x)_n^1 = \sin^2 \pi m_{sx} = \text{const}$, and $((I^x)_n^2 = \cos^2 \pi m_{sx} = \text{const}$ \hfill [6]

These relations represent surfaces of photoelastic isodynes. When the stress state is close to the plane state, the distribution of scattered light intensities represents surfaces of the plane photoelastic isodynes which are identical with the plane elastic isodynes, e.g.:

$p_y = S_s m_{sx} = \text{const}$ \hfill [7]

where the coefficient S_s is time- and wavelength-dependent, i.e., $S_s = S_s(t, \lambda)$.

Figures 4, 5, and 6 illustrate the capabilities of isodyne photoelasticity for the following three cases: experimental isodyne field, Fig. 4, as a measure of reliability of analytical solutions; isodynes in fracture mechanics, Fig. 5; and isodynes in stress analysis of composite structures, Fig. 6.

STRAIN GRADIENT PHOTOELASTICITY

In principle, the path of light propagating through inhomogeneous bodies is curved. It has been shown (Bokshtein 1949; Pindera 1955) that this effect limits the resolution of photoelastic measurements and modifies fields of isochromatics in regions of high birefringence gradients.

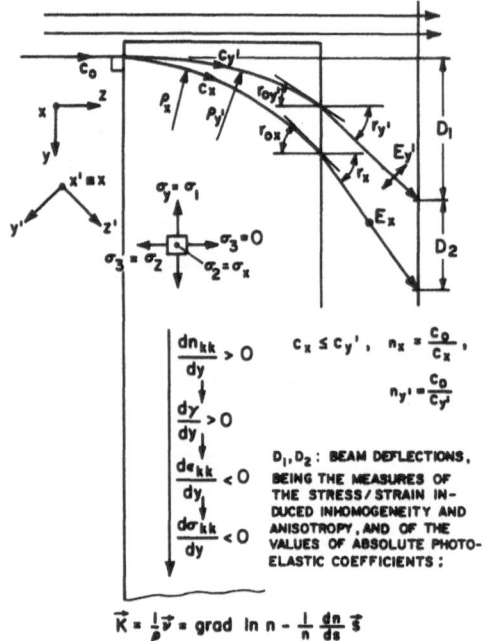

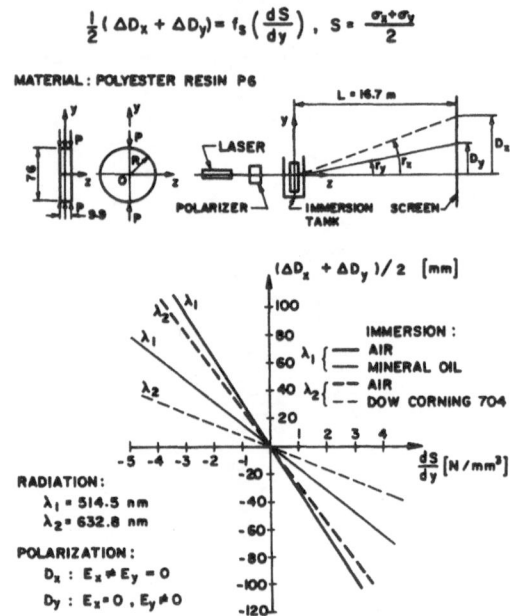

Fig. 7. Light propagation through a plate. Influence of gradients of inhomogeneity and anisotropy caused by strain/stress gradients on the light path

Fig. 8. Strain gradient photoelasticity. Dependence of light deflection on the gradient of the sum of principal stresses

The author has observed that the optical anisotropy caused by an inhomogeneous stress state results in noticeable separation of curved light beams into ordinary and extraordinary beams, Fig. 7. This separation can be quite large, e.g., up to 4 mm for an optical path of 200 mm.

It is common to present the influence of the optical inhomogeneity of a body on the path of a propagating light beam in the form of a simplified relation, Fig. 7:

$$\bar{K} = \frac{1}{\rho}\,\bar{v} = \frac{1}{n}\left(\text{grad } n - \frac{dn}{ds}\,\bar{s}\right) = \text{grad } \ell n\; n - \frac{1}{n}\frac{dn}{ds}\,\bar{s} \qquad [8]$$

The above relation, together with the relations

$$n_1 = n_0 + C_1\sigma_1 + C_2(\sigma_2 + \sigma_3)\;,\;\text{ and }\; n_2 = n_0 + C_1\sigma_2 + C_2(\sigma_3 + \sigma_1) \qquad [9]$$

which follow from the phenomenological model of birefringence (Ramachandran and Ramaseshan 1961), yield simple relations for $\bar{K}_1$ and $\bar{K}_2$. Pindera with co-workers (1978, 1979, 1981, 1982, 1986) established that such relations can be used as a foundation for mathematical models relating the deflections of a light beam traversing a body to the gradients of the sum and difference of principal stresses along the path of the light beam.

Such a model, when applied to plane stress states, yields simple linear equations relating the light beam deflections, stress gradients, and values of the absolute and relative stress-optic coefficients $C_1(\lambda)$, $C_2(\lambda)$, and $C_0(\lambda) = C_1 - C_2$. Figure 8 illustrates the efficacy of the strain gradient photoelasticity.

Two of many possible applications of the gradient photoelasticity are: the development of the more advanced theory of the shadow (caustic) method used in fracture mechanics; and the development of the more advanced thoery of transmission photoelasticity in the presence of high stress gradients (stress concentration problems).

REFERENCES

Alfvén H, Cech F (1985) Space Research and the New Approach to the Mechanics of Fluid Media in Cosmos. In: Niordson FI, Olhoff N (eds) Theoretical and Applied Mechanics. North-Holland, Amsterdam p 1-29
Bokshtein MF (1949) On the Resolving Power of the Polarizing System for Stress Analysis. J. Tekhnicheskoi Fiziki 19:1103-1106
Brillouin L (1964) Scientific Uncertainty and Information. Academic Press, New York
Cloud GL, Pindera JT (1968) Techniques in Infrared Photoelasticity. Exp. Mech. 8:193-201
Doeblin EO (1983) Measurement Systems: Application and Design, 3rd edn, McGraw-Hill, New York
Hausler E (1963) Lichtablenkung durch elastische deformierte durchsichtige Festkörper. Optik, 20:386-390
Hecker FW, Pindera, JT (1978) Influence of Stress Gradient on Direction of Light Propagation in Photoelastic Specimens. VDI-Berichte 313:745-754
Hecker FW, Kepich TY, Pindera JT (1979) Neglected Factor in Photoelasticity: Nonrectilinear Light Propagation in Stressed Bodies and its Significance. In: Aben H (ed) The Eight All-Union Conference on Photoelasticity, Vol 1. Academy of Sciences of the Estonian SSR. Institute of Cybernetics. Tallin p 117-123
Hecker FW, Kepich TY, Pindera JT (1981) Non-Rectilinear Optical Effects in Photoelasticity Caused by Stress Gradients. In: Lagarde A (ed) Optical Methods in Mechanics of Solids. Sijthoff & Noordhoff, Alphen aan den Rijn, The Netherlands p 123-134
Kac M (1969) Some Mathematical Models in Science. Science 166:469-474
Kuhn TS (1970) The Structure of Scientific Revolution, 2nd edn, The University of Chicago Press, Chicago
Mazurkiewicz SB, Pindera, JT (1979) Integrated Plane Photoelastic Method - Application of Photoelastic Isodynes. Exp. Mech. 19:225-234.
Pindera JT (1955) Technique of Photoelastic Investigations of Plane Stress States. Rozprawy Inzynierskie 3:109-176
Pindera JT, Cloud G (1966) On Dispersion of Birefringence of Photoelastic Materials. Exp. Mech. 6:470-480
Pindera JT (1973) Contemporary Trends in Experimental Mechanics: Foundations, Methods, Applications. In: Pindera et al (eds) Experimental Mechanics in Research and Development. Solid Mechanics Division, University of Waterloo. Study No. 9. Waterloo, p 143-168
Pindera JT, Straka P (1973) Responses of the Integrated Polariscope. J. Strain Analysis 8:65-76
Pindera JT (1973-4) Response of Photoelastic Systems. Trans. CSME 2:21-30
Pindera JT, Straka P (1974) On Physical Measures of Rheological Responses of Some Materials in Wide Ranges of Temperature and Spectral Frequency. Rheologica Acta 13:338-351

Pindera JT, Mazurkiewicz SB (1977) Photoelastic Isodynes: A New Type of Stress Modulated Light Intensity Distribution. MRC 4:247-252

Pindera JT (1981) Foundations of Experimental Mechanics: Principles of Modelling, Observation and Experimentation. In: Pindera JT (ed) New Physical Trends in Experimental Mechanics, CISM Courses and Lectures No 264. Springer, New York, p 188-326

Pindera JT, Mazurkiewicz SB (1981) Studies of Contact Problems Using Photoelastic Isodynes. Exp. Mech. 21:448-455

Pindera JT (1981) Analytical Foundations of the Isodyne Photoelasticity. MRC 8:391-397

Pindera JT, Krasnowski BR, (1982) Determination of Stress Intensity Factors in Thin and Thick Plates Using Isodyne Photoelasticity. In: Simpson LA (ed) Fracture Problems and Solutions in the Energy Industry. Pergamon Press, Oxford p 147-156

Pindera JT, Hecker FW, Krasnowski BR (1982) Gradient Photoelasticity. MRC 9:197-204

Pindera JT (1982) New Development in Photoelastic Studies: Isodyne and Gradient Photoelasticity. Opt. Eng. 21:672-678

Pindera JT (1983) Device for Birefringence Measurements Using Three Selected Sheets of Scattered Light (Isodyne Selector, Isodyne Collector, Isodyne Collimator). Canadian Patent No 1153572

Pindera JT (1983) Apparatus for Determination of Elastic Isodynes and the General State of Birefringence Whole Field-Wise, Using the Device for Birefringence Measurements in a Scanning Mode (Isodyne Polariscope). Canadian Patent No 1156488

Pindera JT, Krasnowski BR, Pindera M-J (1984) Analysis of Models of Stress States in the Regions of Cracks Using Isodyne Photoelasticity. In: Pindera JT (ed) Modelling Problems in Crack Tip Mechanics. Martinus Nijhoff Publishers, Dordrecht p 271-286

Pindera JT (1985) Reliability of Analytical and Experimental Methods of Stress Analysis: Influence of Speculative and Physical Methodologies of Modelling in Mechanics. In: Li Chengxiang, Yang Ling, Li Ruqing, Shi Guangyi (eds) Proc. Intern. Conf. on Experimental Mechanics (Beijing, 1985). Science Press, Beijing, China, p 679-686

Pindera JT, Krasnowski BR, Pindera, M-J (1985) Theory of Elastic and Photoelastic Isodynes. Samples of Application in Composite Structures. Exp. Mech. 25:272-281

Pindera JT, Hecker FW, Krasnowski BR (1986) Basic Theories and Experimental Methods of Gradient Photoelasticity. Exp. Mech.: to be published

Popper KR (1968) The Logic of Scientific Discovery, 2nd English edn, Harper & Row, New York

Provan JW (1984) The Micromechanics in Fatigue Crack Initiation. In: Pindera JT (ed) Modelling Problems in Crack Tip Mechanics. Martinus Nijhoff Publishers, Dordrecht p 131-154

Ramachandran GN, Ramaseshan S (1961) Crystal Optics. In: Flügge S (ed) Handbuch der Physik 25. Springer, Berlin p 1-217

Sih GC (1984) The State of Affairs near the Crack Tip. In: Pindera JT (ed) Modelling Problems in Crack Tip Mechanics. Martinus Nijhoff, Dordrecht p 65-90

Sih GC (1985) Mechanics and physics of energy density theory. Theoret. Appl. Fracture Mech. 4:157-173

Smith CW, Epstein JS (1984) Measurements of Near Tip Field Near the Right Angle Intersection of Straight Front Cracks. In: Pindera JT (ed) Modelling Problems in Crack Tip Mechanics. Martinus Nijhoff Publishers, Dordrecht p 325-333

Taya T (1985) Phase Velocity of Longitudinal Wave in Solid Elastic Bar. In manuscript

Thum A, Petersen C, Svenson O (1960) Verformung, Spannung und Kerbvirkung. VDI-Verlag, Düsseldorf

Stresses and Deformations at the Beam-to-Wall Joints of Shear Wall Structures

G. Matsui[1] and Y. Tsuboi[2]

[1] School of Science and Engineering, Waseda University, Okubo, Shinjuku-ku,
Tokyo, 160 Japan
[2] College of Engineering, Hosei University, Koganei, Tokyo, 184 Japan

1. INTRODUCTION

The currently available methods of analysing laterally loaded shear walls with openings are frame analogies taking into account bending, shear, and axial deformations of walls and beams.
In the wide column frame analogy, the beam lengths are taken to be the clear distance between the adjacent walls, and the joints flexibility can be accounted by extending the beam section a distance (one-fourth of the beam depth) into the wall at each end. This idealization is considered as a frame with rigid members in the wall zone in Fig. 1. The value of increasing the effective length of the beams by an amount equal to $(1/4)h$ was obtained from experimental studies of models, and both dimensions of the beams and columns were the same order.[1] The traditional shear wall structures with door high openings may be interconnected by slender beams. Michael[2] suggested the amount equal to half beam depth by considering the wall as a semi-infinite elastic plane. This paper deals with the value of effective length of the beams of the shear wall structures having large wall widths compared with the beam depths. The local deformations of the beam-wall joint can be calculated by considering the walls as infinite-strips subjected to external forces from beams at the beam ends. The external forces can be obtained from photoelastic experiments.

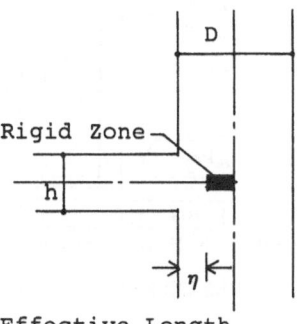

Effective Length

Fig.1 Rigid Zone

2. PHOTOELASTIC EXPERIMENTS OF BEAM-WALL JOINTS

2.1 Symmetric Loading Test

The models were formed from about 5mm thick epoxy resin plates. The dimensions of models are shown in Table 1. Pure bending by a couple applied to the beams is shown in Fig. 2. Using the shear-difference method, normal stresses and shearing stresses at the end of the beams can be obtained from isochromatic and isoclinic fringe patterns. Stress concentrations occurred at the re-entrant corners of the beam-wall intersection. The exact value of stresses cannot be obtained from the results of the photoelastic investigations. The maximum normal stresses are determined from the conditions that the internal resultant moment coincides with the external forces. The maximum shearing stresses can be obtained by summing the shearing stresses acting on the ends of the beams. The resultant of the shearing forces must vanish.

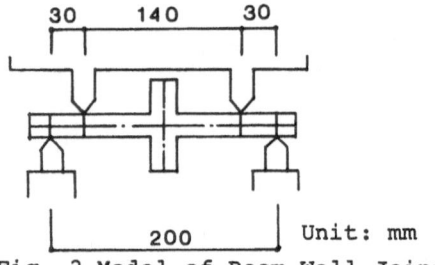

	h	D	D/h	t
Type 1	20	20	1	0.53
Type 2	20	30	1.5	0.54
Type 3	20	40	2	0.54
Type 4	20	60	3	0.54
Type 5	20	80	4	0.55 Unit: mm

Fig. 2 Model of Beam-Wall Joint   Table 1 Variation of Wall Depth

2.2 Antisymmetric Loading Test

Three story beam-wall junction models were formed from about 6mm thick epoxy resin plates as shown in Table 2. Antisymmetric loads applied at four points A, B, C, and D. Normal and shearing stresses can be obtained by the same method in 2.1. The resultant of the shearing force must equal the external force P.

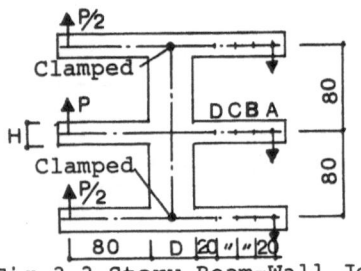

	h	D	t
Type A-1	40	40	5.99
Type A-2	40	60	6.15
Type A-3	40	80	5.43
Type A-4	20	60	5.68
Type A-5	20	80	5.99 Unit: mm

Table 2 Variation of.Beam-Wall Joint

Fig.3 3 Story Beam-Wall Joint

3. EXPERIMENTAL RESULTS

Isocromatic fringe patterns, normal stresses and shearing stresses are shown in Fig. 4. The distribution of the normal stresses at the end of the beams is not linear. The differences of the normal stress patterns are not large for the range of the beam span to depth ratio. The shearing stresses at the end of the beams do not resemble the parabolic distribution given by the elementary theory and there are very large stresses at the tops and bottoms of the beam, while the shearing stresses near the middle of the beams are in the opposite direction of the loading. The shearing stresses vary to some extent with the span-to-depth ratios.
The stress distributions can be assumed as follows:

$$\sigma_x = k_1 \sinh\gamma x/\sinh\gamma a + k_2 y$$

$$\tau_{xy} = c_1(\delta x \sinh\delta x - \delta a \tanh\delta a \cosh\delta x) + c_2(a^2 - x^2) \tag{1}$$

The constants γ, k_1, k_2, δ, c_1, and c_2 can be determined from experimental results. The values of assumed equations and the experimental results are plotted in Fig. 5.

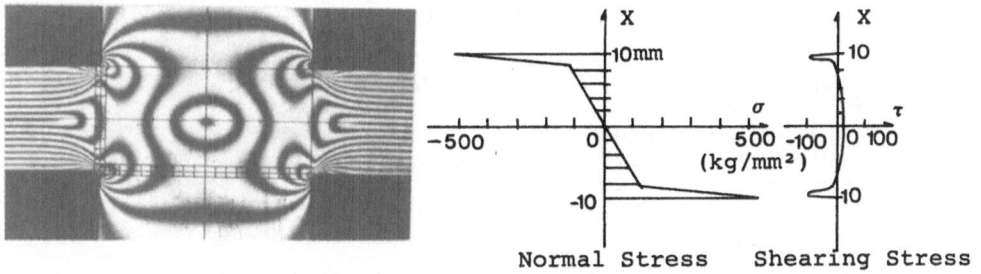

Fig. 4 Isochromatic pattern, Normal and Shearing Stress

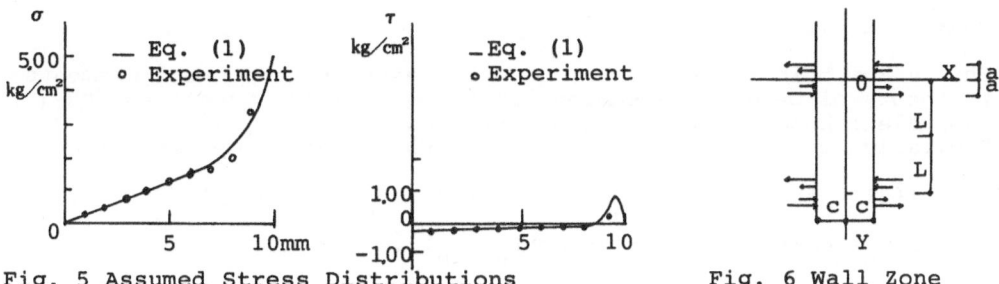

Fig. 5 Assumed Stress Distributions   Fig. 6 Wall Zone

4. DERIVATION OF EQUIVALENT BEAM LENGTH

The wall zone can be taken as an infinitely long strip subjected to the action of repeated normal forces and shearing forces of period 2L on both sides at the beam intersections in Fig. 6. For symmetric loading, the stress patterns can be represented by the Fourier series of the form:

normal stress $\sigma_x = \sum p_n \sin\alpha_n y$ on x=c

shearing stress $\tau_{xy} = -\sum q_n \cos\alpha_n y$ on x=c

(2)

Stresses and displacements of the wall can be obtained from elastic theory. Corresponding stress components are

$$\sigma_x = -\sum [p_n\{-(\beta_n + th\beta_n)\cosh\xi_n + th\beta_n \xi_n \cosh\xi_n\} + q_n\{-\beta_n th\beta_n \cosh\xi_n +$$
$$\xi_n \sinh\xi_n\}] \sin\eta_n / \{\cosh\beta_n \psi(\beta_n)\}$$

$$\sigma_y = \sum [p_n\{(th\beta_n - \beta_n)\cosh\xi_n + th\beta_n \xi_n \sinh\xi_n\} + q_n\{(2 - \beta_n th\beta_n)\cosh\xi_n +$$ (3)
$$\xi_n \sinh\xi_n\}] \sin\eta_n / \{\cosh\beta_n \psi(\beta_n)\}$$

$$\tau_{xy} = -\sum [p_n(-\beta_n \sinh\xi_n + th\beta_n \xi_n \cosh\xi_n) + q_n\{(1 - \beta_n th\beta_n)\sinh\xi_n +$$
$$\xi_n \cosh\xi_n\}] \cos\eta_n / \{\cosh\beta_n \psi(\beta_n)\}$$

where
$\alpha_n = n\pi/L$, $\beta_n = \alpha_n c$, $\xi_n = \alpha_n x$, $\eta_n = \alpha_n y$, $\psi(\beta_n) = th\beta_n + \beta_n(1 - th^2\beta_n)$, $th\beta = \tanh\beta$

206

The complementary energy of this infinite strip in the interval $0 \leq x \leq L$ is given by

$$V = tL/E \sum [(p_n \tanh \beta_n + q_n)^2 / \psi(\beta_n) - (1+\nu) p_n q_n] / \alpha_n \tag{4}$$

where t is the wall thickness.
The complementary energy of "equivalent cantilever" in the wall zone subjected to bending is given by

$$U = M^2 \eta / (2EI). \tag{5}$$

If c is very small in comparison with L, we find, using a triangular stress distribution in (1a),

$$U = M^2 c / (2EI).$$

This value coincides with the quantity given by the usual elementary theory of bending. Assuming that Eq. (4) is equal to Eq. (5), equivalent beam length η can be obtained.
Similarly, for antisymmetric loading, the stress components are

$$\sigma_x = \sum [\{(1 - \beta_n th\beta_n) \sinh \xi_n + \xi_n \cosh \xi_n\} p_n + \{(2th\beta_n - \beta_n) \sinh \xi_n +$$

$$th\beta_n \xi_n \cosh \xi_n\} q_n] \sin \eta_n / \{\cosh \beta_n \psi(\beta_n)\},$$

$$\sigma_y = \sum [\{(1 + \beta_n th\beta_n) \sinh \xi_n - \xi_n \cosh \xi_n\} p_n + \{\beta_n \sinh \xi_n -$$

$$th\beta_n \xi_n \cosh \xi_n\} q_n] \sin \eta_n / \{\cosh \beta_n \psi(\beta_n)\}, \text{ and} \tag{6}$$

$$\tau_{xy} = \sum [(\beta_n th\beta_n \cosh \xi_n - \xi_n \sinh \xi_n) p_n + \{(\beta_n - th\beta_n) \cosh \xi_n -$$

$$th\beta_n \xi_n \sinh \xi_n\} q_n] \cos \eta_n / \{\cosh \beta_n \psi(\beta_n)\} + \tau_0,$$

where $\psi(\beta_n) = th\beta_n + \beta_n (th^2 \beta_n - 1).$

The complementary energy of this infinite strip in the interval $0 \leq x \leq L$ is given by

$$V = tL/E \sum [(p_n + q_n \tanh \beta_n)^2 / \{\alpha_n \psi(\beta_n)\} - (1+\nu) p_n q_n / \alpha_n + 2c(1+\nu) \tau_0^2]. \tag{7}$$

The complementary energy of the wall and equivalent beam given by the elementary theory is

$$U = M^2 L / (4Etc^3) + M^2 \eta / (\tfrac{4}{3} ta^3) \tag{8}$$

From Eqs. (7) and (8), the equivalent beam length η can be obtained. The equivalent beam length can be evaluated for a range of the beam depth h to wall width d ratio as shown in Fig.7.
Assuming that the values of η are the functions of the d/h ratio, the following approximations can be proposed:

$\eta = h/4$; $1 \geq d/h$

$\eta = \sqrt{dh}/4$; $1 \leq d/h \leq 4$ \hfill (9)

$\eta = h/2$; $d/h \geq 4.$

The displacements of the wall are shown in Fig. 8. The displacements of the wall centerline are shown in Fig. 9. The displacements and rotations of the centerline of the equivalent beam are shown in Fig. 10 and 11.

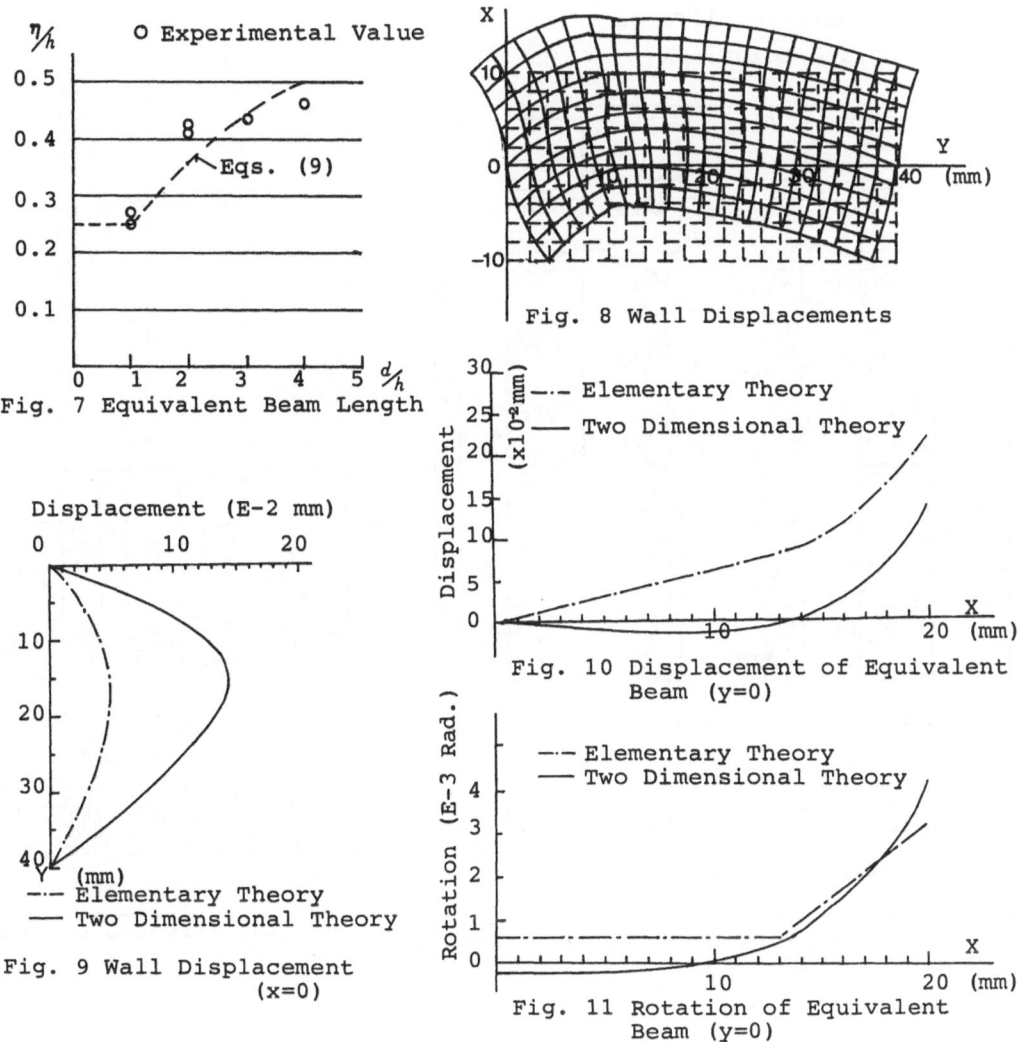

Fig. 7 Equivalent Beam Length

Fig. 8 Wall Displacements

Fig. 9 Wall Displacement (x=0)

Fig. 10 Displacement of Equivalent Beam (y=0)

Fig. 11 Rotation of Equivalent Beam (y=0)

5. SOME EXAMPLES OF PHOTOELASTIC EXPERIMENTS OF SHEAR WALL STRUCTURES

Eight-storied shear walls were formed from 5mm epoxy resin plates. The specimens were clamped at their base and vertical loads applied at each floor. Three main types are considered.

Type A- presents walls with slender and short beams.
Type B- presents walls with large wall width to beam depth ratio.
Type C- presents walls with comparatively same order of dimensions of beams and walls.

Photoelastic experimental result is shown in Fig. 12.
Using the effective length of the beams from the above results, the wide column frame solution appears to give good results for Type C and is a little better than the ordinary method for Type A and Type B.

---- Experimental Value
——— Frame Solution

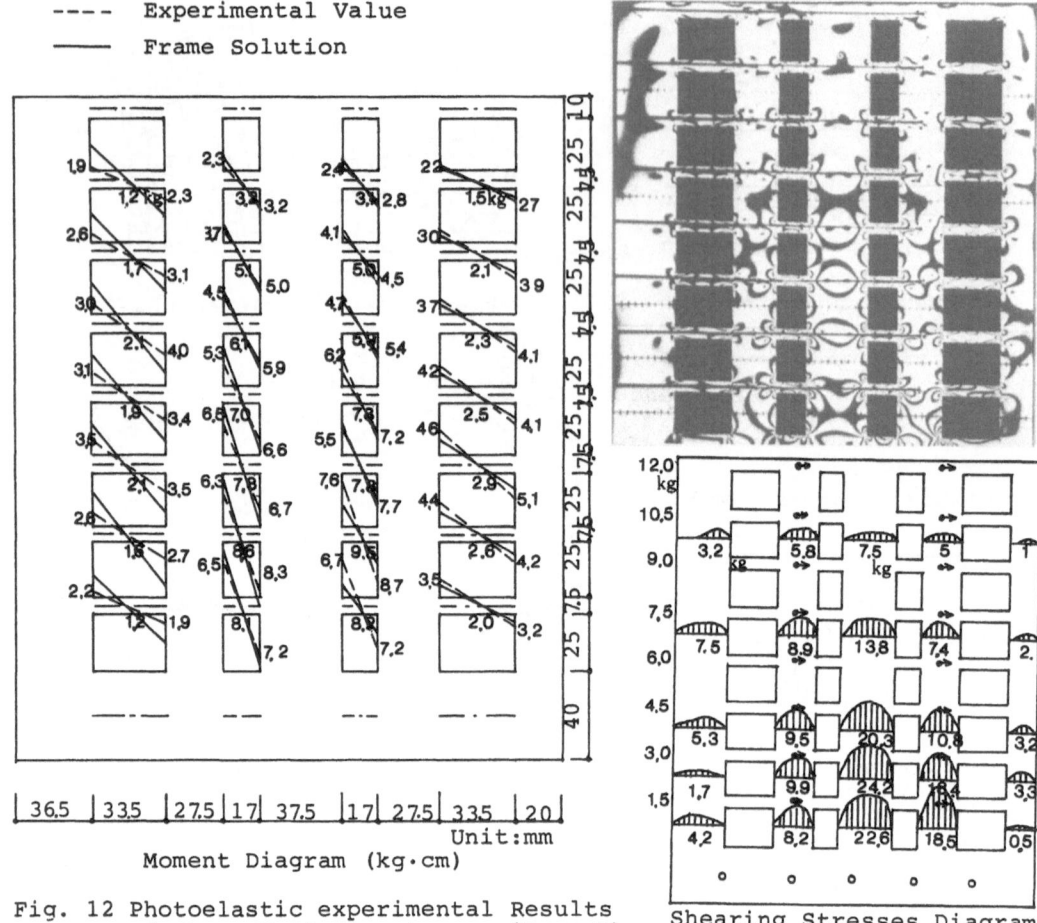

Moment Diagram (kg·cm)

Unit:mm

Shearing Stresses Diagram

Fig. 12 Photoelastic experimental Results
of Shear Wall Structure   (Type B)

6. CONCLUSION

As the wall width becomes wider, the accuracy of the wall centerline displacement decreases when using elementary theory. Large
deviation of the equivalent beam displacement between elementary
theory and two dimensional theory exists.
Elementary theory taking into account the rigid zone can be used to
predict accurately the rotation of the equivalent beam in the wall.

REFERENCES

1) Muto K (1963) Aseismic Design Method, vol. 1, Maruzen, Tokyo
2) Michael D (1966) Tall Buildings, P.253, Pergamon

Three-Dimensional Photoelastic Analysis of Aeroengine Rotary Parts

K. Uchino[1], T. Kamiyama[1], T. Inamura[1], K. Simokohge[1], H. Aono[2] and T. Kawashima[2]

Research Institute[1] and Aero-Engine & Space Operations[2], Ishikawajima-Harima Heavy Industries Co., Ltd., Ohtemachi, Chiyoda-ku, 100 Japan

INTRODUCTION

In the design of aero gas turbine engines, very high reliability and safety are required. The rotary parts in particular are designed with scrupulous care because their failure can cause a serious accident.

Recently, to save weight, designers of aircraft gas turbines attempt to bring the conventional safty factors as close to the actual value as possible. These improvements are achieved on the condition that highly accurate analysis is done. There is also a tendency to adopt 3D FEM analysis, but 3D FEM analysis has the following problems: (1) It takes a long time to model and is very expensive to run on the central processor to obtain a solution; (2) It is very difficult to get correct 3D FEM models of complex gas turbine engine components; and (3) It is risky to use a 3D FEM program whose accuracy has not been verified experimentally. Therefore, during the design stage, it is desirable and realistic to use a conventional 2D FEM program whose accuracy has been verified experimentally and to estimate 3D stress concentration factors using the design data which have been obtained and modified by various experimental techniques.

We have succeeded in offering better design data using 3D photoelastic techniques than was previously available. We have modified design data, verified the stress predictions obtained from a new 3D FEM computer program, and got the stress distribution of three-dimensional structures which are too complex to be modeled numerically.

In this paper, the positioning of 3D photoelasticity in the design flow of aero gas turbines and some examples of 3D photoelasticity applied for gas turbine engine rotary parts are shown. We also show that the combined use of numerical and experimental methods provides a more accurate design tool for modern gas turbine engines.

DESIGN OF GAS TURBINE ENGINE ROTARY PARTS AND ROLE OF 3D PHOTOELASTICITY

The disk is one of the most critical gas turbine engine rotary parts. Fan and compressor disks are designed to stand the centrifugal forces of blades, which are assembled around the disk rim, and the body force of the disk itself. Thermal stress is also considered for turbine disks. In 2D stress analysis, stresses in stress-concentrated sections such as disk/blade attachments and bolt holes are evaluated by multipling the design factors. It was therefore possible to obtain safety disk design, as long as sufficient margins for unknown stress factors including 3D stress concentration were maintained.

As shown in introduction, to save weight, designers attempt to adopt the smallest margin which still guarantees safe design. This requires us to increase the accuracy of 3D stress concentration analysis, because, as the margin has decreased, the possibility of finding fatigure cracks in the engine parts has increased. This is because the fatigure cracks are very sensitive to local stress concentration. Figure 1 shows the design flow which is commonly used to develop jet engine rotary parts. As this flow shows, photoelasticity plays very important roles in verifying the theoretical analysis method and providing the stress concentration design data.

In the next chapter some typical examples of 3D photoelasticity used for gas turbine engine rotary parts are described.

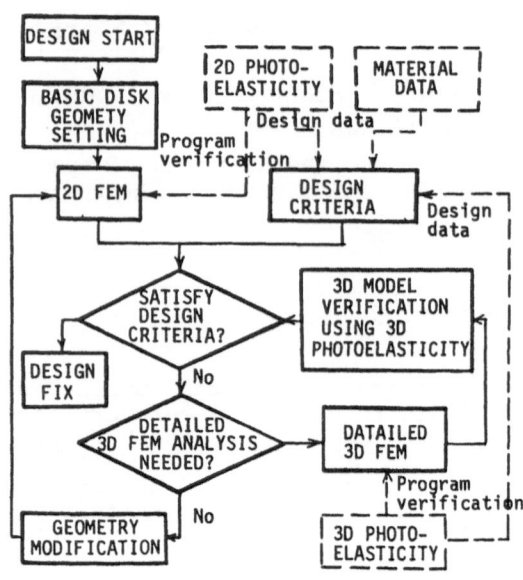

Fig.1.   Design Flow & Role of Photoelasticity

3D PHOTOELASTICITY OF GAS TURBINE ENGINE ROTARY PARTS

3D Stress Concentration of Disk/Blade Attachment.

The design of disk and blade attachment must be very carefully considered in the development of modern high-speed, high-performance gas turbine engines.

2D FEM analysis of a disk and blade attachment region usually shows a high stress concentration in the disk grooves. But actual disk grooves usally have some broach angle, and fatigue test results using actual disks have shown that fatigue cracks were always observed in the acute corner in the disk groove. Figure 2 shows a typical fatigue crack which occured during cyclic spin testing. This means that a double stress concentration is present in the disk/blade attachment section. A series of 3D photoelastic analysis tests was conducted to confirm the stress concentration factor due to the broach angle on some actual disk/blade attachments.

Fig.2.   Fatigue Crack in Dovetail Corner

Modeling: As standard technique, machining is used to make 3D photoelastic models. Specially prepared "formed cutters" which have individual disk groove contours are used to cut disk/blade attachments. Sharp cutting tools are essential and they must be sharpened frequently. In general, relatively high cutting speeds are used with low feed rates. Figure 3 shows a typical example of 3D photoelasticity model with machined disk/blade attachment grooves.

Fig.3. Example of Machined Model (200mm dia x 130mm^t)

When cost prohibits making the model by machining, a mold technique is used exceptionally. We use plaster and silicon-rubber as the mold materials. Various techniques are used to prevent gas formation in the models during molding and curing. When we use a plaster mold, water paint is used as an under coating to prevent gas formation, then silicon-rubber is applied. Water paint must be chosen carefully because normal water paint can't stand the temperature of poured epoxy-resin. When a silicon-rubber mold is used, epoxy-resin must be sufficently cured before pouring. Silicon-rubber material selection is one of the keys to getting a precise 3D photoelastic model. To keep the deformation of the model as small as possible, the material must have a thermal expansion rate as close to the shrinkage rate of epoxy-resin used for model making as possible. Figure 4 shows an example of a silicon-rubber mold technique. Figures 4(a), (b), and (c) show an actual disk, silicon-rubber mold, and 3D photoelastic epoxy model. Figure 4(c) shows an enlarged view of the groove section. Clear and undeformed makings are observed.

(a) Actual Disk (400mm dia)

(b) Silicon-rubber mold

(c) Epoxy Model

Fig.4. Example of Mold Model

Stress Freezing and Slicing: The procedure of stress freezing is quite similar to that of usual 3D photoelasticity. The stress freezing furnance has a rotating axis which rotates with an accuracy of ±0.5%. The model is sliced by a band saw. When the geometry of model is very complex and it is difficult to hold the model mechanically, the model is sometimes cut manually.

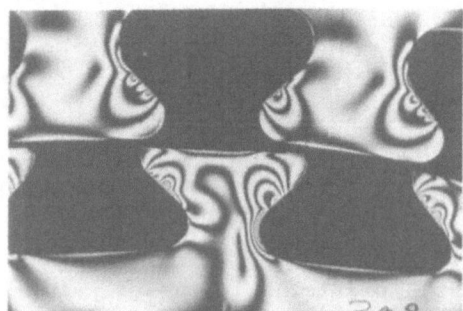

Position 1

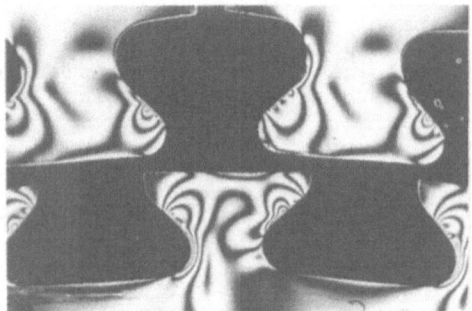

Position 5

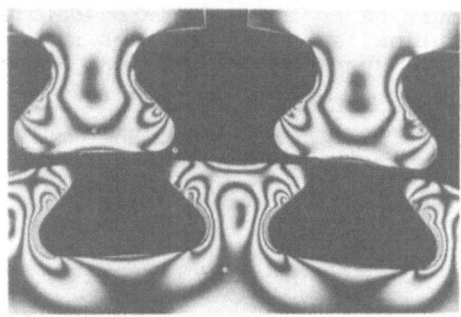

Position 3

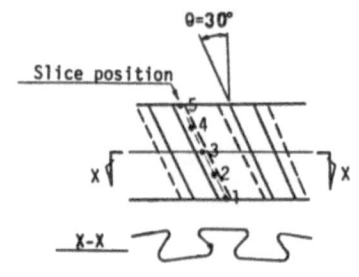

Fig.5. Isochromatic Pattern of 30° Broach Angle Groove

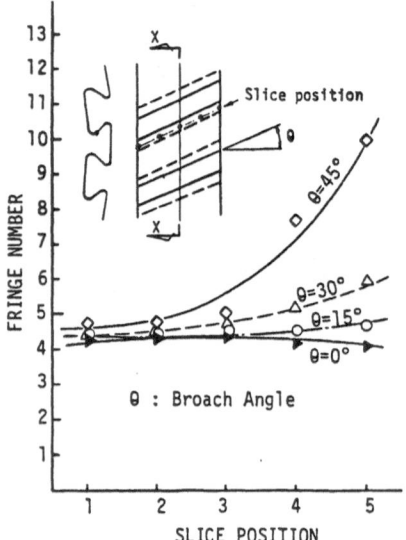

Fig.6. Stress Distribution Angled Groove

Test Results: The slices were taken normal to the axis of rotation from a 3D photoelastic model in which the stress due to centrifugal force was frozen. The typical examples of frozen stress patterns of actual gas turbine engine disk dove-tail grooves, which have 30° broach angles, are shown in figure 5. The figure shows that the highest effective stresses are in the acute corners in the grooves. The stress-concentration factors along the entire free boundary of the fillet were obtained. Figure 6 is the stress distribution along the groove.

A series of tests was carried out for models of various broach angles and groove contours. Figure 7 summarizes the effect of the broach angle. This test result has shown that, if high broach angles are adopted, sufficient margin compared with the stress obtained by 2D FEM should be considered.

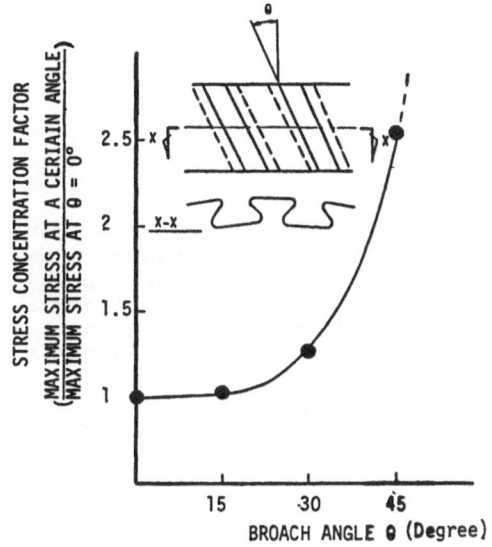

Fig.7. Stress Concentration Factor
vs Broach Angle

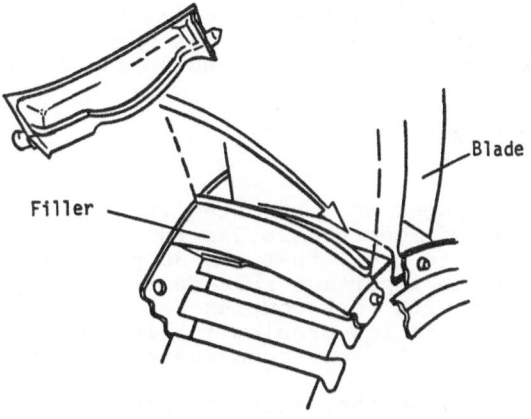

Fig.8. General Idea of Annulus Filler

Annulus Filler

Modern high-bypass ratio engines
tend to have very big fan blades.
Adopting an annulus filler is one
way to avoid the high stress due
to a platform in wide chord fan
blades. These parts form an air
passage and reduce the centrifugal
forces acting on the blade root
section in a conventional platform (Fig.8). This part has very
complex geometry, and it was
difficult to accurately estimate
stress distribution analytically.

Fig.9. Set Up of Stress Freezing

3D Photoelastic Model: We
used a silicon-rubber molding technique. In the first
step, we made a wooden model
based on the drawing. The
3D photoelastic model was
made by pouring the epoxy
resin into the mold. The work
needed great care to prevent
gas formation in the epoxy
model as described before.
Figure 9 shows the stress
freezing set up of the epoxy
model. This model has very
thin walls, so the rotational
speed was limited to about 300
rpm to prevent failure due to
large deformation.

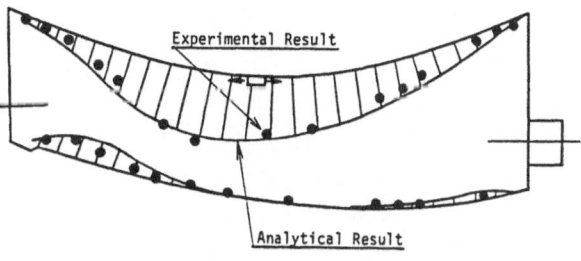

Fig.10. Comparison of Experimental
and Analytical Results

<u>Test Result:</u>   The test results approximately agreed with analysis. Figure 10 compares experimental and analytical stress distribution results.   The 3D test has verified that the 3D FEM program used to analyze the complex annulus filler predicted stress very well.

SUMMARY

This paper has presented the position of 3D photoelasticity in design flow of modern aero gas turbine engine rotary parts.   Two examples are shown to prove that the 3D photoelastic stress freezing technique will play an important role in predicting the stress distribution of rotary parts with very complex geometry.   One example is a double stress concentration problem in blade-disk attachment, the other, an annulus filler with very complicated 3D features.

We believe that the combined use of numerical and experimental photo-elastic methods will provide more accurate design results for modern gas turbine engines.

Stresses in Pin-Loaded Glass-Epoxy Plates Using Transmission Photoelasticity

M.W. Hyer

Department of Mechanical Engineering, University of Maryland, College Park, MD 20742, USA

INTRODUCTION

Materials which exhibit orthotropic elastic behavior, are birefringent, and which can be used for transmission photoelasticity studies are not commonplace. With the increasing usage of fiber-reinforced materials for advanced structural applications, there has been an interest in having such a material available. With a material available, the stress state around holes, notches, and other geometric discontinuities can be evaluated in much the same way they have been evaluated in isotropic materials, namely, by fabricating photoelastic analogues. Horridge (1955) and Hayashi (1958) were the first to suggest the concept of birefringent fiber-reinforced materials. However, Phi and Knight (1969), Sampson (1970), and Dally and Prabhakaran (1971) made significant contributions to the concept by advancing both the fabrication techniques and the characterization of materials which exhibited orthotropic elastic properties and which were transparent. The basis for these materials was the embedding of transparent glass fibers in an transparent epoxy matrix. The indicies of refraction of the glass and epoxy were matched so the composite was indeed transparent. There was considerable difficulty in fabricating the materials. The presence of voids, fiber sizing, index matching, and obtaining sufficient fiber volume fraction were just a few of the problems encountered. However, after considerable effort, sufficient quantities of the material could be fabricated so that specimens representing a fiber-reinforced structural component could be made and photoelastic data obtained. Since then Cernosek (1976), Prabhakaran (1976), Chandrashekhara and Jacob (1977), Agarwal and Chaturvedi (1978), Doyle (1980), and Chaturvedi (1982) have made contributions to the subject with both experiments and with the theory of characterization. Efforts have been made to use the material to study the stress state in problems of interest in the mechanics of composite materials (Voloshin 1980, Prabhakaran 1982). This paper summarizes other work with the material to study the stress state in a problem of interest in the mechanics of composite materials. Specifically this paper examines the stresses around a hole in a plate that is loaded by a pin through the hole. The problem has a direct application to the connecting of composite structures. This paper describes the problem and the experimental set-up used to study it. Typical isochromatic and isoclinic fringe patterns are presented. A scheme to separate the stresses is discussed and then numerical results for the stresses around the hole are presented and interperted. Some of the work has been reported elsewhere (see Hyer and Liu 1984,1985).

SPECIFIC PROBLEM AND EXPERIMENTAL SET-UP

Figure 1 illustrates the problem studied and the experimental apparatus used to study it. The dimensions of the plates, meant to simulate a composite connector, are indicated in the figure. Only the dimensions indicated in the figure were studied, the plate width to hole diameter ratio, W/D, being 4 and the end distance to hole diameter ratio, e/D, being 2. Three different composite plates were tested. Each plate was constructed of 32 layers of glass fibers in an epoxy matrix. The fiber volume fraction of each plate was about 55%. The plates were loaded with a steel pin which passed through the hole and which bore on the bottom half of the hole. The pin fit snugly in the hole, requiring a slight amount of pressure to force it into the hole. The hole was drilled with an ultrasonic core drill. A coolant was used during the drilling operation to insure that excess heat was not generated and introduce unwanted residual stresses at the hole edge. The load was introduced to the pin by steel crossbars which in turn were attached to two U-shaped yokes. The yokes were loaded with in a deadweight loadframe. The load is indicated as P in the figure. The load on the hole was reacted at the top end of the plate by attaching the composite plate to two aluminum plates that extended to the upper end of the loadframe.

The three plates studied had the following lamination sequences: $(0_4/45_4/-45_4/90_4)s$, a quasi-isotropic laminate; 0_{32}, a unidirectional laminate; and $(45_4/-45_4)_{2s}$, an angle-ply laminate. The plates were fabricated by the Illinois Institute of Technology Research Institute (see Daniel et al 1981). The elastic properties of the plates, relative to the x-y coordinate system in Fig. 1, are indicated in the accompanying table. The load level used, 5.92 kN, was selected to be high enough so as to generate a sufficient number of fringes, but not so high that the glass-epoxy material crazed near the areas of high stress concentration. Crazing would cause the material to become opaque and eliminate the possibility of gathering photoelastic data near the regions of high stress concentration, the regions of interest.

A standard split-bench collimated lightsource polariscope was used, the loadframe being positioned between the two halves of the polariscope. The isochromatic fringe patterns were recorded on standard black and white film. High-contrast film was used for recording the isoclines. Because of the presence of residual birefringence in one of the plates, the isoclines had to be measured in the presence of the full load and the corresponding isochromatic fringe pattern.

ISOCHROMATIC AND ISOCLINIC FRINGE PATTERNS

The lightfield isochromatic fringe pattern for the unidirectional plate is shown in Fig. 2. From the figure several features of the experiments are obvious. First, the sensitivity of the fiber-reinforced material was lower than common isotropic photoelastic materials. Second, the material properties of the plate, particularly the birefringent properties, were not spatially uniform. This is evidenced by the lack of symmetry, about the vertical centerline, of the fringe patterns. This lack of symmetry was not due, as first thought, to a lack of symmetry in the loading. Third, the fringes were not sharp and distinct. The transmitted light diffused somewhat as it passed through the material. This no doubt was due to microscopic imperfections in the material and the scattering of light from the interface between the fiber and matrix. Index matching was possible only within a limited range and so the interface still influenced the transmission of light.

The isoclinic fringe pattern for the unidirectional case is shown in Fig. 3. This figure was made by superposing a series of high-contrast photographs taken at the various isocline settings of the polariscope.

It should be noted in Fig. 2 that the isochromatic fringe pattern was skewed in the direction of the fibers. The fringes were concentrated on the centerline below the hole, indicating the stresses did not diffuse toward the sides of the plate. This would suggest that the stress state around the hole would not be influenced by narrowing the plate somewhat. With the angle-ply plate, the isochromatic fringes were biased in the +45° and -45° directions. The isochromatic fringe pattern for the quasi-isotropic plate was similiar to the pattern that would be generated using an isotropic material. The fringes were spread in a semi-circular fashion below the hole.

The isoclinic fringe pattern for the unidirectional plate was also clustered toward the centerline of the plate, below the hole. The clustering of the isoclinic fringes made data reduction difficult. As will be seen, knowing the spatial location of the isoclines was important for separating the stresses. With the 0-20° isoclines clustered together, it was impossible to not introduce errors into this portion of the data reduction.

STRESS-OPTIC LAW

As stated above, much of the past research with elastically orthotropic birefringent materials has dealt with the stress-optic characterization. There has been much debate in the literature regarding the stress-optic laws. However, the one used here simply assumed that the relationship between the dielectric tensor and the applied stress was orthrotopic in the principal material coordinate system. Furthermore, a planar response was assumed and any dielectric effects present before stress was applied were also assumed to be orthotropic. Since the curing of epoxy is an exothermic process, thermally-induced birefringence is possible in these materials. Such an effect would be due mainly to the difference in thermal expansion between the fiber and the epoxy.

Referred to the Cartesian x-y coordinate system of Fig. 1, the relations between the isochromatic and isoclinic fringe numbers and the stresses at a point are:

$$\frac{\sigma_x}{f_x} - \frac{\sigma_y}{f_y} = N_T \cos(2\theta_T) - N_R \cos(2\theta_R)$$

$$\frac{\tau_{xy}}{f_{xy}} = \frac{N_T}{2} \sin(2\theta_T) - \frac{N_R}{2} \sin(2\theta_R)$$

(1)

Here N_T is the isochromatic fringe number at the point, θ_T the isocline, N_R the residual isochromatic fringe number, and θ_R the residual isocline. The term residual, subscript R, applies the the state when no mechanical loads are applied. The subscript T refers to the state when the mechanical loads are applied and stands for 'total', i.e., residual plus mechanically-induced. The values of N_R and θ_R were determined prior to applying the load. They were assumed to be spatially uniform over the plate. The quantities f_x, f_y, and f_{xy} are the material's stress-optic coeffients and they were determined by prior calibration. They were also assumed to be spatially uniform over the plate. The values of f_x, f_y, f_{xy}, N_R and θ_R for each plate are given in the accompanying table. The angles θ_T and θ_R are measured relative to the +x axis. The stress-optic equations referred to the Cartesian x-y system were used to determine the stresses away from the hole. For determining the stresses in the vicinity of the hole, the stress-optic equations were transformed to a cylindrical coordinate system. With the transformation two similiar equations result except that in the cylindrical system the material is not orthropic. The three components of stress appear in each equation. As was seen above, in the Cartesian system the stress-optic equations were uncoupled, the two normal stresses appearing in one equation and the shear stress appearing in the other.

SEPARATING THE STRESSES

Separating the stresses requires a third relation among the three components of stress. Here the plane-stress equilibrium equations were used. Specifically the plane-stress equilibrium equations in finite-difference form were used, the two stress-optic equations and the two plane-stress equilibrium equations forming an overdetermined set of algebraic equations. Two finite-difference grids were used, a rectangular grid away from the hole and a cylindrical grid near the hole. These two grids are shown in Figs. 4 and 5, as is the scheme for subdividing the grids into smaller regions, the regions being denoted as 1 through 11. The lines separating the subdivisions are denoted as AA, BB,....,ab, etc. The subdivision was done to reduce the size of the computational problem. The subdivision scheme will not be discussed here. However, what is important to note is the overlap of the cylindrical grid with the rectangular grid. The main focus of the study was to determine the stresses near the hole. Yet, the boundaries where the stresses were known, namely the traction-free side and bottom edges of the plate, were quite far from the hole edge. The overdetermined set of equations, the photoelastic data, and the boundary conditions were used to compute the stresses in the rectangular grid. Then the results from the rectangular grid along arc 'c-a-d' (see Fig. 5) were used as boundary conditions to compute the stresses in the cylindrical grid.

The development and check-out of the stress separation scheme amounted to at least 50% of the effort in this study. The scheme was first checked by generating the isochromatic and isoclinic data for a hole loaded by known edge tractions and in a infinite plate. Exact solutions, and thus 'exact' fringe patterns, for this problem are known. The generated photoelastic data and the overdetermined set of equations, along with psuedo-boundary conditions (psuedo because the plate was really infinite in extent), were used to re-calculate the stresses. Agreement between the stresses from the exact solution and the stresses calculated with the stress separation scheme was very good. Thus there was confidence in the approach.

The other major portion of the stress separation scheme was determining the photoelastic data at each grid point. With the low fringe density and lack of sharpness of the fringes, and with the clustering of the isoclinic fringes, the task was difficult and susceptible to error. To determine the photoelastic data at each grid point, Tardy compensation was used where the isoclinic was known with certainty, e.g., along the centerline of the specimen. Also, because of the lack of centerline symmetry of the fringe pattern, data from the left and right sides of the plates were averaged and only the right half of the plate analysed. The spatial locations of the whole and half-order isochromatic fringe data , and the spatial location of the isoclines in 5° increments, were used with a least-squares data fit to determine the fringe data at the grid points.

NUMERICAL RESULTS

The distribution of stresses around the hole-edge of each plate is shown in Figures 6, 7, and 8. The stresses are shown as a function of the circumferential location around the hole, $\theta = 0°$ corresponding to the bottom of the hole and $\theta = 90°$ corresponding to the net section. The stresses have been nondimensionalized by bearing stress, S, the bearing stress being defined as $S = P/tD$ (refer to Fig. 1 for the definition of t and D). The three components of stress shown are: the circumferential stress, σ_θ; the radial, or bearing, stress, σ_r; and the friction-induced shear stress, $\tau_{r\theta}$. Also shown on each figure is the cosinusoidal distribution often assumed in analyses.

From the figures it is clear that while the radial stress distributions and magnitudes were somewhat similiar for the three cases, the circumferential stress distributions and magnitudes were different. Also, among the three cases the shear stresses differed by sign and in the character of the distribution. The stresses for the quasi-isotropic case were similiar to the situation that would occur for an isotropic plate. The circumferential stress increased gradually from the bottom of the hole, reaching a maximum at the net section. Conversely, the radial stress was large at the bottom of the hole, directly on the centerline, and then decreased as the net section was approached. That the radial stress was not a maximum on the centerline, rather the maximum was off the centerline, is felt to be due to friction. Such effects have been observed in analytical studies of pin-loaded holes (see Hyer and Klang 1985). In contrast to the steadily increasing circumferential stress with increasing θ for the quasi-isotropic case, for the unidirectional case the circumferential stress remained small until $\theta = 60°$. At $\theta = 60°$ the circumferential stresses increased rapidly. The peak circumferential stress was significantly larger for the unidirectional case than for the quasi-isotropic or angle-ply cases. For the angle-ply case the circumferential stress reached a maximum value at $\theta = 70°$ and the stress remained constant in value to the net section. For the angle-ply case the radial stress reached a maximum value at $\theta = 45°$.

Because of their low magnitudes, the friction-induced shear stresses themselves were not significant. However, the effects of friction were felt to be significant. Besides shifting the peak radial stress from the centerline, directly beneath the hole, friction is felt to be responsible for the general flattening of the radial distribution at the bottom of the hole. In addition, as has been shown analytically, friction does influence the peak value of the circumferential stress. Finally, it is seen that the often-assumed cosinusoidal distribution for the radial stress is inaccurate directly below the hole.

ACKNOWLEDGEMENT

The research effort which led to the results presented here was financially supported by the Structures Laboratory, USARTL (AVSCOM). The
technical monitor was Donald J. Baker, NASA Langley Research Center.

REFERENCES

Agarwal, BD and Chaturvedi, SK (1978) Development and Character-
    ization of Optically Superior Photoelastic Composite Materials.
    Int J Mechanical Science 20:407-414

Cernosek, J (1976) On Photoelastic Response of Composites.
    Experimental Mechanics 16:354-357

Chandrashekhara, K and Jacob, KA (1977) Experimental-Numerical
    Hybrid Technique for Stress Analysis of Orthotropic Composites.
    In: Holister, GS (ed) Developments in Composite Materials,
    1. Applied Science, London, p 67

Chaturvedi, SK (1982) Fundamental Concepts of Photoelasticity
    for Anisotropic Composite Materials. Int J Engineering Science
    20:145-157

Dally, JW and Prabhakaran, R (1971) Photo-orthotropic-elasticity.
    Experimental Mechanics 11:346-356

Daniel, IM, Niro, T and Koller, GM (1981) Development of Orthotropic
    Birefringent Materials for Photoelastic Stress Analysis. Cont-
    ractor Report, National Aeronautics and Space Administration,
    CR-165709

Doyle, JF (1980) Constitutive Relations in Photomechanics. Int
    J Mechanical Science 22:1-8

Hayashi, T (1958) Photoelastische Untersuchungen der Spannungs-
    verteilung in der Durch Fasern Verstarken Platte. Proceedings
    Int Union on Theo and Appl Mech-Warsaw Symp:501-511

Horridge, GA (1955) A Polarized Light Study of Glass Fibre
    Laminates. British J of Applied Physics 6:314-319

Hyer, MW and Liu, D (1984) Stresses in a Quasi-Isotropic Pin-Loaded
    Connector Using Photoelasticity. Experimental Mechanics 24:48-53

Hyer, MW and Liu, D (1984) Stresses in Pin-Loaded Orthotropic
    Plates Using Photoelasticity. Contractor Report, National
    Aeronautics and Space Administration, CR-172498

Hyer, MW and Liu, D (1985) Stresses in Pin-Loaded Orthotropic
    Plates:Photoelastic Results. J Composite Materials 19:138-153

Hyer, MW and Klang, EC (1985) Contact Stresses in Pin-Loaded
    Orthotropic Plates. Int J Solids and Structures 21:957-975

Pih, H and Knight CE (1969) Photoelastic Analysis of Anisotropic
    Fiber-Reinforced Composites. J Composite Materials 3:94-107

Prabhakaran, R (1976) The Interpertation of Isoclinics in Photo-orthotropic-elasticity. Experimental Mechanics 16:6-10

Sampson, RC (1970) A Stress-Optic Law for Photoelastic Analysis of Orthotropic Composites. Experimental Mechanics 10:210-215

Voloshin, A (1980) Stress Field Evaluation in Photoelastic Anisotropic Materials: Experimental Numerical Technique. J Composite Materials 14:342-350

TABLE

Mechanical and Optical Properties of the Plates

plate property	quasi-isotropic	unidirectional	angle-ply
E_x, GPa	19.7	12.3	12.4
E_y, GPa	19.7	37.2	12.4
G_{xy}, GPa	7.38	3.93	10.9
ν_{xy}	0.328	0.300	0.577
f_x, kPa/fringe/m	96.8	68	674
f_y, kPa/fringe/m	96.8	129	674
f_{xy}, kPa/fringe/m	96.8	57	115
N_R, fringe	0	0.12	0
θ_R, degrees	0	0	0

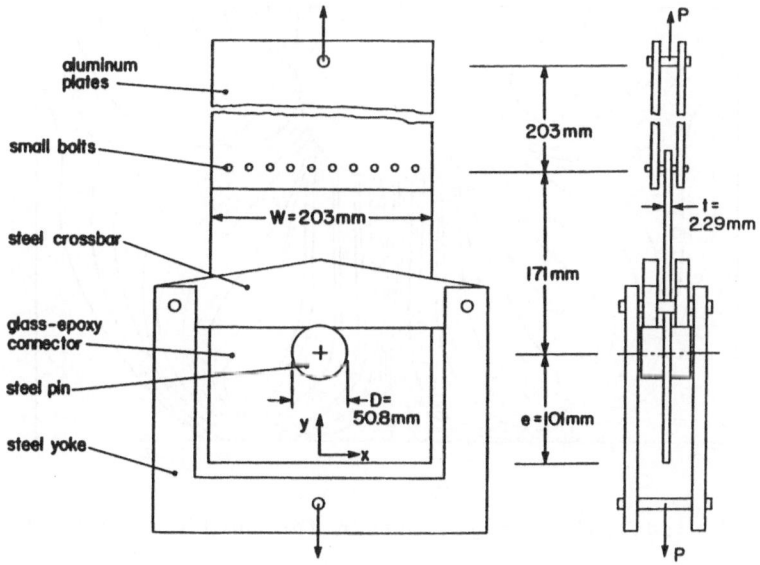

Fig. 1   Problem Studied and Apparatus Used.

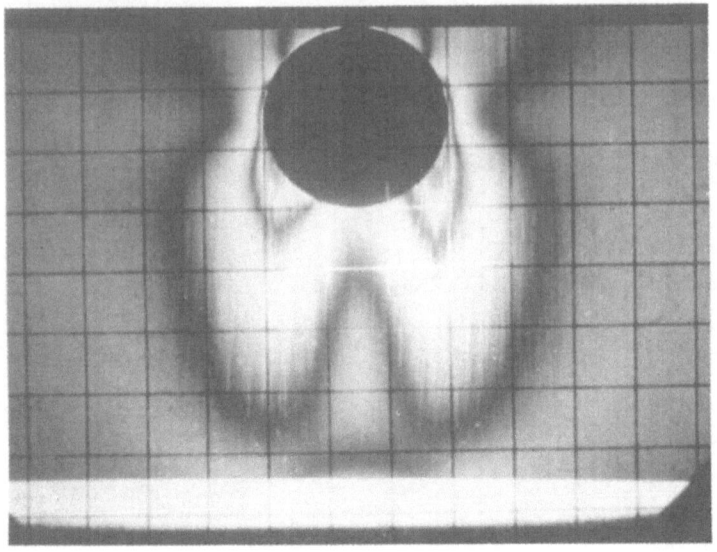

Fig. 2 Isochromatic Fringe Pattern for Unidirectional Plate.

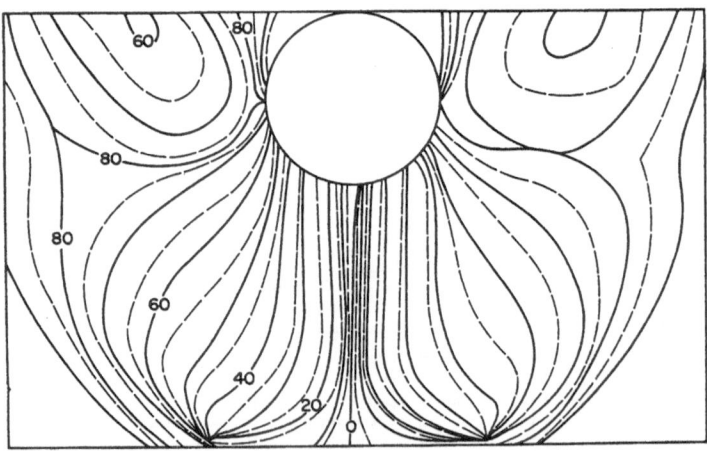

Fig. 3 Isoclinic Fringe Pattern for Unidirectional Plate.

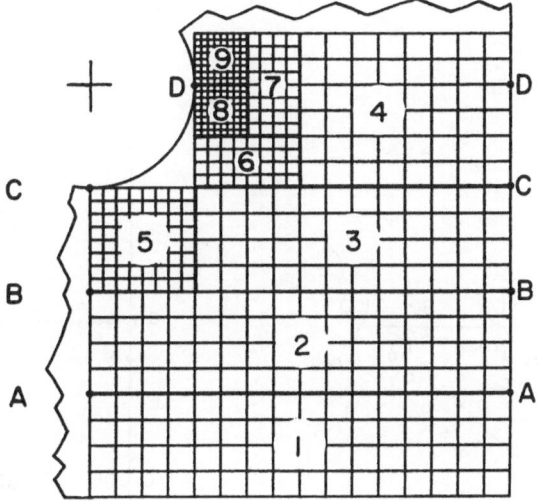

Fig. 4 Rectangular Finite-Difference Grid.

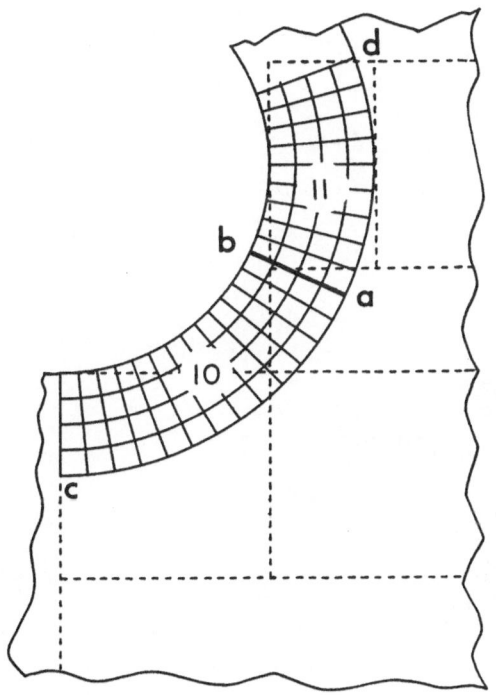

Fig. 5 Cylindrical Finite-Difference Grid.

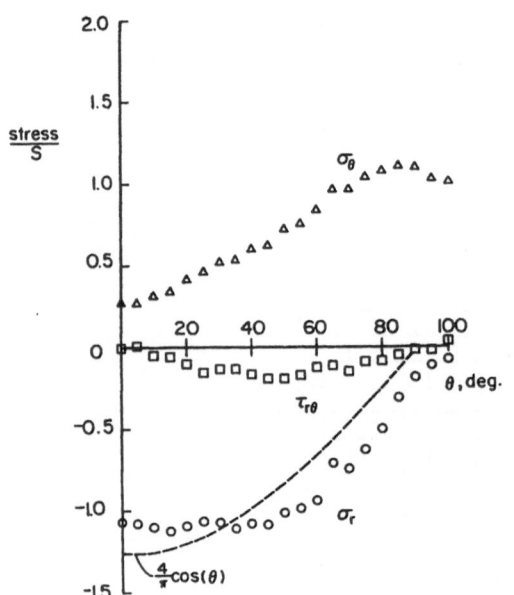

Fig. 6   Hole-Edge Stresses in
Quasi-Isotropic Plate.

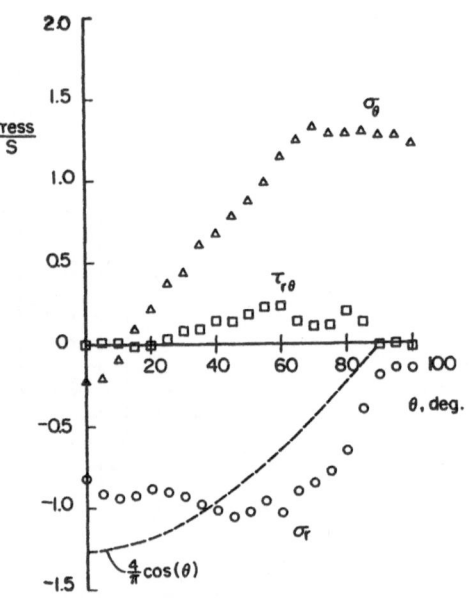

Fig. 7   Hole-Edge Stresses in
Unidirectional Plate.

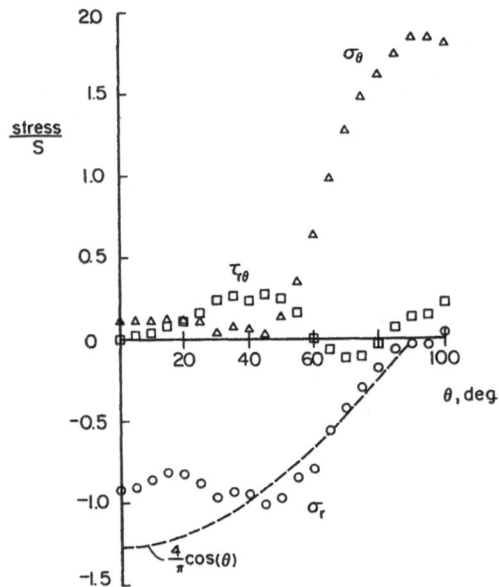

Fig. 8   Hole-Edge Stresses in
Angle-Ply Plate.

Photoelastic Analysis of Stresses in Composites

Vincent J. Parks

Civil Engineering Department, The Catholic University of America, Washington, District of Columbia, 20064, USA

INTRODUCTION

The term "composite" in this paper will be used in a somewhat broader sense than in the usual definition. Consider a series of materials and structures; first, a typical composite material, random fibers of glass embedded in a plastic matrix; second, fibers oriented in one direction in the plastic matrix; third, the fibers oriented in varied directions, layer by layer, in the plastic. Fourth, consider single layers of uni-directional fibers in plastic that are cast and subsequently bonded to each other with the fibers in varied directions, to give a system very similar to the one previously described. Fifth, consider fiber rein-forced layers bonded alternately to layers of a homogeneous material such as a metal, ceramic or glass. Sixth, consider layers of two dif-ferent homogeneous materials bonded together. Finally, consider two irregular shapes of different materials bonded together. All the above examples are here termed composites.

Every body has a shape, however in the first examples above, the term composite emphasized the character of the material. By implication, whatever shape the body had could be called a composite structure. The final example is not composed of composite materials, but will here be called a composite structure, and the term composite, in this paper, will henceforth refer to a composite material or composite structure.

To be complete in the definition of composites, it will be stipulated that the mechanical properties of the composite component materials will be different, especially Young's modulus or Poisson's ratio. If all properties were the same, there would be little reason to combine the materials, and there is no such known composite.

As examples, composites will be taken to include bricks made with straw, concrete, reinforced concrete, fiberglas, metal and carbon fiber reinforced materials, carbon-carbon materials, plywood, fiberglas cylinders, modern layered skis, systems with plastic and metal bonded to bone or teeth, ceramic-metal insulators, printed circuit boards, steel and plastic reinforced tires and so on.

Composites as defined above, including all the examples given, have the common feature that when subjected to load they exhibit a buildup of stress (a stress concentration) at the interfaces between the different materials. This stress concentration is most severe at the edges of the interfaces, where the interfaces meet the surface of the body. Even if there is no intersection of interface and surface, stress buildup occurs on the interfaces.

Muskhelishvili (1953) considered the problem of bonded homogeneous
materials, and solved several shapes where the interface did not intersect the boundary. Muki and Sternberg (1969) determined the transfer
of forces in the pin-pull problem. Hein (1971) analyzed several homogeneous layer problems for stresses, showing that the stress at the intersection of the interface and boundary was singular. A similar conclusion was obtained for the interface between two layers of an anisotropic material (representing a fiber reinforced matrix) by Wang (1982).

THE PRIMARY DIFFICULTY IN PHOTOELASTIC ANALYSIS OF COMPOSITES

There is an inherent difficulty in photoelastic analysis if the composite itself cannot be analyzed. If the composite is made of one or
more birefringent materials, the structure itself can be loaded and
possibly analyzed, or if only surface stresses are needed, it may be
possible to apply a photoelastic coating to the surface and analyze it.
Beyond these two special situations, the usual method of modeling the
composite structure must be employed. To properly model the composite,
the elastic moduli of the model materials should be in the same ratio
as those of the composite, and in the addition the Poisson's ratios of
the model materials should be equal to those of the corresponding composite component materials. Obtaining model materials with the proper
mechanical properties is the inherent and primary difficulty of photoelastic analysis of composites.

Poisson's ratios are very seldom matched. The main emphasis in choos-
ing the photoelasticity materials is in obtaining the best ratios of
elastic moduli. Glass, hard plastics such as epoxy, and soft plastics
such as polyurethane are the three primary groups of photoelastic
materials. Glass is 20 times more rigid than epoxy and epoxy is about
1000 times more rigid than polyurethane. At the higher temperature
used in stress-freezing, materials soften, but the high ratios remain,
and the problem is further complicated by various viscoelastic effects.
In general, both three- and two-dimensional photoelastic analyses have
large ratios of elastic moduli. This is often true even where non-
birefringent materials are used to model some of the materials.

The following six analyses avoid or partially treat the difficulty in
various ways, and emphasize the current limitations of the photoelastic
method applied to composites.

RESIDUAL STRESSES IN GLASS RINGS

Figure 1 shows the photoelastic patterns in electrical insulating units
made by bonding molten glass between metal rings and a central pin.
Residual stresses build up in the glass as it cools and hardens. The
analysis was aimed at comparing residual stresses for various glasses
and heat treatments. The metal rings and pins in several samples were
dissolved with acid, and it was seen that the photoelastic patterns
completely disappeared. This insured that the residual stresses were
completely elastic at ambient temperature, and not "frozen-in" during
cooling. Small holes were drilled in the glass to separate the principal stresses. Using the axisymetry to specify the isostatics, the
Lané-Maxwell equations were used to separate stresses along the radial
lines. The units were one-third the thickness of the outer diameter,
and so the stresses obtained were the average values of the stresses
through the thickness. By use of the actual composite structure the
primary difficulty of photoelastic analysis in composites is avoided.

STRESSES IN THE MIRROR OF A PASSIVE SOLAR ENERGY COLLECTOR

Figure 2 shows a photoelastic pattern in a millimeter thick curved
mirror held to a 10 meter radius, by bonding to a hexagonal honeycomb-

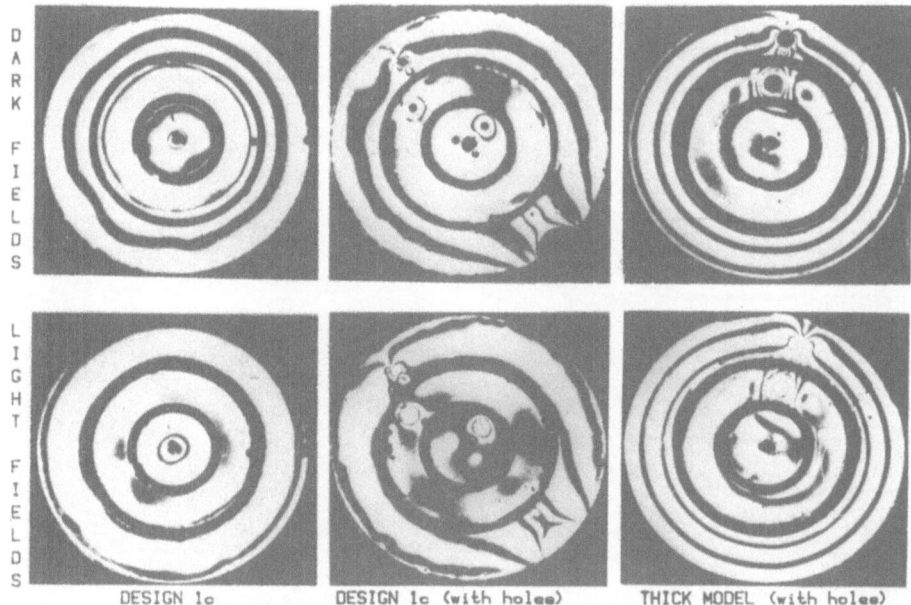

DARK FIELDS

LIGHT FIELDS

DESIGN 1c   DESIGN 1c (with holes)   THICK MODEL (with holes)

Fig. 1. Dark and light field patterns of two insulators. Each has 3 rings of glass. The one on the left is as-manufactured. The middle photos are of the same insulator, after a hole was drilled in each of the 3 glass rings to help separate stresses. The insulator on the right is a thicker model with holes drilled in the outer two rings.

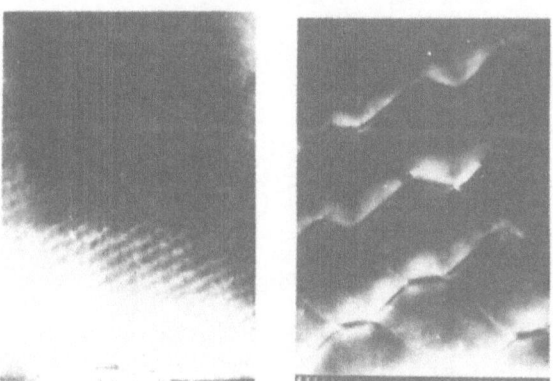

Fig. 2. The left photo is a 400 mm wide region of a curved mirror. The enlargement is 32 mm wide. The hexagon patterns in the photo correspond to the bonded mesh on the back of the mirror.

mesh, which in turn was held in place by a curved steel sheet. The pattern was obtained by a 10° rotation of the analyzer of the polariscope (Tardy's method). Although no variation in the mirror surface could be detected by surface measurements, or reflection measurements, it is clear in the photoelastic patterns that the mirror is being deformed along the bonding lines of the hexagon backing. The analysis was conducted because of mirrors cracking during the passage of the sun overhead, and the differential heating of the structure. A redesign to distribute the bonding should reduce the chances of failure.

This is a second example of avoiding the primary difficulty of photoelastic analysis by use of the actual structure.

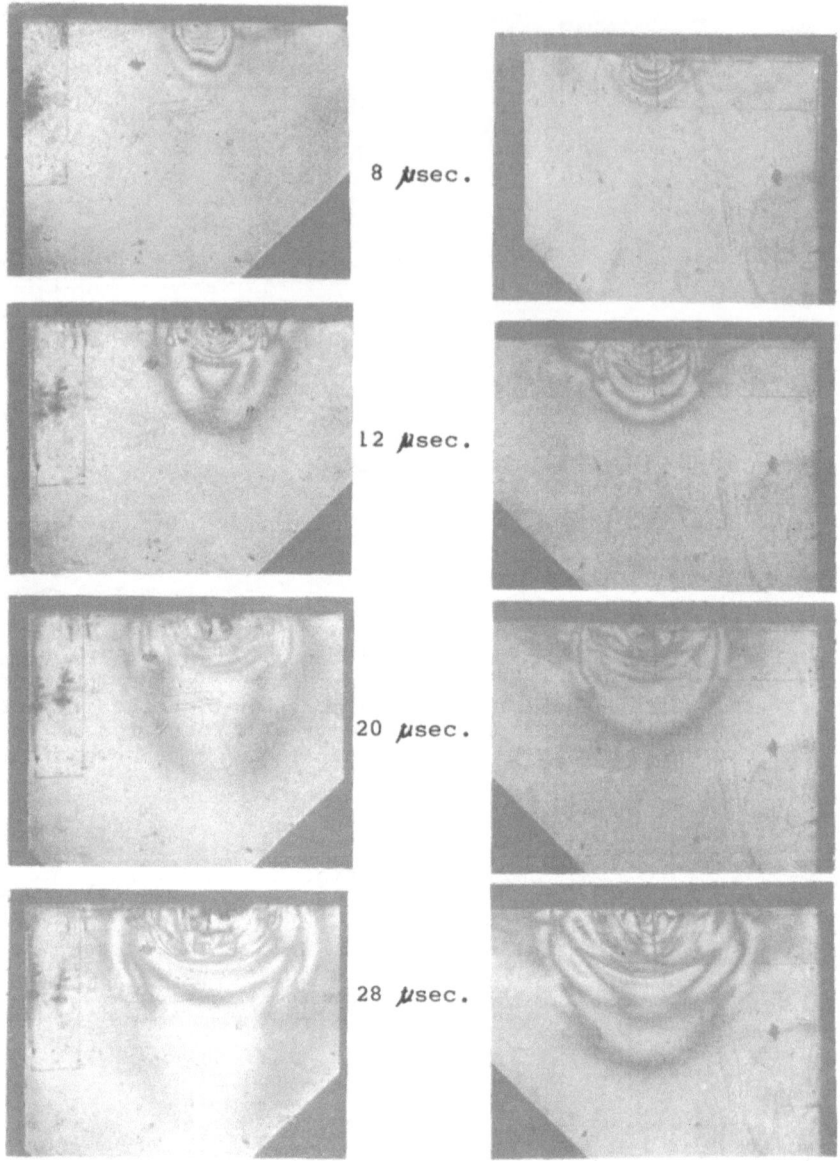

8 μsec.

12 μsec.

20 μsec.

28 μsec.

Fig. 3. The series of photographs on the left are with the reinforcing fibers running vertically, parallel to the direction of the load. The photographs on the right are of the same model loaded a second time with the fibers turned horizontally. The damage from the previous load can be seen on the right of the model.

DYNAMIC WAVES IN A LAYERED COMPOSITE

Figure 3 shows the photoelastic patterns in a 1.5 mm thick photoelastic coating bonded on a 6 mm sheet of epoxy, reinforced with glass fibers in one direction. The patterns were recorded with a camera of the Cranz-Schardin type, using a bank of lasers for light. The camera was one recently developed by Dally and Sanford (1982). The optical system was laid out similar to Kawada (1984) with a mirror backed coating and a 122 cm field lens. Pockel cells allowed shutter speeds of 1.5 nsec., and times between frames controlled to within 0.1 microseconds. The sheet was loaded by 100 mg lead azite explosive charges, on both an edge parallel to the reinforcing fibers, and on an edge perpendicular to the reinforcing fibers. The patterns show a stress wave speed of 6500 m/sec in the fiber direction, and 2100 m/sec perpendicular to the fibers.

Since photoelastic coating was used here on the actual composite, there is no photoelastic model, and the primary difficulty is once again avoided.

LOADED EMBEDDED PIN IN A LARGE MASS

Figure 4 shows the meridian slice from a three-dimensional photoelastic model of a metal pin in a glass mass. The integrity of the insulator is tested by pulling the pin from the mass, with the mass supported on a rigid ring, several diameters larger than the pin diameter, that encircles the pin. The glass mass is modeled with a conventional photoelastic epoxy with an elastic modulus of 14 MPa at 130°C. The model shown in Fig. 4 has a pin made of extruded acrylic with an elastic modulus of about 5 MPa. This is the best combination of materials found for the analysis.

The ratio of steel-to-glass elastic moduli is 3. The best ratio of model materials was one-third, and here illustrates the primary difficulty.

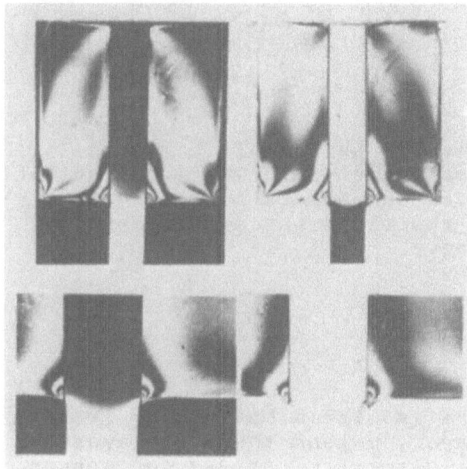

Fig. 4. Meridian slice of a pin embedded in a large mass and pulled. The lower photographs are of the same slice thinned down by 1/2 and slightly enlarged.

ASTM F19-64 TENSION SPECIMEN

Figure 5 shows a set of meridian slices from three-dimensional models
of the American Society of Testing Materials F19-64 tension specimen.
The F19-64 is designed to test bonding of cermaic with metal. Thus the
specimen is composed of two like halves of ceramic material, separated
and bonded with a metal film at the horizontal mid-plane. Depending on
the ceramic, the modulus may be greater or less than the bonding metal.
To study the specimen with the materials available, it was possible to
make models of the extreme cases; high ratio of metal-to-ceramic
moduli, and high ratio of ceramic-to-metal moduli, and to make a model
of ceramic and metal with equal moduli. The model with equal moduli is
just a one material model. The high ratio models both used a conventional three-dimensional epoxy model material for the ceramic. For the
high metal-to-ceramic model, an actual metal shim was used. For the
high ceramic-to-metal model, a silicone rubber was used to represent
the metal.

 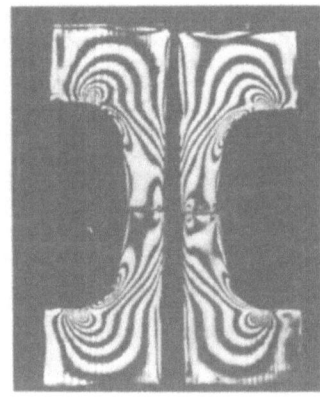 

NO LAYER   SOFT LAYER   HARD LAYER

Fig. 5. These meridian slices from three 3-D axisymmetric models
loaded in tension and the stress patterns frozen-in. The model at left
was in one piece. The model in the middle has a soft rubber layer on
the horizontal mid-plane. The model on the right has a thin sheet of
aluminum foil on the horizontal mid-plane.

The photoelastic patterns show that there is little buildup of stresses
in the hard-ceramic with a soft metal layer, unless the soft layer has
a surface void, in which case, a typical crack-like stress concentration forms. Surface voids in the hard metal layer may also occur, but
even without voids, the hard layer produces a stress concentration by
"pinching" the softer ceramic material. The strategy here in confront-
ing the primary difficulty is to "bracket" the actual moduli ratio with
two extreme cases.

THE BONDED SLAB

Figure 6, taken from Durelli and Parks (1967), shows two- and three-dimensional models of a soft slab (e.g., rubber) bonded to a hard base (e.g., metal). The photoelastic materials are polyurethane (2-D) and epoxy at 275° (3-D) each bonded to a steel base. The actual loading was thermal shrinkage. The loading of the models was the skrinkage which occurs in casting the materials. Since the prototype has a large ratio of elastic moduli (more than ten), it was possible to represent the stress field in the soft material accurately with a photoelastic material.

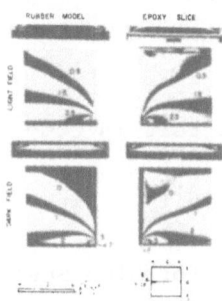

Fig. 6. The photographs at the left show the pattern at the corner of a rubber strip bonded to a steel strip and shrunk in curing. The photographs on the right show the pattern on a central slice cut from a slab bonded on a steel plate and shrunk in curing.

MACROANALYSIS AND MICROANALYSIS

There is a class of composites that lends itself to photoelastic analysis. The composites are those of glass fibers embedded in plastics. Both glass and plastics are transparent and birefringent. Still regular glass fiber reinforced plastics are not sufficiently transparent to permit analysis, primarily due to differences in indices of refraction of the glass and the plastic. However, if the indices of glass and plastic are carefully matched the composites can be "modeled" and successfully analyzed. Dally and Prabhakaran (1971) report a number of orthotrophic models prepared in this way. Since the resultant stresses are the average of stresses in the two materials (glass and plastic) this analysis is termed macroanalysis. Marloff and Daniel (1969) followed a similar approach except that the individual fibers were modeled with large polycarbonate straight rods in an epoxy plastic. By heating the model to the softening temperature of the epoxy a moduli ratio of 175 was obtained. The model is a magnification of the fiber geometries and allows the stresses in the plastic matrix to be analyzed apart from the stresses in the fibers. This is termed microstress analysis.

CONCLUSION

The primary difficulty of photoelastic stress analysis of composites restricts its application to the following areas.

1. Composites in which one or more of the composite materials are birefringent materials and the area of interest is accessible to analysis.
2. Composites in which stresses or strains are only required on the surface of the loaded body, and the surface is accessible to view and can be bonded to a photoelastic coating.
3. Composites in which the materials have properties that are proportional to corresponding properties in photoelastic materials (Fiberglas is a notable example), and the average stresses are required.
4. Composites in which the stresses are required in the soft material, and the soft material has a Young's modulus of one-tenth or less than that of the hard material.

ACKNOWLEDGEMENTS

The support of The Catholic University of America and the U.S. Naval Research Laboratory in preparing this paper is gratefully acknowledged. Several of the models were analyzed for Sandia National Research Laboratories. Their support was very helpful. Miss Madeline Sapienza is given special thanks for her careful preparation and editing of the manuscript.

REFERENCES

Beyer AH, Solakin AG (1934) Photoelastic Analysis of Stress in Composite Materials. Trans ASCE 99:1196-1212

Dally JW, Prabhakaran R (1971) Photo-orthotropic-elasticity. Experimental Mechanics 11:346-356

Dally JW, Sanford RJ (1982) Multiple Ruby Laser System for High Speed Photography. Optical Engineering 21:704-708

Durelli AJ, Parks VJ (1967) Photoelastic Stress Analysis on the Bonded Interface of a Strip with Different End Configurations. Am Ceramic Soc Bull 46:582-586

Hein, VL, Erdogan F (1971) Stress Singularities in a Two-Material Wedge. Int Journ of Fracture Mech 7:317-330

Kawata K, Takeda N, Hashimoto S (1984) Photoelastic Analysis of Dynamic Stress Concentration in Composite Strips. Experimental Mechanics 24:316-326

Marloff RH, Daniel IM (1969) Three-dimensional Photoelastic Analysis of a Fiber-reinforced Composite Model. Experimental Mechanics 9:156-162

Muki R, Sternberg E (1970) Elastostatic Load-Transfer to a Half-Space from a Partially Embedded Loaded Pin. Int J Solids Structures 6:69-90

Muskhelishvili NI (1953) Some Basic Problems of the Mathematical Theory of Elasticity, 2nd ed, Noordhoff, Leyden

Wang S, Choi I (1982) Boundary-Layer Effects in Composite Laminates. J Applied Mech 49:541-560

Photostress and New Photo-elastig Coating Technique by in Situ Testing of Bridges

T. Jávor

Research Institute of Civil Egnineering–VÚIS, Lamačská 8, Bratislava, Czechoslovakia

INTRODUCTION

The Photostress and the photo-elastig coating technique weapply the Research Institute of Civil Engineering, Bratislava besides other experimental methods not only for the experimental stress analysis of models in laboratory but also during the construction and testing of bridges in situ. While the photoelastic model is still the only method for three-dimensional analysis, the surface coating technique permits the measurement of surface strains in the elastic or plastic ranges on structures, joints anchoring areas of prestressed concrete beams, atc., previously inaccessible to photoelasticity. The use of experimental stress analysis techniques in situ is expanded in such fields as: establishment of design criteria, reduction of weight and cost of similar new structures or structural elements and improvement of product reliability. The photoelastic coatings, the photostress method combines the best features of strain gages and classical photoelasticity by providing a visible picture of the surface stress distribution of the component and stress distribution which is accurately readable at any point for both direction and magnitude.

BASIC ANALYSER AND DATA OF THE USED PHOTOELASTIC MEASUREMENTS.

For the photostress analysis in situ we are using the Photolastic Reflection Polariscope made by Photolastic, Inc., USA and developed by the Measurements Group Vishay (Dr. F. Zandman). The basic Analyser, Model 031, consists of two ballbearing mounted Polarizer-Quarter Wawe Plate Assemblies attached to a common frame, and mechanically connected so that they rotate in unison. The assembly is equiped to receive the special ligth source, and the assembly is provided with measurement scales. The instrument is also equipped to accept many new accessories which greatly increases the versatility of the instrument and permits any photoelastic coating task to be accurately performed.

The Basic Analyzer measures three major pieces of data:
- The magnitude of the difference of the principal strains or stress in biaxial state
- The magnitude and sign of the tangential stress at free boundaries, or in any region of uniaxial stress condition and
- The directions of the principal strains or stress.

The instrument may be hand-held or mounted on a tripod. The hand-
-held feature is used to inspect stress areas for possible detailed
analysis by quickyscanning the entire test part. The portable operation is also used when a large number of point by point measurements are to be made on a structure, and for analysing hard-to-see
areas where a tripod would be awkward. In other cases, when attention is concentrated on only a few areas, or when a test is being
conducted on small parts, the instrument will usually be mounted on
the tripod. Experimental stress analysis is not always reduced to
measuring the magnitude of stress. If the part being stress analysed is being done so because of actual service failures, the display of the complete stress distribution on the part will usually
offer suggestions on how to modify designs to prevent failures. The
photoelastic pattern also yields valuable design information, on
how to modify the part to make it lighter, and at the same time
less stressed. The visual stress display shows the relative importance of various load modes applied. Often such design information
is reealed not by highly stressed areas, but by low stressed areas
where material could be removed. In many practical applications of
bridges, one of the principal stresses is zero. In this case we ha-
ve the stress $\delta = N \dfrac{fE}{1 + \mu}$, where the fringe value $f = \dfrac{\lambda}{2\,tK}$

where λ is the wawe length, N is called Fringe order which we are
measuring, E is the Young modulus, μ Poisson ratio, t is the thickness of coating and K is the sensitivity of plastic.

The newest compensator developed by Photolastic (Model 332) provides a direct digital readout of the strain when the Null Balance is
achievedat the point of measurement. This unit is also available
with a printer system (Model 432) which "prints out" the Point Number (2 digits), strain direction angle (2 digits), and the strain
magnitude (4 digits).

In certain cases, however, a more complete analysis is required that
necessitates separation of the two principal strains, and obtaining
the individual values of each. To accomplish this a second reading
is required, and that reading must be taken with oblique incidence
light. The Oblique Incidence Adapter for use with the Photolastic
Analyzer has a fixed mirror angle which provides for simplified data
reduction. By oblique incidence, we mean that the light from the polarizer traverses the photoelastic coating at an angle and the bire-
frigence measured depends on the secondary principal strain in the
plane perpendicular to the light path. Thus, an oblique incidence
reading combined with a normal incidence reading provides us with
the necessary information for determining the separate values and
directions of the principal strains $\mathcal{E}_x$ and $\mathcal{E}_y$. When two readings
are obtained, one in normal incidence and the other in oblique incidence, the values of the two principal strains are given by:
$$\mathcal{E}_x = f\,(1.5\,N_o - N_{normal}); \quad \mathcal{E}_y = f\,(1.5\,N_\varphi - 2N_{normal}),$$
where the numerical values are coefficients derived from the deve-
lopment equations for oblique incidence measurements.

The use of monochromatic viewing in photoelastic coatings falls into
two distinct categories: black and white photography and identification of fringes when the colors washout at higher fringe orders. A
truly monochromatic light source provides a very low light intensity
only usable in a dark room or inside of the box-girder bridge. Semi-
monochromatic lights used with standard photographic filters cause
a shift in the fringe positions from color to black and white. The
most efficient and economical solution is to use standard high intensity light and filter the desired band of wave lengths. The Photo-

lastic Monochromator is a narrow band interferential filter. It provides a band pass of 100 A° atthe wave length of the tint of passage producing a black fringe at every location where the tint of passage is observed in white light. It can be used in-hand or attached to the lens of the camera. Photographic recording of the observed fringe patterns provides the simplest and most accurate method of recording data without transcription errors of forgotten details. A very convenient method of recording isoclienics is to use color transparency film. While the analyzer-polarizer assembly is rotated successively to different positions, obtain photographs without changing the position of the camera. The resulting slides can then projected later on white paper for tracing the isoclinics. When black and white film is used, the isoclinics should be recorded at a low strain level to avoid confusion with isochromatic fringes.

Correction Factors

If the photoelastic cating exhibits an initial color pattern prior to loading, we are speaking about the parasitic birefringence. There are cases where permanent birefringence will occur due to mishandling of the plastic or to yielding of the part after it has been coated. In these cases, the directions of principal strains of the parasitic birefringences may not necessarily coincide with the directions of principal strains due to load. It is important to underline that subtraction of the two states of stress is only permissible when the directions of the principal stresses coincide for both states of stress.

In other cases, when the temperature is changing during a test, a system of stresses will develop in the photoelastic coating due to the difference in the coefficient of thermal expansion between the test part and coating. If this happens, a correctio factor must also be used to compensate for the effect or temperature change on the readings. In regions not located on boundaries, normal incidence reading is not affected by the change of temperature, and the pattern observed is directly due to the thermal stresses to be measured. In oblique incidence a "zero shift" will result due to change of temperature only. In normal incidence however, fringes will appear on the edges due to a change in temperature. The most convenient procedure for analysis in this case is to prepare a dummy specimen not subjected to the same stresses in the part, but to the same changes of temperature as the investigated part, and then to take comparative readings.

In the case of plane stress problems where the bending action is negligible, there is some reinforcing effect although it is very small. When thin beams or plates are subjected to bending, the plastic coating reinforces the part and the measured strain must be corrected for this reinforcing effect. The correction factor applied must take into consideration three different effects as follows:
- The coating increases the stiffness of the plate,
- The neutral axis of the coated structure shifts,
- The reading is an average strain through the coating thickness, and corresponds to the middle plane of the coating, which is located further from the neutral plane than the surface of the structure.

Photoelastic Coating Materials

The photoelastic coating materials are produced in three forms:
sheets for coating flat parts, liquids for coating complex-shaped
parts and special photoelastic sprays. The materials are grouped
into high, medium and low modulus materials. Pre-manufactured
sheets are most economical for testing flat parts as well as for
testing of bridge structures in situ. The Photolastic Division of
Measurements Group, Inc. the PS-1 sheets manufactures with a reflective backing. All sheets are supplied with a protective paper
coating. Liquid plastic materials are used for making coatings for
structures with complex contours which cannot be coated satisfactorily with flat sheets.

PHOTOELASTIC MATERIALS

For the choice of photoelastic sensitive surface materials there
is necessary to evaluate the kind of the surface, the sensitivity
of the photoelastic material, the stiffness effect of the surface
sheet, the maximum measurable strain, the environment temperature,
the location and the selection of sheet adhesive. According to the
kind and form of the surface it is possible to apply pre-manufactured thin photoelastic plates or sheets for stress analysis of
bridge structures in-situ. The liquid and sprayed material did not
prove successful for measurements in-situ. There were mostly plates
or wall surfaces, and the thin plates, delivered by manufacturer,
with guaranteed uniform physical and photoelastic properties constant thickness with a tolerance of 0,07 up to 0,02 mm, with simple
working resp. cutting into strips, were fully satisfying. The de-
cisive criterion for the choice of photoelastic sheet is its optical sensitivity, where the relation for the difference of principal
strain is:

$$\mathcal{E}_1 - \mathcal{E}_2 = N \cdot \frac{\lambda}{2 t k} = N \cdot f \cdot$$

It follows from this relation, that the measurement results by the
reflection photoelastic method depend on the order of the value f,
i. e. on the sheet sensitivity and on the sensitivity of the used
apparatus resp. on the photoelastic distinction by which the value
N is measured, i. e. the number of the observed orders. It is evident that the value f depends on the sheet thickness and on the
factor K. The plastics with low sensitivity (factor K = 0,005 and
modulus E = 0,7 kp/mm^2) issue in the case of high values f
(μm/m/order). For plastics with medium sensitivity K = 0,02,
E-modulus = 21 kp/mm^2, while for high sensitivity plastics, where
f is low, K = 0,10 up to 0,16 and E = 250 up to 350 kp/mm^2. The
emergence of the stiffness effect of the surface sheet is connected
with the above mentioned especially in the case of thicker sheets.
Naturally this effect does not appear at concrete bridges, but at
thin wall metal structures it must be considered and the measured
strain must be corrected, for a transfer of the neutral axis, on
the sections under bending stress, should not occur, due to the
stiffening of thin wall cross section by thick surface sheet. This
effect of stiffening is dropped for extremely thin sheets, as well
as for thin wall metal structures, too. At the photostress analysis of concrete structures occurs the effect of the concrete surface influence on the sheet and the aggregates point influence on the
photoelastic material. We have made the analysis of this phenomena
in laboratory on prisms with various concrete surfaces, on which
photoelastic sheets of various thicknesses were sticked on. Then

the prisms were tested for establishing their E-moduli by centric pressure. At the same time the strain was analyzed on the prism surfaces with regard to the optimum choice of the photoelastic sheet. We used 3 kinds of sheets for these tests; namely PS 1A, PS 1B and PS 2, delivered by the firm VISHAY. In all cases there was necessary to grind the concrete surfaces and fill them up by adhesives, and only then the photoelastic sheets and the electric strain gauges for measurement checking were sticked on. For these given cases we considered as optimum the material PS 1B, 1 mm thick, with a constant sensitivity 300, optical ciefficient K = = 0,150, E = 2,5 GPa and maximum elongation 10 %.

The choice of optically sensitive sheets depends on their maximum elongation in the linear area resp. on the diagram strain/elongation, too. The firm VISHAY recommends for measurements in elastic and viscoelastic area of solid materials, as metalls, glass or concrete, with maximum elongation up to 3 %, the sheets PS 2, PS 8, PL 1 and PL 8. For relatively elastic and soft materials, as rubber, wood and plastics, with elongation from 3 up to 30 %, the sheets PS 3 are recommended, and for elongations up to 100 %, the sheets PL 2, PL 3 and PS 4 are suitable. This problem is critical at measuring in plastic area with great strain, where the adhesive, under the sheet, must follow the strain in the same way as the sheet. During our measurements this problem occured only at observing the strain of fissures and cracks. In these cases it is necessary to choose a thicker sheet e. g. PS 1A, 3 mm thick.

The adhesives used for the photoelastic sheets must have low viscosity, short hardening time, small shrinkage and E-modulus similar to the E-modulus of the applied sheet. For the photostress method at sticking the reflection property of the sheet mus be kept and the constant of the optical sensitivity of the adhesive as well as its thickness must be considered, as the adhesive becomes a part of the photoelastic active system. So the best advantage is to use adhesives delivered by the same manufacturer as the sheets are. At short measurements in situ it is necessary to consider the influence of the environment temperature changes through the proper values of the factor K do not change in the range of -20° up to +60°C. During the measurement under thermal changes the so-called edge--effect occurs in the places of sticked on sheet edges, due to the different coefficients of thermal expansivity of the sheet and the observed material, e. g. concrete. At sticking on the sheets in-situ under lower temperatures it is necessary to take in account a longer setting time of the adhesives.

APPLICATION OF THE PHOTOSTRESS METHOD

At our Institute we apply the photostress method for investigation of more complicated parts of concrete as well as steel bridges, even for evaluation of the stress state of the so-called long stands for manufacture of big pretension prestressed concrete bridge beams. There are always measurements in situ under relatively bad weather conditions and often on complicated structures and places with bad access.

Photoelastic measurements of the stress state of gusset plate of steel framwork bridges

On the highway-railway bridge over the river Danube in Bratislava (Fig. 1) there were made stress state measurements of the gusset

plate on the main beam over a bridge support, in 6 places, according to the scheme in Fig. 2. The measurements were made by the
photostress method by means of optically sensitive sheets, 3 mm
thick, type PS - 1A in places 1, 2, 3 and of the sheets type PS 1B,
2 mm thick, in places 5 and 6. The sheets were square-shaped,

Fig. 1. The highway-railway bridge over the river Danube in Bratislava, where we investigated the gussets by the photostress method
during the bridge loading test

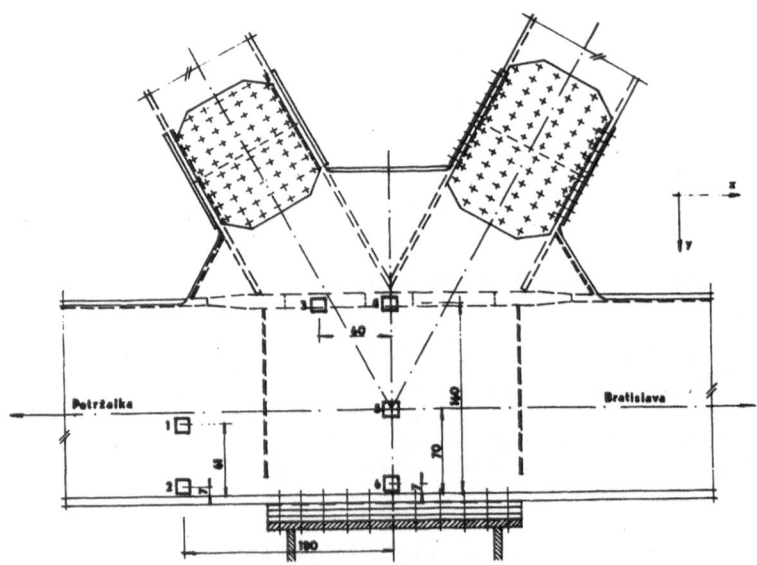

Fig. 2. Location scheme of measurement places on the gusset plate
of the Danube bridge at investigation by the photostress method

12,5 x 12,5 cm. This method was chosen because of extremely bad
weather conditions and with record to the sufficient resistance
of sheets, also during more days lasting measurements. We used
a portable reflection polariscope VISAHY 031, a compensator 032
and an automatic reader, model 532. The measurements were concentrated on the analysis of various stages of the bridge loading
test. The measurement sheets were photographed on colour slides,
so that, after enlargement by the slide projector, it was possible
to investigate the observed part in detail. As according to the designer analysis the strain value of 500.10^{-6} should not be exceeded
in the joint of the gusset plate, and the maximum strain measured
in place Nr. 4 was 310.10^{-6} in longitudinal "x" direction, the gusset plate complied fully. In place Nr. 5 the values $\varepsilon_x = -270.10^{-6}$
resp. $\varepsilon_y = 70.10^{-6}$ were reached.

Stress measurements on a high way concrete bridge

The prestressed concrete bridge over the river Sázava, near the locality Hvězdonice, on the highway Prague - Bratislava, is one of
the most important bridge structures in Czechoslovakia. It consists
of 11 fileds, each one 54 m long, with a two-beam cross section.
The bridge was concreted in-situ by means of the so-called sliding
centering. During the construction and during the demanding loading
tests the bridge was investigated by the photostress method, besides extensive tensometric measurements, by built-in acoustic rea-
ders (300 pieces). There were the measurements strips, (2 x 20 cm),
of an optically sensitive material Umapolar, Czechoslovak production, with optical sensitivity k = 11,8 kp/cm, sticked on by the
adhesive PL-1 Bi-Pax, on several fields, in the center of span, and
over the supports, in the level of the upper and lower fibres. With
regard to the cold weather the setting of the adhesive was retarded,
and so it was necessary to plaster temporily the strip ends, to the
bridge structure. There were applied the same equipments of the
firm VISHAY as in the above mentioned case. We used a compensator
with zero balance for direct evaluation and by means of an interconnected digital evaluation apparatus, model 532, we obtained directly the strain values in μ strain (10^{-6}). The values were recorded by a printer which registered the angle of the stress direction, too. The scheme of the sheet strips location of the chosen
fields is given in Fig. 3 and the measured values were in good accor-

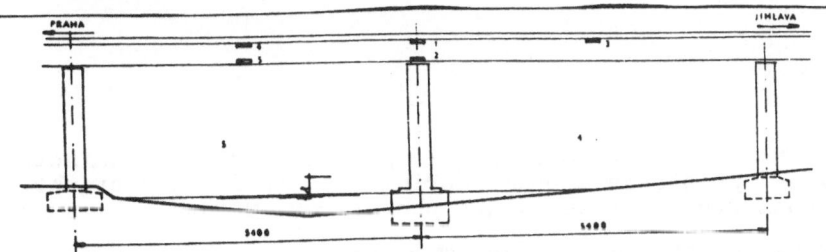

Fig. 3. Location scheme of the photoelastic sensitive Umapolar
strips at stress measurements by the photostress method in two chosen bridge fields on the highway bridge in Hvězdonice near Prague

dance with the theoretical assumption (examples: field Nr. 5, under
loading, the longitudinal stress measured in the span center, in the
upper fibre was - 1,11 MPa, in the low fibre of the same cross section + 2,70 MPa, under the same load, the stress measured over the

support in the upper fibre was + 1,48 MPa and under the same load, in the same place of the adjacent field Nr. 4 the measured stress was + 1,85 MPa.

The photostress method gives a good account as a fast method, the high humidity over the river having any essential influence; at usage of electric resistance strain gauges this problem represents a great disadvantage.

The analysis of concrete anchoring blocs of long stands for the manufacture of pretension prestressed concrete bridge beams

The need of increase of safety and serviceability of concrete bridges lead in Czechoslovakia to the production of pretensioned prestressed concrete beams in long stands. As the long stands are built as concrete moulds with concrete anchoring blocs it was necessary to evaluate the stress state of these concrete equipments; for instance for the manufacture of railway bridge beams there are anchored up to 52 pairs of strands tensioned by a force of 52 x 320 kN. The successive tensioning of these strands, Ø 15,5 m x 2, was observed in detail by the photostress method on the anchoring wall of the manufacturing stand, as it is seen in Fig. 4. The Polariscope VISHAY 031 as well as the sheets PS - 1A of the same firm, sticked on

Fig. 4. The anchoring wall of a long stand for the manufacture of pretensioned prestressed beams during the observation of the three--dimensional stress state by the photostress method

by the adhesive PC-1 were used. The measurements were made at successive tensioning the cable pairs, namely at first 12 pieces, then 18, 24, 30 etc. up to 52 to the full tension. We determined the maximum stress in concrete for 12 pairs of strands - 40 MPa under full tension of all cables, the E-modulus of concrete being 36 000 MPa.

Similarly we proceeded at checking of the anchoring chamber of the manufacturing plant for road bridge as well as at investigation of the beam surfaces manufactured there. We applied optically sensiti-

ve strips PS - 2B VISHAY (8,2,5 cm) and we sticked them on the pe-
rimeter of the anchoring chamber and on the surface, in the middle
of the beam length. The measurements were made at various stages
of cable tensioning, after their total tension and after the pretension transfer into the beams. There was determined a maximum
tension of 4,84 MPa in the long stand and of - 3,50 MPa in beams;
this represented a good agreement with the analysis requirements
and proved again the advantages of the photostress method also for
measurement of the stress state of complicated concrete structures.

The surface photoelastic method was employed for strain measurement of beams at pretension transfer for the analysis of cable anchoring in concrete by band. In spite of disadvantageous influence
of aggregates, laying near the concrete surface (which causes un-
sufficiently exact limits of interference fringes and decreases so
the measurement accurancy)the course of evaluated strain in Fig. 5
shows, that the anchoring, i. e. the full taking over the preten-

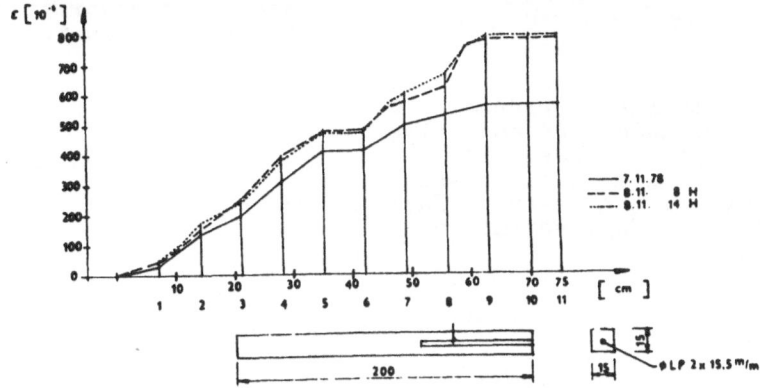

Fig. 5. The strain course during the prestress transfer into the
test beam determined by the photostress method

sion sets in a length of cca 62 m, which corresponds with international standards and regulations. The optically sensitive sheet PS -12
was sticked on the whole surface of the anchoring area of a beam
and the measurements were made immediately after the introduction
of pretension and then after 8 resp. 14 hours, as it is shown on
the diagram of measured strain in relation with the length i. e.
in the position of checked places.

We investigated the anchoring area of post-tensioned concrete beams
on photoelastic model in laboratory, too (Fig. 6), namely for various cable placements, i. e. for various excentrically introduced
forces. In these cases the prestressing force is introduced into
the beam by an anchor supported on the beam face; it is an expres-
sively simplier case than modelling the force introduced into the
model by band of the reinforcement and concrete. In this case we
realized the introduction of shear forces by thin metallic strips
sticked into the model of optically sensitive material. Every strip
was loaded separately and so it was possible to model the shear
course (Fig. 7).

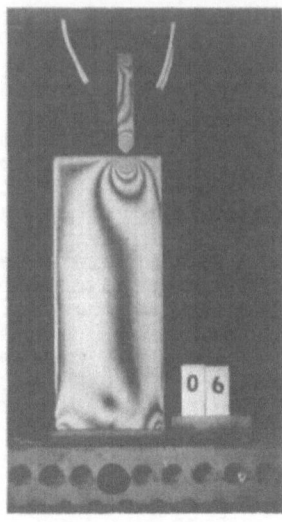

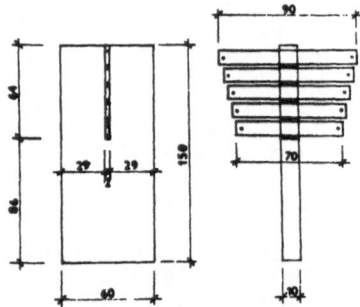

Fig. 7. Scheme of a photoelastic model with sticked-in
metallic strips enabling the
band analysis of the reinforcement and concrete by the
successive transfer of shear
forces

Fig. 6. Photoelastic stress
investigation under the anchoring
area of the postensioned prestressed beam on a model in the
laboratories at the Research Institute of Civil Engineering in
Bratislava

CONCLUSION

In the contribution we pointed out new possibilities of the usage
of the reflection photoelastic method, i. e. of the photostress
method, for analysis of the stress state not only of the exposed
places of steel bridges by measurements in situ, but mainly of
prestressed concrete bridges, their anchoring areas as well as their
characteristic strain and stress. The contribution brings the basic
description of the applied apparatuses, the choice of optically
sensitive sheets incl. relevant correction factors, too.

Integrated Photoelasticity as Tensor Field Tomography

H.K. Aben

Institute of Cybernetics, Academy of Sciences of the Estonian SSR, Akadeemia tee 21, Tallinn 200108, Estonia, USSR

INTRODUCTION

Tomography is a powerful method of determining the internal structure of various objects (Herman 1980). In tomography a certain radiation (X-rays, protons, acoustic waves, light rays, etc.) is passed through a section of the object in many directions, and properties of the radiation (intensity, phase, deflection, etc.) after passing the object are recorded for many rays. Experimental data for a certain value of the angle θ (Fig. 1) are named projection.

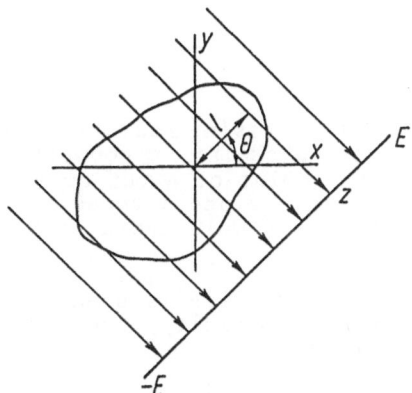

Fig. 1. Data collection in tomography

If $f(r,\varphi)$ is the function that determines the field, experimental data for a real number pair $1, \theta$ can be expressed by the Radon transform

$$\int_{-\infty}^{\infty} f(r,\varphi)dz = g(1,\theta) \tag{1}$$

If projections for many values of θ have been recorded, the function $f(r,\varphi)$ is determined from the Radon inversion

$$f(r,\varphi) = \frac{1}{2\pi^2} \int_0^\pi d\theta \int_{-E}^{E} \frac{\partial g(1,\theta)}{\partial 1} \frac{d1}{r\cos(\theta-\varphi) - 1} \tag{2}$$

Orthodox tomography considers only determination of scalar fields, i.e.
every point of the field is characterized by a single number - the coefficient of attenuation of X rays, acoustical or optical index of refraction, etc. (Sweeney, Vest 1973; Herman 1980; Vishnyakov, Levin
1982). In integrated photoelasticity polarized light is passed through
the object and transformation of the polarization of light is measured
for many light rays. If the object has certain properties of symmetry,
the stress distribution in the object can be determined. Since determination of stress in integrated photoelasticity usually is also carried out by sections, integrated photoelasticity can be considered a
kind of optical tomography. However, in this case one has to determine
a tensor field (stress tensor or dielectric tensor), i.e. every point
of the field is characterized by six numbers (since stress and dielectric tensors are symmetric).

In several papers (Davin 1969; Zimin, Shakhurdin 1978; Kubo, Nagata
1979) determining of a three dimensional tensor field on the basis of
integrated optical measurements has been considered. However, the methods proposed require passing of light through the object in many directions which are not in the same plane. Therefore, these methods are cumbersome both experimentally and numerically, and they cannot be considered tomographic. There is no information available about application
of these methods to practice.

In this paper some specific features of tensor field tomography will be
outlined and a new method of determining an axisymmetric state of
stress by integrated photoelasticity will be described.

PARTICULAR FEATURES OF TENSOR FIELD TOMOGRAPHY

The influence of a point of a scalar field on the passing radiation does
not depend on the direction of the latter, since a scalar is geometrically represented by a sphere. The influence of a point of a tensor field
depends on the direction of the passing radiation, since a tensor is
geometrically represented by an ellipsoid (this is a well known fact of
crystal optics).

In scalar field tomography usually nonpolarized radiation is used. In
tensor field tomography (e.g. in integrated photoelasticity) one has to
use polarized radiation since it is more informative.

Scalar field tomography is based on line integrals of the field (Eq. 1)
for many rays. In the general case when the principal directions of the
tensor are not constant on a light ray, line integral through a tensor
field does not have any physical meaning and it cannot be determined experimentally. Therefore, what can be measured in tensor field tomography demands investigation.

Since a scalar field is uniquely determined by the Radon inversion
(Eq. 2), there is no need for *a priori* information about the field. In
the case of a tensor field the number of unknowns is much larger and
therefore there is obviously a need for *a priori* information. In the case
of a stress field one has such kind of information in the form of equations of equilibrium and compatibility, macrostatic and boundary conditions.

The possibility to determine a scalar field on the basis of line integrals through it follows from the existence of the Radon inversion (Eq.2).
The problem whether the experimental information obtained is sufficient

to determine a tensor field demands in each particular case special investigation.

THEORY OF CHARACTERISTIC DIRECTIONS

Let us consider what can be measured on each light ray when polarized light passes a three dimensional photoelastic model and rotation of the principal stress axes is present. In this case the light vector is transformed by a unitary unimodular matrix U as follows (Aben 1966, 1979)

$$\begin{pmatrix} E_{1*} \\ E_{2*} \end{pmatrix} = U \begin{pmatrix} E_{10} \\ E_{20} \end{pmatrix} \qquad (3)$$

where E_{j0} are components of the incident light vector and E_{i*} are components of the emergent light vector. The most general expression for U is

$$U = \begin{pmatrix} e^{i\xi} \cos\theta & e^{i\zeta} \sin\theta \\ -\bar{e}^{i\zeta} \sin\theta & \bar{e}^{i\xi} \cos\theta \end{pmatrix} \qquad (4)$$

where parameters ξ, ζ and θ depend on the stress distribution between the points of entrance and emergence of light. Besides, these parameters also depend on the wavelength and on the choice of coordinate axes.

Due to the property of unitarity, it is possible to show that for a light ray in a nonhomogeneous photoelastic medium there always exist two perpendicular directions of the polarizer by which the light emerging from the model is also linearly polarized. The corresponding directions of the light vector at the entrance of light are named primary characteristic directions, and at the emergence of light, secondary characteristic directions. They are determined by primary (α_0) and secondary (α_*) characteristic angles (Fig. 2).

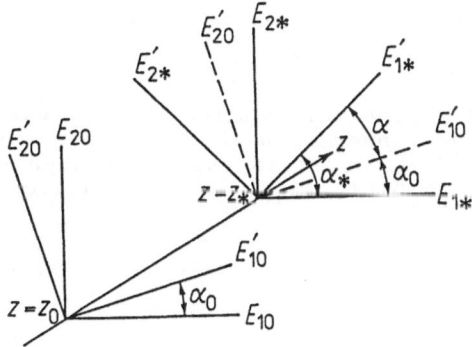

Fig. 2. Primary (E'_{10}, E'_{20}) and secondary (E'_{1*}, E'_{2*}) characteristic directions

Characteristic directions are determined by formulas

$$\tan 2\alpha_0 = \frac{\sin(\zeta + \xi)\sin 2\theta}{\sin 2\xi \cos^2\theta - \sin 2\zeta \sin^2\theta} \tag{5}$$

$$\tan 2\alpha_* = \frac{\sin(\zeta - \xi)\sin 2\theta}{\sin 2\xi \cos^2\theta - \sin 2\zeta \sin^2\theta} \tag{6}$$

The seeming "rotation" of the plane of polarization is determined by the characteristic angle α

$$\tan 2\alpha = \tan 2(\alpha_* - \alpha_0) = \frac{2\sin 2\theta\cos\xi\cos\zeta}{\sin^2\xi - \sin^2\zeta - \cos 2\theta(\cos^2\xi + \cos^2\zeta)} \tag{7}$$

The relative phase retardation between vibrations along the secondary characteristic directions is named the characteristic phase retardation 2γ

$$\cos 2\gamma = \cos 2\xi \cos^2\theta + \cos 2\zeta \sin^2\theta \tag{8}$$

If the primary and secondary characteristic angles α_0 and α_* , and the characteristic phase retardation 2γ are measured experimentally, it is possible to determine the parameters ξ, ζ and θ

$$\tan\xi = \frac{\cos(\alpha_0 + \alpha_*)}{\cos(\alpha_0 - \alpha_*)} \tan\gamma \tag{9}$$

$$\tan\zeta = \frac{\sin(\alpha_0 + \alpha_*)}{\sin(\alpha_0 - \alpha_*)} \tan\gamma \tag{10}$$

$$\tan\theta = \frac{\cos\xi}{\cos\zeta} \tan(\alpha_0 - \alpha_*) = \frac{\sin\xi}{\sin\zeta} \tan(\alpha_0 + \alpha_*) \tag{11}$$

Thus, in tensor field tomography one can determine three parameters ξ, ζ and θ of the transformation matrix U (Eq. 4) on every light ray. These parameters are in a complicated way determined through the distribution of components of the tensor on the ray.

CONSTANT PRINCIPAL AXES

If the section under investigation is a plane of symmetry x,y , there is no rotation of principal axes of the tensor on a ray. In this case simple integral relationships are valid. If light is passed through the body parallel to the y direction, the absolute (Δ_x, Δ_z) and relative (Δ) optical retardations are

$$\Delta_x = \int n_x dy \quad , \quad \Delta_z = \int n_z dy \tag{12}$$

$$\Delta = \Delta_x - \Delta_z = \int (n_x - n_z) dy \tag{13}$$

Here index denotes the direction of the light vector. Angles of deflection (φ_x, φ_z) of the light rays can be expressed in the following way (Aben, Krasnowski, Pindera 1984)

$$\varphi_x = \frac{1}{n_0} \int \frac{\partial n_x}{\partial x} \, dy \quad , \quad \varphi_z = \frac{1}{n_0} \int \frac{\partial n_z}{\partial x} \, dy \tag{14}$$

Here n_0 denotes the index of refraction of the nonstressed material. Only two of the five relationships (12) to (14) give independent information about the distribution of the refractive index tensor.

DETERMINATION OF STRESS IN A PLANE OF SYMMETRY OF AN AXISYMMETRIC BODY

Let us express the stress components in cylindrical coordinates as follows:

$$\sigma_r = \sum_{k=0}^{m} a_{2k} \, \rho^{2k} \quad , \quad \sigma_\theta = \sum_{k=0}^{m} b_{2k} \, \rho^{2k} \quad , \quad \sigma_z = \sum_{k=0}^{m} c_{2k} \, \rho^{2k} \tag{15}$$

Here a_{2k}, b_{2k} and c_{2k} are coefficients to be determined on the basis of experimental data, and ρ is dimensionless radius.

The refractive index tensor has the following components

$$n_r = n_0 + C_1 \sigma_r + C_2 (\sigma_\theta + \sigma_z)$$

$$n_\theta = n_0 + C_1 \sigma_\theta + C_2 (\sigma_z + \sigma_r) \tag{16}$$

$$n_z = n_0 + C_1 \sigma_z + C_2 (\sigma_r + \sigma_\theta)$$

If we pass polarized light through the model parallel to y axis (Fig.3), the integral Wertheim law (13) yields

$$\Delta(\xi) = 2C_0 R \int_0^{\sqrt{1-\xi^2}} (\sigma_z - \sigma_r \cos^2\theta - \sigma_\theta \sin^2\theta) d\eta \tag{17}$$

where $C_0 = C_1 - C_2$, $\xi = x/R$, $\eta = y/R$. Introducing polynomials (15) into (17) yields

$$\frac{\Delta(\xi)}{2C_0 R} = \sum_{k=0}^{m} (c_{2k} G_{2k} - a_{2k} F_{2k} - b_{2k} H_{2k}) \tag{18}$$

where

$$G_0 = \sqrt{1-\xi^2} \quad , \quad F_0 = \xi \arccos \xi \quad , \quad H_0 = G_0 - F_0$$

$$G_{2k} = \frac{\sqrt{1-\xi^2}}{2k+1} + \frac{2k}{2k+1} \, \xi^2 G_{2k-2} \tag{19}$$

$$F_{2k} = \xi^2 G_{2k-2} \quad , \quad H_{2k} = G_{2k} - \xi^2 G_{2k-2}$$

Let us assume that besides $\Delta(\xi)$ also the distribution of the angle of deflection $\varphi_z(\xi)$ of the light rays will be measured experimentally. Taking into account Eqs. (14), (15) and (16), we have

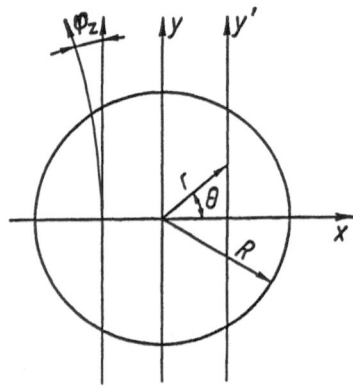

Fig. 3. Passing of light through a plane of symmetry of an axisymmetric body

$$\frac{n_0 \varphi_z(\xi)}{4C_1} = \sum_{k=1}^{m} (qa_{2k} + qb_{2k} + c_{2k})K_{2k} \tag{20}$$

where $q = C_2/C_1$, $K_{2k} = k\xi G_{2k-2}$.

Two sets of equations (17) and (20) are not sufficient to determine three sets of unknown coefficients in polynomials (15). However, we may use *a priori* information in the form of the compatibility equation

$$\frac{\partial}{\partial\rho} [\sigma_\theta - \mu(\sigma_z + \sigma_r)] - (1+\mu)\frac{\sigma_r - \sigma_\theta}{\rho} = 0 \tag{21}$$

where μ is the Poisson ratio. Introducing polynomials (15) into Eq.(21) we can eliminate one set of unknown coefficients:

$$c_{2k} = -\frac{1+(2k+1)\mu}{2k\mu} a_{2k} + \frac{2k+1+\mu}{2k\mu} b_{2k} \tag{22}$$

Now relationships (17) and (20) yield a system of equations from which the coefficients a_{2k} and b_{2k} can be determined. Thus a particular kind of tensor field can be determined in a tomographic way.

It is opportune to compare the described method with methods which are used to determine an axisymmetric refraction index scalar field (Maruyama, Iwata, Nagata 1976; Ugniewski 1980). In the latter case the distribution of the scalar refractive index is determined on the basis of *one* set of experimental data by the aid of the Abel inversion which is a special case of Radon inversion. No *a priori* information is used.

In our case we have to determine *three* scalar fields (15). For that we have to make on each light ray *two* experimental measurements. In addition, only the use of *a priori* information in the form of the compatibility equation (21) permits us to solve the problem.

EXPERIMENTAL EXAMPLE

The method described was used to determine stresses in a plate loaded as shown in Fig. 4. The plate was made of plexiglas. Experimental distributions of the relative optical retardation $\Delta(\xi)$ and of the angle of deflection $\varphi_z(\xi)$ are also shown in Fig. 4. Distribution of the stress σ_z for various values of m in polynomials (15) is shown in Fig. 5. Analytical solution (Sneddon 1951) is shown by the dotted curve.

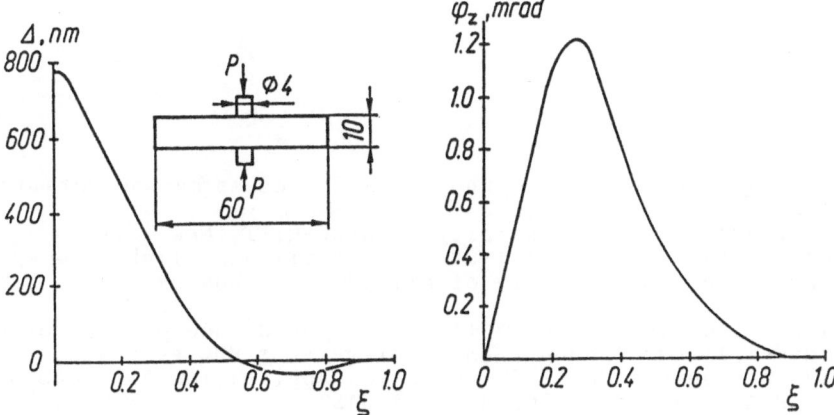

Fig. 4. Distribution of integral phase retardation $\Delta(\xi)$ and of angle of deflection $\varphi_z(\xi)$

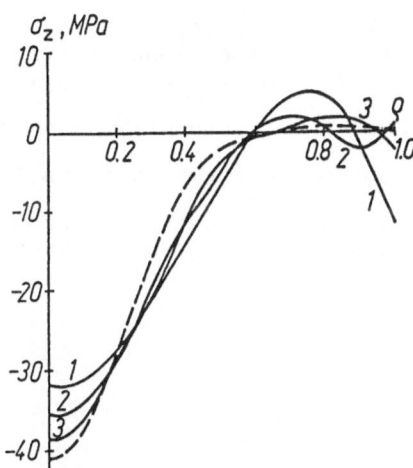

Fig. 5. Distribution of σ_z for different values of m in Eqs. (15): 1 - m=2 , 2 - m=3 , 3 - m=4

CONCLUSIONS

It has been shown that integrated photoelasticity can be considered a particular kind of tomography. Its main specific feature is that a tensor field is to be determined. In the general case in tensor field tomography line integrals of the field cannot be used. One has to measure the characteristic parameters instead of that. However, if the principal axes of the tensor are constant on light rays, one may use integral relationships (12) to (14). A new method of determining stress distribution in a plane of symmetry of an axisymmetric body has been developed. Practical application of the method has been described.

The author expresses his gratitude to K.J.Kell for carrying out the experiment.

REFERENCES

Aben HK (1966) Optical phenomena in photoelastic models by the rotation of principal axes. Exp Mech 6:13-22
Aben H (1979) Integrated photoelasticity. McGraw-Hill, New York
Aben HK, Krasnowski BR, Pindera JT (1984) Nonrectilinear light propagation in integrated photoelasticity of axisymmetric bodies. Trans CSME 8:195-200
Davin M (1969) Sur l'exploitation de l'information donnée par un "effet résultant" recueilli sur chaque sécante traversant une éprouvette et appartenant à un ensemble donné de sécantes. Application à la photo-élasticité. C R Acad Sc Paris 269:A 543-A 545
Herman GT (1980) Image reconstruction from projections. Academic Press, New York
Kubo H, Nagata R (1979) Determination of dielectric tensor fields in weakly inhomogeneous anisotropic media. J Opt Soc Am 69:604-610
Maruyama Y, Iwata K, Nagata R (1976) A method for measuring axially symmetrical refractive index distribution using eikonal approximation. Jap J Appl Phys 15:1921-1927
Sneddon IN (1951) Fourier transforms. McGraw-Hill, New York
Sweeney DW, Vest CM (1973) Reconstruction of three-dimensional refractive index fields from multidirectional interferometric data. Appl Opt 12:2649-2664
Ugniewski S (1980) Analysis of schlierengrams in refractometry of axisymmtric objects. Appl Opt 19:3421-3422
Вишняков ГН, Левин ГГ (1982) Оптическая томография фазовых объектов. Опт и спектроск 53:731-735
Зимин ВД, Шахурдин ВИ (1978) Соотношения для теневых и интерференционных методов исследования напряженно-деформированного состояния твердых тел. Приклад мех 14:№5,25-29

Stress Analysis for Axi-Symmetrical Problems by Scattered Light Photoelasticity

Makoto Kuramoto

The Institute of Vocational Training, Aihara, Sagamihara, Kanagawa, 229 Japan

PREFACE

Formerly in the photoelastic experiments, indistinct isoclinic lines had to be applied in order to resolve the stresses in axi-symmetrical problem. The problem of axi-symmetrical stresses can be analyzed by the scattered light photoelasticity. This method need to obtain a stress equilibrium equation and three scattered isochromatic fringe patterns by three different incidences of polarized light beam. For the purpose of checking the accuracy of this experimental method, the author performed the stress analysis of a round shaft with a semicircular ring groove under the room temperature tensile loading, and also the author developed the epoxy resin of a new photoelastic material. This material had almost free stress initiation and photoelastic stress sensitivity of this material was higher value. As a result, it was proved that this technique led to a higher accuracy of the experimental stress analysis.

THEORY

In the scattered light photoelasticty, the incidence of a polarzed light beam provides the relation, given below, between scattered isochromatic fringe order Ns and secondary principal stress difference ($\sigma_1' - \sigma_2'$) on the assumption that the principal stress axis does not rotate significantly during incidence.

$$Ns = \alpha \int_0^s (\sigma_1' - \sigma_2') \, dt \tag{1}$$

where α= Photoelastic stress sensitivity and S= Measuring point.
For stress analysis in axi-symmetrical problems in the cylindrical coordinates, two shearing stress components $\tau_{r\theta}$ and $\tau_{\theta z}$ are 0 at any point assuming that the z axis is a symmetry axis. Assuming the volumetric forces K_r, K_θ and K_z are negligible to simplify the problem, four stress components (σ_r, σ_θ, σ_z and τ_{rz}) may be obtained. The integral stress equilibrium equations are given below.

$$\sigma_r = (\sigma_r)_{r_0} - \int_{r_0}^r \frac{\partial \tau_{rz}}{\partial z} \, dr - \int_{r_0}^r \frac{\sigma_r - \sigma_\theta}{r} \, dr \tag{2}$$

$$\sigma_z = (\sigma_z)_{z_0} - \int_{z_0}^z \frac{\partial \tau_{rz}}{\partial r} \, dz - \int_{z_0}^z \frac{\tau_{rz}}{r} \, dz \tag{3}$$

If the polarized light beam is given at any angle against the symmetry axis (z axis) along the principal section(r z plane) (see Fig. 1), one of the secondary principal stresses along the principal section line in the plane is perpendicular to the incoming beam; and the principal section is principal stress σ_θ, whose direction is perpendicular to the principal section. As $\tau_{\theta z}$ and $\tau_{r\theta}$ are 0 on the principal section line, the other two principal stresses are in the principal plane and are expressed in three stress components(σ_r, σ_z and τ_{rz}). The stress (another secondary principal stress) coordinate transformation equations to the ξ η coordinate system, which is obtained by transforming the coordinates by an angle of ϕ from the r axis at measuring point 0' in the principal section, are obtained from the Mohr's stress circle as follows (see Fig. 2):

$$\begin{aligned}
\sigma_\xi &= \sigma_r \cos^2\phi + \sigma_z \sin^2\phi + \tau_{rz} \sin 2\phi \\
\sigma_\eta &= \sigma_r \sin^2\phi + \sigma_z \cos^2\phi - \tau_{rz} \sin 2\phi \\
\tau_{\xi\eta} &= -(1/2)(\sigma_r - \sigma_z)\sin 2\phi + \tau_{rz} \cos 2\phi
\end{aligned} \right\} \tag{4}$$

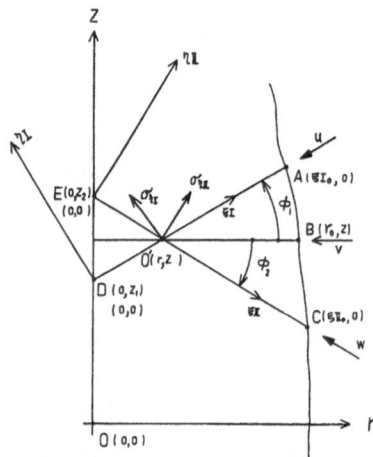

Fig. 1 Principal section of the model and direction of incidence of the polarized light beam.

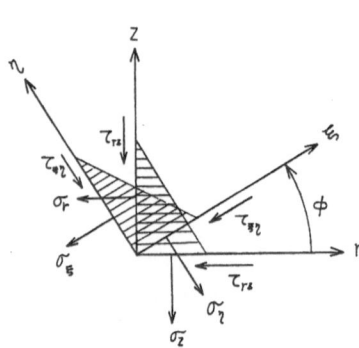

Fig. 2 Transformation of stresses by the coordinates.

Stresses $\sigma_{\eta I}$ and $\sigma_{\eta II}$ perpendicular to the ξI and ξII axes in Fig. 1 obtained by transforming angle ϕ into ϕ_1 and ϕ_2 in the above equations are as follows:

$$\sigma_{\eta I} = \sigma_r \sin^2\phi_1 + \sigma_z \cos^2\phi_1 - \tau_{rz} \sin 2\phi_1 \qquad (5)$$

$$\sigma_{\eta II} = \sigma_r \sin^2\phi_2 + \sigma_z \cos^2\phi_2 - \tau_{rz} \sin 2\phi_2 \qquad (6)$$

Assuming that points $A(\xi I_0$, $0)$, $B(r_0$, $z)$ and $C(\xi II_0$, $0)$ are on the free boundary, which is represented by three coordinate systems (ξI , ηI),(r ,z) and (ξII , ηII) whose origins are D, O and E give the polarized light beam in the directions u, v and w opposite to the ξI, r and ξII axes. The u, v and w values are 0 at points A, B and C. The following relations are obtained between all these coordinate systems:

$$u = \xi I_0 - \xi I \qquad\qquad v = r_0 - r \qquad\qquad w = \xi II_0 - \xi II \qquad (7)$$

Assuming fringe orders (obtained by incidence of the polarized light beam in the directions u, v and w) to be N_u, N_v and N_w for points O' or the measuring point, relational expressions between the fringe order and the secondary principal stress difference are obtained as follows by changing (1) to a differential equation and substituting (5) and (6):

$$\sigma_{\xi I} - \sigma_\theta = -\frac{1}{\alpha} \cdot \frac{dN_u}{du} \qquad (8) \qquad\qquad \sigma_z - \sigma_\theta = \frac{1}{\alpha} \cdot \frac{dN_v}{dv} \qquad (9) \qquad\qquad \sigma_{II} - \sigma_\theta = \frac{1}{\alpha} \cdot \frac{dN_w}{dw} \qquad (10)$$

If these basic measurements are obtained, values τ_{rz}, $\sigma_r - \sigma_z$ and $\sigma_r - \sigma_\theta$ may be obtained easily. Assume that incident angles of the polarized light beam ϕ_1 and ϕ_2 are equal (constant) in magnitude but opposite in direction, that is, $\phi_1 = -\phi_2 = \phi$. By τ_{rz} from (8) minus (10), eliminating σ_θ or σ_z from (8) plus (10), and using (9), $(\sigma_r - \sigma_z)$ and $(\sigma_r - \sigma_\theta)$ are obtained as follows:

$$\tau_{rz} = \frac{-1}{2\alpha \sin 2\phi} \left(\frac{dN_u}{du} - \frac{dN_w}{dw} \right) \qquad (11) \qquad\qquad \overline{} \quad \sigma_r - \sigma_\theta = \frac{1}{2\alpha \sin^2\phi}$$

$$\sigma_r - \sigma_z = \frac{1}{2\alpha \sin^2\phi} \left(\frac{dN_u}{du} + \frac{dN_w}{dw} - 2\frac{dN_v}{dv} \right) \qquad (12) \qquad\qquad \times \left(\frac{dN_u}{du} + \frac{dN_w}{dw} - 2\cos^2\phi \frac{dN_v}{dv} \right) \qquad (13)$$

These values may be obtained at any point, but normal stresses σ_r, σ_θ and σ_z are not obtained independently. If any one of these normal stress components is obtained, the other components may be obtained accordingly. For that purpose, stress equilibrium integrel equation (2) or (3) should be used. In the case of (2), $(\sigma_r)_{r0}$ is σ_r at the integration start point. If the σ_r value at the integration start point is not known, no values at any point may be obtained. If the integrstion start point is specified on the free boundary, stress component $(\sigma_n)_{r0}$ perpendicular to the boundary is 0, and $(\sigma_r)_{r0}$ at the start point is necessarily obtained from (14) as shown below (see Fig. 3).

$$(\sigma_n)_{r_0} = (\sigma_r)_{r_0} \cos^2\beta + (\sigma_z)_{r_0} \sin^2\beta + (\tau_{rz})_{r_0} \sin 2\beta = 0 \qquad (14)$$

where β is an angle between the r axis and the normal to the free boundary at point r_0. Using the value of $(\sigma_z)_{r0}$ obtained from equation (12), σ_r at point r_0 or $(\sigma_r)_{r0}$ may be obtained as follows:

$$(\sigma_r)_{r_0} = \frac{1}{2\alpha} \left\{ \frac{\sin^2\beta}{\sin^2\phi} \left(\frac{dN_u}{du} + \frac{dN_w}{dw} - 2\frac{dN_v}{dv} \right)_{r_0} + \frac{\sin 2\beta}{\sin 2\phi} \left(\frac{dN_u}{du} - \frac{dN_w}{dw} \right)_{r_0} \right\} \qquad (15)$$

Hence, three scattered isochromatic fringe pattern differential values and one stress equilibrium equation may give all stress components.

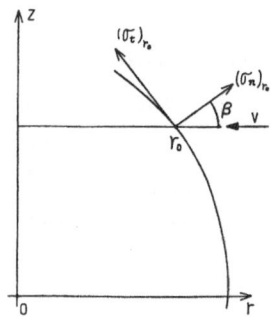

Fig. 3 Relation of stresses to coordinate at r_0 free boundary point.

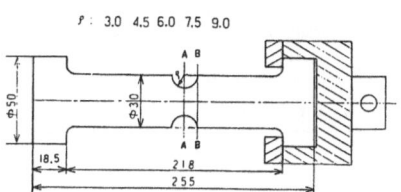

Fig. 4 Specimen geometry and the tensile jig.

PHOTOELASTIC MATERIAL

In the photoelastic experiments, epoxy resin composed of Ciba Araldite B and hardening agent H.T.901 reacted in the ratio 10 : 3 is used. A large block of this resin has the demerit of initial internal stress. Miyata[1~3] and others developed a new resin in which Araldite F is hardend with H.T.905 at room temperatures to avoid the demerit. The resin is free from spots which are special to the scattered light method and initial stresses.
The author applied a new method in which hardening agent H.T.901 is replaced with H.T.903. Araldite B was conditioned at approximately 90°C for more than 24 hours, and then filtered through Millipore microfiber glass prefilter at 120 through 130°C and 2 kgf/cm^2 of pressure difference to remove impurities of more than 1μm. In the same manner, hardening agent H.T.903 was filtered at 90 through 100°C and atmosphere pressure to remove impurities. The two were mixed in a ratio of 10 : 3 at approximately 90°C and stirred for approximately two hours. Silicone oil of more than 0.02% was added during stirring to improve the scattering power. After vacuum degassing at 140 through 145°C, the mixture was poured into a mold (60φ x 300 mm), and conditioned in a furnance at 90°C for 48 through 72 hours, and removed from the mold after hardening. The molded résin was conditioned at 130°C for 72 through 96 hours again to be hardened, and slowly cooled to room temperature at a cooling rate of 1.2°C/hr.

TEST SPECIMENS

The test specimens geometry is shown in Fig. 4. Five specimen types were subjected to the test. Each specimen is a round shaft 30 mm in diameter (2R) with a semicircular ring groove 3.0, 4.5, 6.0 and 9.0 mm in radius (ρ). Each bar was machined approximately 0.05 mm longer than specified, then filed with emery paper #400 through #1200, and polished to mirror finish with buffing cloth (red oxide), then annealed and finished with buffing cloth again. The bars were tensed at both ends using jigs as shown in Fig. 4 with no bending moment applied. Load P, nominal stress σ_n' for the minimum section, photoelastic stress sensitivity α (wave length: 6328 Å), and its error are given in Table 1. The photoelastic stress sensitivity was calculated from values obtained by the transmission light (from a 5461 Å mercury vapor lamp) method using a disk (50 mm). The table shows that $\rho=4.5$ and 9.0 mm specimens have similar stress sensitivity (their errors are within ±0.005 mm·fr/kgf). This is because both were molded simultaneously. The other three specimens were molded independently.

The photoelastic stress sensitivity of resin containing hardening agent H.T.903 ranges from 0.76 to 0.79 mm·fr/kgf, while that of H.T.901 ranges from 0.81 to 0.89 mm·fr/kgf (from the author's results)[4~6]. Vales for H.T.903 are lower than those for H.T.901 by 6 to 11%. The material used by Miyata and others showed $\alpha=0.40$ mm·fr/kgf (very low) of photoelastic stress sensitivity. This reveals that the material composed of Araldite B and hardening agent H.T.903 is useful in scattered light photoelasticity experiments.

Table 1 Load, photoelastic stress sencitivity and kind of the specimens.

ρ(mm)	3.0	4.5	6.0	7.5	9.0
P(kgf)	537	470	374	250	120
σ_n'(kgf/mm)	1.19	1.36	1.47	1.42	1.06
α(mm·fr/kgf)	0.787	0.761	0.776	0.781	0.759
Error of α	±0.002	±0.002	±0.002	±0.003	±0.005

EXPERIMENTAL APPARATUS

The experimental optical system is shown in Fig. 5.[7] Light source d is a TOSHIBA 60 mW He-Ne gas laser (wave length: 6328Å, single mode), which sends a laser beam into the photoelastic model a, moving up and down by the beam scanning apparatus c. The beam scanning apparatus is designed so that the laser beam from the light source through three total reflection prisms M_1, M_2 and M_3 may not change polarization plane direction after passing through prism M_3. Prisms M_1 and M_2 are mounted on separate blocks to allow the two to rotate on their individual optical paths. This causes slicing in the normal direction (z axis) to the model as well as in any oblique direction to the principal section (r z plane).

The plane polarized light beam passing through prism M_3 may be circularly polarized by 1/4 wave length plate Q_1 (first), then changed to a plane polarized light beam any direction by 1/4 wave length plate Q_2 (second)[8]. This causes four types of photographs (dark field, 1/4 forward pattern, light field, and 1/4 backward pattern[9~10].

The beam from the light source is 1.4 mm in diameter. To minimize the slice thickness for each model, the model may be set at the focus of a long lens (f=400mm). The minimum slice thickness is approximately 0.6 mm.

View camera (b) is a Toyo View 75M, which uses a 180 mm focal length lens. The view angle is approximately 30°, and the angle in liquid is approximately 18°. Cut films of 4×5 inch (Tri-x) and developer solution D-19 were used.

A vertical incidence (direction v) immersion fluid bath and an oblique incidence (directions u and w) immersion fluid bath were used individually. The oblique incident angle is 45°, and the oblique incidence immersion fluid bath is capable of taking fringe photographs in both the u and w directions. The immersion fluid solution is a mixture of carbon disulfide and benzene.

Fig. 6 is a photograph of the u direction incidence immersion fluid bath installed. The photo shows that the dial gauge indicates deflection of the loop gauge to give applied loads.

EXPERIMENTAL RESULT

Light field scattered isochromatic fringe patterns of models (groove diameter: $\rho=9.0$ mm) without load and under a 120 kgf load for vertical incidence (direction v) and oblique incidence (direction u) of the laser beam are given in Figs. 7 to 10.

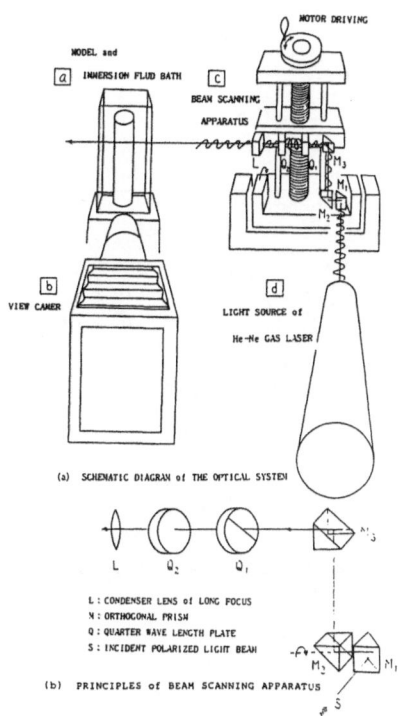

(a) SCHEMATIC DIAGRAM of THE OPTICAL SYSTEM

L : CONDENSER LENS of LONG FOCUS
N : ORTHOGONAL PRISM
Q : QUARTER WAVE LENGTH PLATE
S : INCIDENT POLARIZED LIGHT BEAM

(b) PRINCIPLES of BEAM SCANNING APPARATUS

Fig. 5 Schematic diagram of the scattered-light
photoelasticity.

Fig. 6 Photograph of loading apparatus and
system for the measurement of *V*-
direction incidence.

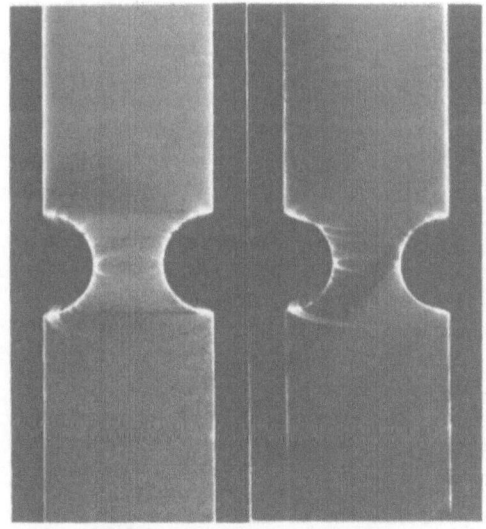

Fig. 7   Fig. 8

Fig. 7 Scattered isochromatic-fringe pattern
under no loading in vertical incidence
of the laser beam.

Fig. 8 Scattered isochromatic-fringe pattern
under no loading in oblique incidence
of the laser beam.

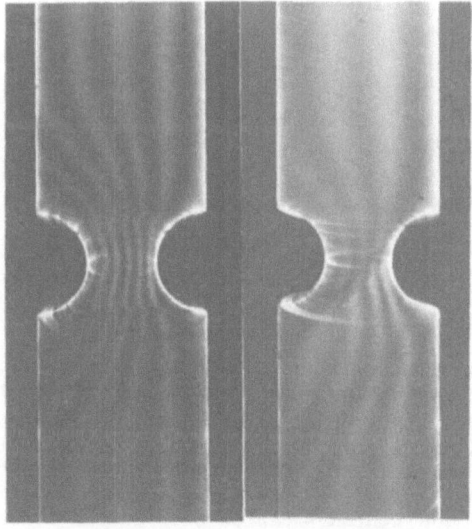

Fig. 9   Fig. 10

Fig. 9 Scattered isochromatic-fringe pattern
under 120 kgf loading in vertical incidence of the laser beam.

Fig. 10 Scattered isochromatic-fringe pattern
under 120 kgf loading in oblique incidence of the laser beam.

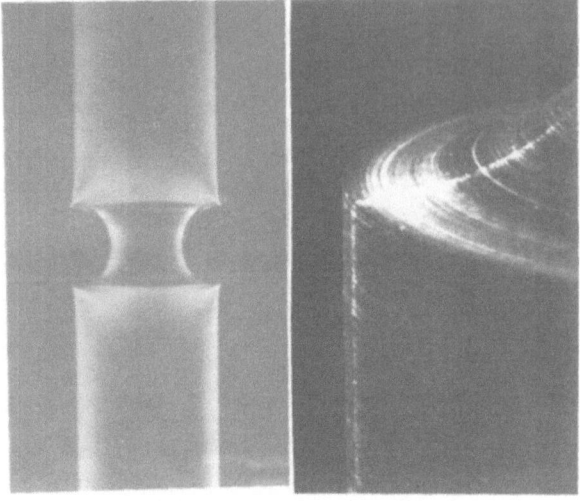

Fig. 11   Fig. 12

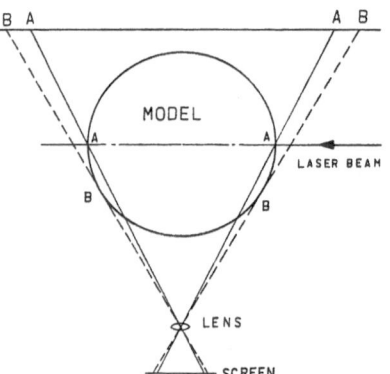

Fig. 11 Isochromatic-fringe pattern under no
loading by the transmission photoelasticity.

Fig. 12 Expansile photograph of a part of Fig.
12.

Fig. 13 Difference between real and apparent
boundaries.

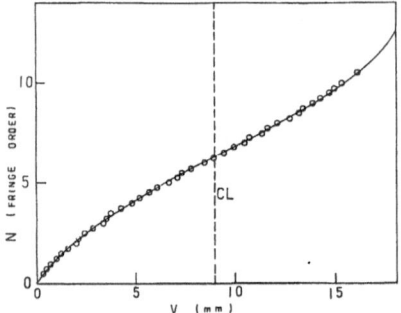

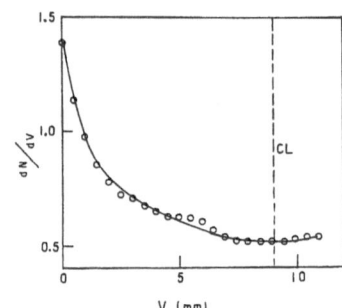

Fig. 14 Scattered isochromatic fringe order in
vertical incidence at the model along
the r-axis (A-A).

Fig. 15 Fringe spacing in vertical incidence at
the model along the r-axis (A-A).

Fig. 11 gives an isochromatic fringe pattern (taken in a transmission light dark
field) of the model, which was heat-treated after completion of the experiment
(machining) then immersed in the immersion fluid solution without load approximately
7 days later. These photographs show distance fringe patterns with little initial
stresses. Fig. 11 shows that this material is slightly subject to the time-edge
effect.
Slight contours can be seen at the laser beam position and outside of it on the
boundary of the model in Fig. 10. Fig. 12 is an enlarged photograph of a part of
Fig. 10. This is because point B (in Fig. 13) is taken in the photograph in
Fig. 10. Misreading of points A and B may cause significant errors in later
analyses. Great care was exercised in taking photographs. To prevent errors,
careful buffing is necessary. Rough buffing may cause defects such that the upper
and lower parts of photographs differ in brightness because their distances to
the film surface are different. Isochromatic fringe photographs taken in the way
mentioned above were enlarged five times the original for analysis.

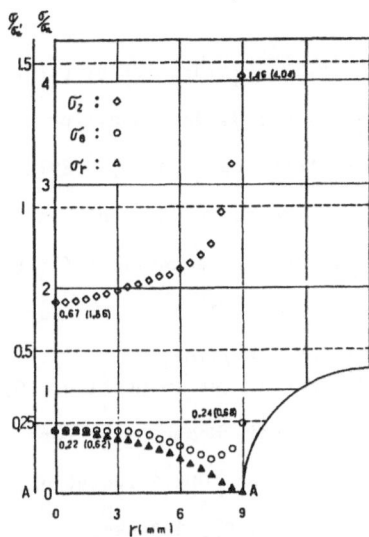

Fig. 16 Distribution of stresses at the model
along the r-axis (A-A).

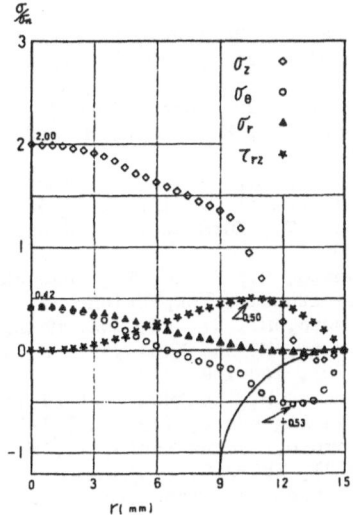

Fig. 17 Distribution of stresses at the model
along the B-B line.

ANALYTICAL RESULT

Only analyses of experimental values (obtained above) for groove radius ρ=6 mm
are described below.
Fig. 14 gives the relation between fringe order N and incident light path length
v(the polarized light beam has the smallest section A-A in the vertical direction
(v)). To create this figure, fringe patterns of orders 1/4, 1/2, 3/4 and 1 were
analyzed. Fig. 15 gives fringe spacing which is obtained by differentiating fringe
orders in Fig. 14 by incident light path length. The same processing was carried
out in the u direction. Test specimens used were symmetrical in the longitudinal
direction, so the processing in the w direction was omitted. The stress
distribution is given in Fig. 16, which was obtained by using those experimental
values in equations in Paragraph 2. The stress distribution in section B-B is
given in Fig. 17. In these figures, σ and σ' are nominal stresses on the largest
and smallest sections, respectively. Stress concentration factor K at the stress
concentrated point is 1.46, which meets Nishitani's theoretical value (1.45) quite
well[11]. The value of σ_θ at this position is somewhat higher than those at other
positions. Values at other positions possibly increase with σ_z because those
positions are stress concentrated points. In section B-B, compressive stress σ_θ
is highest at a position approximately 2.5 mm from the groove corner end. Values
σ_r and σ_z are negative around t here. The value τ_{rz} is highest near the position
of r=10.5 mm. The value σ_z at the point of r=0 in the smallest section (z=0) is
1.86, while σ_z=2.00 at z=9 mm. Values σ_θ and σ_r are 0.62 and 0.42 at the positions
z=0 and z=9 mm respectively.
The method may cause errors near incident points because differential values are
used. The inner value of σ_r calculated from (3) is not large. This means that
inner stress values are reliable. The groove corner end in section B-B is
unreliable. The differentiation in Fig. 13 was carried out directly on the paper.
A computer will analyze data and give more correct values. Analysis for other
types and numerical-analytical detailed procedures will be reported later.

In conclusion, I wish to tender my cordial thanks to Mr. T. Kunio, Professor of
Keio University, for his kind guidance in this experiment; the staff of Mizojiri
Optical Co., Ltd. who produced the beam scanning equipment; Mr. S. Ishino of
National Research Laboratory of Metrology for his kind cooperation in the design
and calibration of the loading apparatus; and Mr. T. Nabeshima, Mr. M. Okimoto,
and Mr. S. Ueno, University students, for their kind cooperation in the experiment.

REFERENCES

1. Miyata, T., Horisawa, H., Hashimoto, M. and Ogata, M., Trans. 1st symposium of Photoelasticity, (1972), 315–325.
2. Miyata, T., Horisawa, H., Hashimoto, M. and Ogata, M., J. Jap. Soc. for Nondestructive Inspection, 22, 6(1973), 344–349.
3. Kikuchi, S., Hashimoto, M. and Ogata, M., Nissan technical review, 9(1974), 39–50.
4. Kuramoto, M., Takada, T., Kuno, H. and Kunio, T., Trans. 2nd Symposium of Photoelasticity, (1974), 43–50.
5. Takada, T., Kuramoto, M., Kuno, H. and Kunio, T., Powder Technol., 14(1976), 51–60
6. Kuramoto, M., Kunio, T., Takeuchi, H. and Matsumoto, M., J. Jap. Soc. for Nondestructive Inspection, 23, 5(1974), 228–239.
7. Kuramoto, M., J. Jap. Soc. Mech. Eng., 77, 668(1974), 697–702.
8. Kuramoto, M. and Kunio, T., J. Jap. Soc. For Nondestructive Inspection, 19, 4(1970), 160–168.
9. Kuramoto, M., Lecture Text of Technical Course of Strain Measurement for Design Engineers, (1968), 151–164.
10. Kuramoto, M. and Kunio, T., Manual of Stress Measurment, Ohm Corporation, (1972), 189–249.
11. Nishitani, H., Trans. Jap. Soc. Mech. Eng., 26, 167(1960), 983–987.

Applications of Coherent Optics
to Experimental Mechanics

C.E. Taylor

Engineering Sciences, University of Florida, Gainesville, FL 32611, USA

When the invention of the first laser was announced, it was immediately clear that all branches of science would be drastically affected by this new kind of light. However, at the beginning it was not also obvious that all facets of modern life would be so greatly influenced. Few could have anticipated the enormous impact on everything from long distance communications and recording devices to credit cards and cash registers, and the list goes on. Since lasers produce light, it was only natural that they be considered as potential sources of light for all optical instruments and optical methods.

Before lasers, optical methods which require only conventional light sources had already been well established in experimental mechanics because they incorporate many intrinsically attractive features. For example: (1) Optical methods generally give full field information, thereby giving a visual description of the deformation of an entire member and do not require prior knowledge of the location of the critical areas, (2) They have extremely fast response times and no inertia effects, thus even the fastest moving physical phenomena may be observed, (3) Optical methods usually do not require physical contact, thus they do not alter the behavior that they are employed to study, (4) Many optical methods are incredibly sensitive, and can readily detect the change in optical path length of the order of a half wave length of light, a few millionths of an inch, and (5) They can be used on real existing structures under live loads; actual boundary conditions are automatically applied and it is not necessary to make assumptions based upon incomplete information.

In 1960 researchers in the field of dynamic photoelasticity needed light that was: (1) plane polarized, (2) monochromatic enough for good fringe visibility, (3) very intense so that extremely short exposure times may be employed, and (4) collimated so that the light may be directed efficiently through a polariscope. Ruby lasers which were invented at that time provided almost the specific answer to the photoelastician's prayers. (Green light would have been nicer, but that was to come later). Early applications of lasers to dynamic photoelasticity were reported by Fourney (1963), North (1965), and the others (1965). The intensity of the available light was so greatly improved that dynamic scattered light photoelastic studies were made by Hemann (1967). A technique which had required long time exposures before the advent of lasers could now be employed with exposure times less than 0.2 microseconds, and was feasible for studying stress waves in solids.

In addition to the above-mentioned characteristics of laser light was the property called coherence. This often brought about the creation of interesting fringe patterns due to interference from nearly parallel surfaces several inches apart, and also resulted in a grainy textured appearance for surfaces when observed or photographed. The fringes could be removed easily by misalignment, but the latter phenomenon, now called speckle, was more difficult to control and was eventually recognized to be not noise but unwanted information. This will be treated later.

It was this elusive property, namely coherence, that was destined to play an important role in the optical revolution. Although Gabor (1949) had published his pioneering work on holography in 1949, it was the introduction of lasers to holography by Leith and Upatnieks (1962) in 1962 that caused interest in the subject to mushroom. Whereas photography records the intensity distribution across the image of an object, holography records both intensity and phase information. The latter is just what is needed for interferometry. This inevitably led to the development of techniques for holographic interferometry by Horman and others. By double exposure it was possible to record two holograms on the same photosensitive plate, and the recorded information could be reconstructed simultaneously. When the two holograms depicted the same object under two slightly different loading conditions and interference fringe patterns are proportional to the change in optical path length (from laser to object to hologram) between the first and second exposures. In cases where the initial and final configuration differ by only a few wave lengths, interference fringe patterns are clearly visible on a diffusely reflecting object. These patterns are often aesthetically beautiful and contain a wealth of information. In theory if one knows the displacement at each point on the surface, they can compute the resulting strains. In practice this is a lengthy process and the accuracy obtained is poor when differentiating experimentally obtained data (to compute strains). In addition, the exact interpretation of the fringes depends upon the procedures and the geometry of the optical system by which they are created. The fringes are related to the displacements and the refractive index changes along the light path. So even if all the fringe orders were determined the problem is far from being solved.

The speckle pattern was found to contain much useful information and methods were developed to use double exposure speckle photographs (specklegrams) to measure in-plane displacements. If one considers a tiny bright spot or speckle to be a small aperture in a photographic transparency, then in a doubly exposed photograph it will appear as two apertures. If illuminated by coherent light then Young's fringes would be expected to be formed. For each speckle pair the Young's fringes would be very dim, but their orientation would be perpendicular to the displacement direction and their spacing would be inversely proportional to the magnitude of the displacement. Within each locality, say a millimeter in diameter, there would be hundreds of speckle cells but the displacement would be reasonably uniform so the resulting Young's fringes should reinforce each other, and consequently be visible if an unexpanded laser beam is shined on a specklegram. The Young's fringes formed by two unrelated speckles cells would be randomly oriented and spaced and would thus show up as background noise. The fringes obtained in speckle method are therefore characteristically poorly defined because of the low signal-to-noise ratio.

Much activity in the field of optical methods of stress analysis led to the development of combinations of the various methods. Papers on "holographic photoelasticity", "holo-moire", etc. soon appeared. The one common thread was that all the methods involved the interpretation of interference fringes, and more data processing was necessary to obtain the desired information from the directly obtained data.

With all of the favorable attributes, one may ask why optical methods are not more widely used in research and in practice. A partial answer is that most optical methods yield a wealth of information about the state of stress or strain, but not in a directly usable form. Even for a trained observer, the reduction of the available information to a useful form is often tedious or impractical.

This brings us to the inevitable conclusion that some help from computers is not only desirable, it is necessary if optical methods are to remain a viable approach to the solution of problems in stress analysis. The idea was not new, for Purse and Allison (1972) had developed a computer controlled system to analyze photoelastic models. Now however, technological advances have opened up new horizons. One can now envision an optical system intergated with its own computer which can take the information provided by nature, process that information, and compute the quantity that is required by the engineer or designer.

The whole field of optical stress analysis is now well into a new revolution that was made possible by an explosion in usage of computers. Digital image processing is at the center of revolution. It can now be applied to advantage to all the traditional branches of optical stress analysis; photoelasticity, moire, holographic interferometry, speckle interferometry, and many combinations of them. At the present time, there are at least three distinguishable approaches to the solution of these problems, namely: (1) analysis of fringes, (2) analysis of intensity levels, and (3) mapping of small sub-images of an object as it is deformed.

Perhaps the most obvious way to utilize optical digital image processing is to analyze fringe patterns, whether those fringes were generated by photoelasticity, moire, or any other interferometric process. The computer must first be "taught" how to recognize and analyze the part of a scene that is of interest. After reflecting a short time on the problems involved in teaching a computer how to specify the extent of a model and identify each fringe order, one finds that the task is surprisingly difficult. The human eye and brain are an incredibly complicated optical-computer system that can not be duplicated easily. Certain functions, such as recognizing objects and fringes we can perform instantly, but at processing data we are quite slow. On the other hand, computers are "dumb" at recognition, but are incredibly fast at performing other operations. One conclusion that may be reached is that an interactive system (i.e., one where human intervention in the process is required) may be the best that can be expected at present. T. Chen (1985) developed such a system which also incorporated the capability of various filtering techniques for fringe enhancement and could be applied to photoelasticity, speckle interferometry, and holographic interferometry. The computer can then reduce the data to a form that is directly usable. For example, the beautiful fringe patterns obtained by double exposure holographic interferometry are contours showing the change in optical path length between the first and second exposures. In most cases the rigid body motion is the dominant contributor in the formation of those patterns, and the deformation contributes only in a minor way. Unfortunately, it is the latter which is of primary interest to the stress analyst. The desired information is available, but much tedious work is required in order to separate it away from the unwanted information. The ultimate goal of the stress analyst is to provide in real time that information which is required by the engineer or designer. If one is interested in determining the maximum tensile stress wherever it occurs in a structure, a computer generated image displaying contours of the maximum principal stress at each point would be invaluable. Without interacting computers (and admittedly over-simplifying) one could say the photoelasticity yields directly only the difference in principal stresses at each point,

moire yields the displacements, and holographic interferometry yields
beautiful fringe patterns which contain a lot of information.

A completely different approach for using processing of images makes
use of the optical system's ability to recognize 256 different gray
levels or intensity levels. Burger (1984) has applied this idea to
photoelasticity and called it "half fringe photoelasticity" because
the load magnitudes are intentionally maintained low enough so that
the maximum fringe order to occur anywhere in the model is less than
one half. Under those conditions each intensity level may be uniquely
associated with a (partial) fringe order, and the analysis is
immensely simplified. The same approach has been applied to the
analysis of moire fringe patterns.

The third approach may be used for objects which exhibit some random
but distinguishable characteristic, such a laser speckle pattern or
naturally occurring grains. In these cases it is possible to by-pass
the fringe formation step. Small sub-images may be identified and
these may be mapped as the object is deformed. An algorithm has been
developed by Ranson (1984) and his colleagues which will give (a) local displacement, (b) rotation, and (c) deformation of each small
area. From these data the strain may be calculated. Applications have
included studies in micromechanics and biomechanics.

In conclusion which one may say that optical methods of stress analysis still play an important role in engineering,and will continue to
do so in the future. Whereas computers may now perform some of the
functions that were previously done by optics, the same computers have
opened up many new and exciting areas. Hardware has not been discussed it is in a state of rapid change where dramatic improvements
continue. In addition, almost every organization has some computers
already available with which the optical system can be made
compatible.

REFERENCES

Burger, CP (1984) Automated Moiré and Photoelastic Analysis with
Digital Image Processing. In: Proceedings of the 1984 SESA Fall
Conference, Nov 1984
Chen, TYF (1985) Application of Digital Image Processing and Computer
Graphics Techniques to Photomechanics. In: Ph.D. Thesis,
University of Florida, Gainesville, Florida
Fourney, ME (1963) On the Application of a Laser to High Speed Photo-
graphy. In: Thesis, California Institute of Technology, Pasadena,
California
Gabor, D (1949) Proceedings of the Royal Society, 197
Hemann, JH (1967) An Application of Scattered Light Photoelasticity to
Dynamic Stress Analysis. In: Ph.D. Thesis, University of Illinois.
Horman, MH (1965) An Application of Wavefront Reconstruction to
Interferometry. Appl Opt 4: 333-336
Leith, EN, Upatnieks, J (1962) Journal of the Optical Society of
America, 52
North, WPT (1965) A Laser Light Source in Dynamic Photoelasticity.
In: Ph.D. Thesis, University of Illinois
Nurse,P, Allison, IM (1972) Automatic Acquisition of Photoelastic
Data. Proc JBCSA, Conference on Recording and Interpretation of
Engineering Measurements, Inst Mar Eng, London, 203-207
Ranson, WF, (1984) Digital Image Processing. In: Proceedings of the
1984 SESA Fall Conference, Nov 1984
Taylor, CE, Bowman, CE, North, WP, and Swinson, WF (1965) Applications
of Lasers to Photoelasticity. Exp Mech 6: 289-296

Separation of Principal Stresses Using Boundary Element Method

Yasushi Mitsui and Shun-ya Yoshida

Department of Civil Engineering, Faculty of Engineering, Shinshu University, Wakasato, Nagano, 380 Japan

INTRODUCTION

The experimental method of photoelasticity has been widely applied to predict stress distribution in a model structure since the entire distribution can be observed visually[8]. However, many elaborate techniques and much experience are required to obtain experimental results with a high degree of accuracy. Furthermore, it is fairly troublesome and difficult to determine principal stresses throughout the model. Thus, the photoelastic method is not so commonly used, while numerical procedures have been broadly adopted for stress analysis.

In this paper, we propose an effective method to determine principal stresses using a hybrid of the photoelastic experiment and the numerical procedure of boundary elements which solves the Laplace equation[4-6].

CALCULATING THE SUM OF PRINCIPAL STRESSES

In the two-dimensional continuous body shown in Fig.1, it is well-known that the sum u of principal stresses p and q satisfies the following Laplace equation when the body forces are assumed to be constant or absent[3];

$$\nabla^2 u(x) = 0 \qquad \text{in } \Omega \qquad (x \in \Omega) \tag{1a}$$

with the Dirichlet boundary conditions i.e.,

$$u(y) = \bar{u}(y) \quad \text{on } \Gamma_1 \quad (y \in \Gamma_1) \tag{1b}$$

or Neumann boundaly condtions

$$g(y) = \partial \bar{u}(y) / \partial n = \bar{g}(y) \qquad \text{on } \Gamma_2 \qquad (y \in \Gamma_2) \tag{1c}$$

where n is the unit outward normal to the surface Γ, and $\bar{u}$ and $\bar{g}$ are specified values of the function and its normal derivative.

Suppose that $u*(x,y)$ is the weighting function of a weak formulaton of Eq.1. Let us choose $u*(x,y)$ from the fundamental solution of Eq.1. Then, we have the following equation [1]:

$$u(x) + \int_{\Gamma_2} u(y)g*(x,y)d\Gamma + \int_{\Gamma_1} \bar{u}(y)g*(x,y)d\Gamma$$
$$= \int_{\Gamma_2} \bar{g}(y)u*(x,y)d\Gamma + \int_{\Gamma_1} g(y)u*(x,y)d\Gamma \tag{2}$$

in which u(x) indicates the value of u at the point x located in the interior of the domain Ω, and g*=∂u*/∂n. For a two dimensional problem, we can set

$$u*(x,y) = (1/2\pi)\ln(1/r) \qquad (3)$$

in which r is the distance between the internal point and the boundary one, i.e., r=|x-y|.

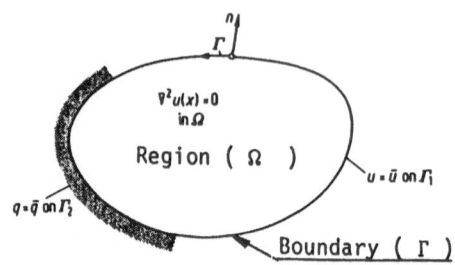

Fig.1. Notation for potential problems

Taking interior point x infinitely close to the boundary Γ in Eq. 2 yields the following boundary integral equation;

$$c(y)u(y) + \int_{\Gamma} u(y)\dot{g}*(x,y)d\Gamma = \int_{\Gamma} g(y)u*(x,y)d\Gamma \qquad (4)$$

$$\Gamma = \Gamma_1 + \Gamma_2$$

where constants c(y) are determined depending on the angle at the point y (c(y)=0.5 for a smooth boundary).

Now we discretize Eq.4 on the boundary. Let Γ be divided into 1 linear elements. As well as the finite element method, the shape function N is introduced interpolating the function u and g by the nodal values at the ends of each element. That is, u_j and g_j in the jth element can be expressed using the nodal values u_j^a, g_j^a as

$$\{ u_j \} = [N_j]\{ u_j^a \} \quad , \quad \{ g_j \} = [N_j]\{ g_j^a \} \qquad (5)$$

Substituting these into Eq.4, the following matrix equation can be derived;

$$[C_j][N_j]\{u_j^a\} = \sum_{j=1}^{1} (\int_{\Gamma} [N_j][u*]d\Gamma_j) \{g_j^a\} - \sum_{j=1}^{1} (\int_{\Gamma} [N_j][g*]d\Gamma_j) \{u_j^a\} \qquad (6)$$

When all the nodes are considered, Eq.6 produces a [1X1] system of equations which can be represented in matrix form as

$$[H] \{ u^a \} = [G] \{ g^a \} \qquad (7)$$

By solving this simultaneous equation, the nodal values u^a and g^a on the boundary are determined. Using Eq.2 , we can compute the sum of principal stresses u at any point in the interior region.

If numerous internal points should be treated or the boundary is rather broad, it may be convenient to separate the domain Ω into

several blocks introducing artificial boundaries in the interior of Ω.
Then we can apply Eq.7 to each blocked region considering continuous
conditions and calculate u values at any internal point in each block.
This procedure is called the blocked boundary element
technique[1,4,5,7].

EXAMPLE

To verify the validity and accuracy of the proposed method, an example
founded in Frocht [3] is treated here. In the following analysis, the
functions u and g assumed to have linear in the element,and the
integration are performed numerically using the four-point Gauss
quadrature rule.

Tension Problem of a Band Plate with U-Shaped Deep Grooves

Fig.2 shows the dark-field isochromatic pattern of a 30.81-mm wide,
4.37-mm thick plate model with deep grooves. The radius of the tip of
a groove is 2.48 mm and the depth of the groove is four times the
radius. The fringe value of the model is 3.445 KPa. The pure tension
P=949.5 N acts in vertically . The maximum stress at the tip of the
groove is observed as 12 fringes. From this photograph, principal
stresses and the sum of principal stresses are easily obtained on the
boundary line as shown in Fig. 3.

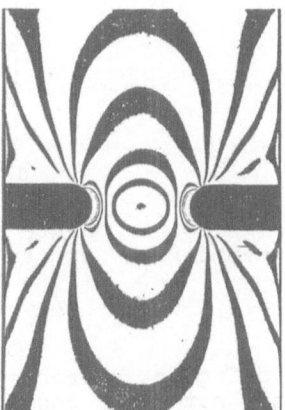

Fig.2. Dark-field isochromatic fringe pattern
in band plate with deep grooves

   Referring to these values, the boundary is divided into a series of
linear elements(Fig.4). Owing to the geometrical symmetry and the
applied force, we may treat only a quater of this model for stress
analysis, but here we treat the whole region in order to compare with
Frocht's data which is indicated by * in the figure. The boundary is
divided into 112 elements, and 18 internal points are specified. The
mesh division of the grooves is irregularly fine since experimental
data reported by Frocht is used directly in the numerical analysis. In
Fig.5 the curves of the sum of principal stresses u along the axis of
symmetry are plotted. It is noticed that the experimental results [3]
coincide with the proposed analysis fairly closely.

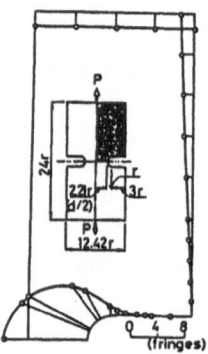

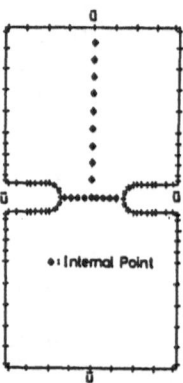

Fig.3. Principal stresses on the boundary for deep groove problem (Experimental results)

Fig.4. Mesh division of boundary elements for deep groove problem

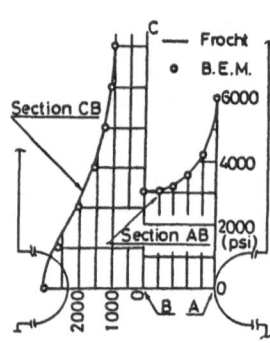

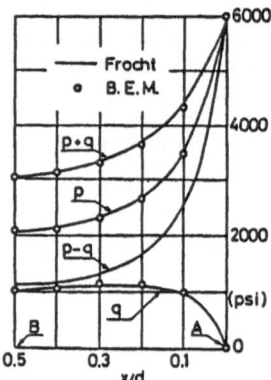

Fig.5 Curves of sum of principal stresses across section of symmetry (1 psi=6.89 kPa)

Fig.6 Separation of principal stresses p and q across A-B (p-q from experiment,p+q from calculation)

Since the difference of principal stresses (p-q) is experimentally observed in the isochromatic pattern, and the sum u is calculated by the boundary element method as stated before, principal stresses can be separated at any point in the interior of the domain. The separated stresses on section A-B are plotted in Fig.6 and compared with the data given by Frocht.

Comparison with the Results by FEM

Next, we compare these results by boundary element analysis with a finite element one by first-order triangular elements. The mesh and internal points for the boundary element analysis are shown on the left-hand side in Fig.7 where + marks indicate the interior points at which the sum of principal stresses u is to be calculated. Symbol $\bar{g}$ denotes the natural boundary condition due to the model symmetry. The net of finite elements is shown on the right-hand side of the same figure. The mesh division is determined according to the boundary element mesh and internal points of the same figure. It is noteworthy that the mesh pattern of boundary elements need not change if more interior points are added while the finite element mesh must be altered if more points are included.

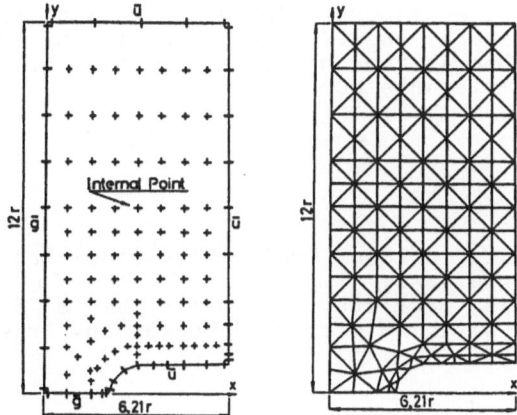

Fig.7 Mesh division in BEM and FEM for deep groove problem

Table 1 Computational conditions (HITAC M-240H computer)

Method (1)	Node (2)	Element (3)	Internal point (4)	CPU, in seconds (5)
(a) Finite element method	144	240	—	0.21
(b) Boundary element method	39	39	85	0.37
(a)/(b)	3.69	6.15	—	0.57

The values calculated by these two methods give almost the same computational results. Table 1 illustrates the computational conditions of both analyses. The nodes and elements in boundary element technique are remarkably reduced, so that less computer resources are needed. This is the greatest advantage of the boundary element method.

CONCLUSION

In this paper, a method of determining the distribution of principal sresses in a photoelastic model by hibridizing the photoelastic experiments and the numerical procedure by boundary elements is proposed. The results are briefly as follows:
1. The accurate experimental data and the boundary element analysis provide highly accurate stress distribution in the photoelastic model structure.
2. The boundary element technique remarkably reduces the computer resources in comparison with "domain type" methods , such as finite element or finite difference methods[2,4-6].
3. The mesh pattern of boundary elements is adequately determined observing the order of the fringe.

4. The boundary element method does not need to redefine the mesh pattern of elements if any number of interior points are added to calculate stresses. Thus, the boundary element scheme is more flexible than "domain type" methods.

5. Once isochromatic figures in experiments are obtained, the principal stresses are easily determined at any time if needed. Therefore, this method is convenient in time limited experiments which use time-sensitive materials such as polymer resin.

ACKNOWLEDGEMENTS

The authors would like to thank Prof. Nishida of Sci. Univ. of Tokyo for his helpful suggestions.

This research is partially supported by a Grant-in-Aid from Scientific Research of the Ministry of Education, Science, and Culture of Japan.

REFERENCES

1. Brebbia, C.A. and Waiker, S., Boundary Element Techniques in Engineering, Newnes-Butterworths, Boston, Mass., 1980.

2. Connor, J. J. and Brebbia, C. A., Finite Element Techniques for Fluid Flow, Butterworth & Co., London, England, 1976.

3. Frocht, M. M. , Photoelasticity, Vol.II, John Wiley & Sons, Inc., New York, N.Y., 1941.

4. Mitsui, Y. and Yoshida, S., Boundary Element Method Applied to Photoelastic Experiments, Jour. JSCE, Vol.66, No.5, 1981, pp.62-66 (in Japanese).

5. Mitsui, Y. and Yoshida, S., Separation of Principal Stresses Using Blocked Boundary Element Method, Proc. Japan Soc. Photoelasticity, Vol.3, No1, 1981, pp.15-23 (in Japanese).

6. Mitsui, Y. and Yoshida, S., Boundary Element Method Applied to photoelastic Analysis, Proc. ASCE, EM Div., Vol.109, No.2, 1983, pp.619-631.

7. Mitsui, Y.,Ichikawa, Y.,Obara, U. and Kawamoto, T., A Coupling Scheme for Boundary and Finite Elements Using A Joint Element, Int. Jour. Anal. Meth. Geo., Vol.9, 1985, pp.161-172.

8. Tsuji, J., Nishida, M. and Kawata, K., Experimental Method for Photoelasticity, Nikkan-Kohgyoh Newspaper LTD., 1965 (in Japanese).

Non-Destructive Three Dimensional Photoelasticity
Finite Strains Application

A. Lagarde

Laboratoire de Mécanique des Solides, Unité de Recherche Associée au CNRS,
40, avenue du Recteur Pineau, 86022 Poitiers Cedex, France

I. Introduction

Photoelasticity always remain widely used especially the technique of freezing and slicing for the investigation on a three-dimensional model.

The subject of this paper is to present two optical slicing methods using the scattering light phenomenon, one point wise, the other whole-field.

The main advantage of these methods lies in their ability to be carried out without mechanical slicing of the model. Furthermore, study of static elasticity problems become possible in finite strains.

From a fundamental point of vew, these methods have the interest of taking into acount fine representation the thin or thick photoelactic medium, in the general case where the secondary principal directions of the stresses rotating along the beam. These methods also have the interest of an accurate measurement of optical parameters. It is then possible to determine the stress tensor by integrating the equilibrium relationships.

First we expose how to represent the photoelastic medium and give the analysis conditions of a thin slice.

2. The current scheme

First recall a basic hypothesis that light is propagating in a photoelastic medium and assume the medium to be isotropic (indeed current photoelastic materials are slightly anisotropic). It follows that for a ray of light propagating along the $\vec{z}$ direction, the direction planes for the component waves (x, y) are orthogonal to $\vec{z}$.

For a photoelastic medium it can be shown that the principal and secondary principal directions of the index of light refraction tensor and the stress tensor coincide and that the change of index is related to stress in the following form (referred to a principal coordinate system)

$$n' - n_o = c_1 \, \sigma' + c_2 (\sigma'' + \sigma_z)$$

$$n'' - n_o = c_1 \, \sigma'' + c_2 (\sigma' + \sigma_z)$$

where n' and n'' denote the secondary principal secondary indices in the wave-plane (x, y) and σ', σ'' are corresponding secondary principal stresses ; c_1, c_2 are constants for a photoelastic material.

In three-dimensional photoelasticity it is usually assumed that the directions of secondary principal stresses and their values are constant through the thickness dz of a slice having its parallel faces normal to $\vec{z}$. This assumption allows to consider this slice as a birefringent plate characterized with the two following parameters

- secondary principal angle $\alpha = (x, \sigma')$ - angular birefringence $\phi = \dfrac{2\pi\delta}{\lambda}$,

$$\delta = dz(n' - n'') = C(\sigma' - \sigma'')dz \qquad C = c_1 - c_2 \quad \text{(C being a photoelastic constant).}$$

The two quantities namely α and $\sigma' - \sigma''$ are of interest to an engineer as σ_x σ_y and τ_{xy} may be determined from the well known relationships.

When   and   are measured for three series of three mutually orthogonal planes it is possible to integrate the equilibrium equations using the finite difference calculus.

3. Propagation of light waves through photoelastic medium

The problem of propagation of light waves was studied by many authors [1, 2, 3, 4]. Here, the essential concepts from Aben's works will be introduced.

3.1 Generalization of the concept of "isoclinic"

Aben in 1966 showed that when rotation of secondary principal axes was present, there were always two pairs of perpendicular conjugate "characteristic directions" (fig. 1). He distinguished the primary characteristic directions at the entrance of light (Δ'_e, Δ''_e), and the secondary characteristic directions (Δ'_s, Δ''_s) for the light emerging from the medium. The light linearly polarized at the entrance along one of the primary directions emerges as linearly polarized along the conjugate secondary direction. We will denote by R the angle determined by two such directions, $R = (\Delta'_e, \Delta'_s)$ and by α^* the angle (x, Δ'_e).

The characteristic directions are generally different from the secondary principal directions of the stress tensor (or those of index tensor). At the entrance we have (σ'_e, σ''_e) and at emergence we have (σ'_s, σ''_s) from the medium. We denote $\alpha_0 = (\sigma'_e, \sigma'_s)$.

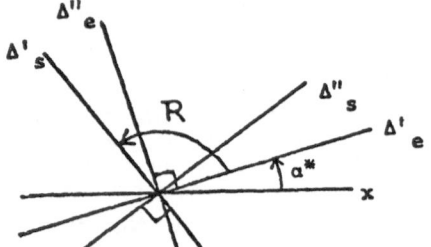

Figure 1 : Orientation of the "characteristic directions" at the entrance and at the emergence from the medium

3.2 Admissible hypothesis for a thin slice

For the case where $d\alpha/dz$ and $\sigma' - \sigma''$ are constant through a thickness, important conclusions follow (Aben) :

- The bisecting lines for the angles formed by two associated "characteristic directions" coincide with those for the angles formed by the associated secondary principal directions at the entrance and at the exit, (fig. 2).

Remark : The bisectors mentioned correspond to the secondary principal directions (mechanical or optical) at mid thickness ; so their directions are defined by the angles $\pm$ R/2 from the characteristic directions :

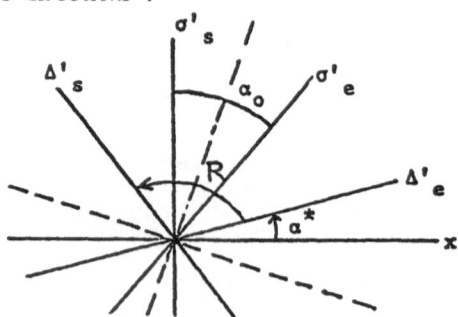

Figure 2 : Orientation of the characteristic Δ' directions and the σ' directions at the entrance and the emergence

- The phase difference ϕ^* characteristic to the medium traverses by the light wave along two characteristic orthogonal directions is generally different from the angular birefringence which would result in the absence of rotation R.

- The quantities R, ϕ^*, ϕ, and α_o obey the following relationships

$$\text{tg } R = \frac{\text{tg }\alpha_o - \dfrac{\alpha_o}{X}\text{ tg }X}{1 + \dfrac{\alpha_o}{X}\text{ tg }\alpha_o\text{ tg }X} \qquad \cos\phi^* = 1 - \frac{\phi^2}{2X^2}\sin^2 X \qquad \text{with } X = \frac{\sqrt{\phi^2 + 4\alpha_o^2}}{2}$$

Note that for R = 0, $\alpha = \alpha_o$ and $\phi = \phi^*$, one finds again the classical scheme of birefringent plate representing a slice dissected from a model.

Using the expressions above, nomograms have been drawn which give ϕ and α_o versus physical quantities ϕ^* and R. One can thus determine the difference of secondary principal stresses $\sigma' - \sigma''$ and secondary principal directions (through the angle α^*) if the angle α describing the orientation of the characteristic directions at the entrance is known.

It should be noted that a computer program which evaluates the two mechanical parameters $\sigma' - \sigma''$ and α as a function of three optical parameters α^*, R, ϕ^* requires approximatly 2 seconds on a programmable hand computer. The optical parameters under consideration can be accurately determined by the point wise method with linear detection (see sect. 5-1).

The nomogram illustrate the following particular situations :

- ϕ^* exceeds ϕ if $\phi > \pi$. This is consistent with the result of Drucker and Midlin [2] who indicated that "the rotation of secondary principal directions increased the number of isochromatic fringes".

- ϕ^* is less than ϕ if $\phi < \pi$ and in the neighbourhood of this value the error becomes important (for $\alpha_o = 30°$, $\phi^* = 140°$ for $\phi = 180°$).

- If the ratio α_o/ϕ is small, $\phi = \phi^*$ and there is coïncidence of the secondary principal directions and the characteristic ones.

Thus we conclude that the medium under consideration can be represented in general as a birefringent (having the axes (Δ'_e, Δ''_e) located by $\alpha^* = (x, \Delta'_e)$ and characterized by an angular birefringence ϕ^* followed by a rotatory power R.

3.3 Representation of the photoelastic medium

Recall the work of Aben, which allowed the determination of two parameters of interest from the three physical parameters and note that this result is based on the hypothesis that the rotation rate of principal secondary axes and the shear-stresses are constant through the thickness. This hypothesis is perfectly admissible for a thin slice.

To our knowledge our group was the first to represent a thin slice schematically by a birefringent plate followed by a rotatory power (or inversely with a birefringent plate whose orientation is shifted of the value of the rotatory power). Thus we were able to take Aben's results into account in the new methods that were developed (see sect. 5).

The hypothesis above is, however, very restrictive for representing the behaviour of a slightly anisotropic medium with a large thickness traversed by light.

In this case, one can resort to some discretization of the body into the series of thin slices, each one being represented by a birefringent plate followed by a rotatory power. In this context, the matrix formalism by Aben insurs that the entire body can be represented as the combination of a birefringent plate followed by a rotatory power. In this way he generalized Poincare's theorem established for series of birefringent plates. This theorem has been employed by Robert and Guillemet [5] and Robert and Royer [6] to give the same representation based on a set of discrete birefrigent plates.

4. Analysis of a thin slice

4.1 Point-wise analysis

The analysis of a slice removed from the stress-frozen photoelastic model by a point-wise plane photoelastic method where the slice is regarded as a birefringent plate can lead to considerable errors if rotation of secondary principal directions takes place. To detect rotation one can use an ellipsometry technique which employs circular incident light and permits one to determine ϕ. It is known now (see sect. 2) that errors made can reach 40° for the rotation $\alpha_0 = 30°$ when ϕ is close to π. The same thing happens for the orientation of secondary principal stresses which do not correspond with the obtained directions. Consequently, large errors may result for the stress components when integrating the equilibrium equations. This is a well known situation in plane photoelasticity.

4.2 Whole-field analysis with a plane polariscope [7, 8, 9]

Here, the analysis with a light-field polariscope is presented as it corresponds to the whole-field method of optical slicing. One can conduct an analogous study for a dark polariscope.

Let us examine a slice (which should be obtained by freezing and slicing) in a plane light-field (rectilinear) polariscope. The slice is represented by a birefringent plate and a rotatory power. Let I_0 designate uniform light-field illumination and x the polarizing axis of the polarizer. Then the light intensity is :

$$I = I_0 \left(\cos^2 R - \sin 2\,\alpha^* \sin^2 (\alpha^* + R) \sin^2 \frac{\phi^*}{2} \right)$$

The extremum values for intensity distribution correspond to :

$$\alpha^* = \frac{R}{2} + k\,\frac{\pi}{4} \quad ; \quad k = 0,\ 1,\ 2\ \dots$$

In order to specify the condition of analysis of fringe patterns we plotted the variations of I max and I min versus ϕ for different values of α_0 obtained following the relationships given in sect. 3.2. As an example, curves were plotted for $\alpha_0 = \frac{\pi}{9}$ in fig. 3.

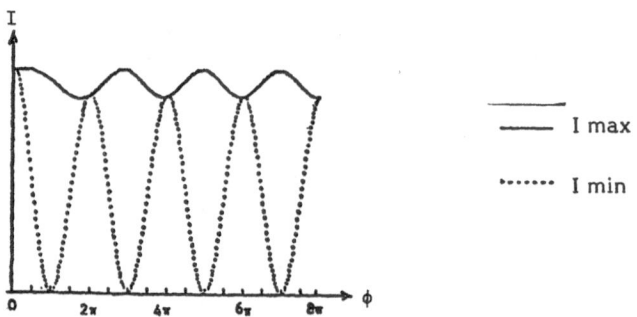

Figure 3 : Variation of I max and I min as a function of ϕ $(\alpha_0 < \frac{\pi}{9})$

The foregoing analysis indicates that for small values of α $(\alpha \leqslant \frac{\pi}{9})$ the maximum intensity I max shows a reduced modulation. Thus, it characterizes an isoclinic zone which permits one to locate the secondary principal stress directions (or those of the indices) in the median plane of a slice. This zone corresponds to $\alpha^* = \frac{R}{2}$ mod $\frac{\pi}{2}$ (see remark sect. 3). The orientation of the polarizer then coincides with one of the secondary principal directions in the median plane. This interesting result is analogous to the one established in 1957 by Hickson [10] for a dark-field polariscope.

It should be emphasized, that in order to avoid errors during the numerical integration procedure, the discretization points should lie on the median planes of the slices.

As α_o increases, the I max modulation increases and it becomes very pronounced for $\alpha_o = \frac{\pi}{3}$. In this case the isoclinic zone disappears although it should be noticed that the isoclinics are discernible up to the α_o value of $\frac{\pi}{6}$.

The term I min which is strongly modulated for α_o close to $\frac{\pi}{6}$, characterizes the isochromatic pattern. The extremum values occur for $\phi = k\pi$ ($k = 1, 2 \,..$) and it follows that localization of fringes is practically independent of the rotation of secondary principal axes.

We can now conclude by noting the following result : investigation of a slice within the plane (rectilinear) polariscope allows one to determinate the secondary principal stress directions in the median plane (without resorting to rotatory power measurements) and the angular beirefringence ϕ for the multiple π-values when the rotation of the secondary principal axes is less than $\frac{\pi}{6}$.

We should point out that the condition on the rotation of the secondary principal axes is not very limiting since one is able to choose the slice-thickness for the non-destructive optical slicing method.

5. Optical slicing methods

The point wise and whole-field methods which we are going to present are using the light scattering phenomenon as polarizer. The loaded photoelastic model is placed in an immersion tank with a liquid of the same refractive index.

5.1 Point wise method with linear detection [11, 12, 13]

Principle

This method has been described as an application of a linear detection ellipsometer. A more synthetic approach is proposed here. It must be pointed out also that it can use the scattering phenomenon as analyser. To set things out clearly, let us consider a solid photoelastic model placed in an immersion tank filled with index liquid and illuminated by a light beam propagating along the $\vec{U}$ direction (fig. 4). The observation of the scattered beam at point M is effected along the direction $\vec{u}$ orthogonal to $\vec{U}$. Besides, though the principal beam is still passing through point M, it can be orientable within a plane orthogonal to $\vec{u}$. Using the law of Rayleigh, we have at point M of the model an orientable polarizer P along the direction $\vec{u} \wedge \vec{U}$.

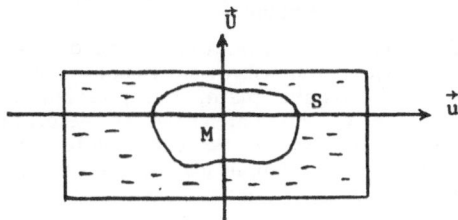

Figure 4 : Solid in an immersion tank

At point S, on the scattered beam issued from the model - the active beam - we have, at the exit of the tank, an ellipsometer. It includes an orientable composite plate and a rotating analyser A. The composite plate consists of an annular polarizer linked with a quarter wave plate Q so that the direction of polarization coïncides with the fast axis X_1 of the quarter wave plate. The reference beam goes through the annular polarizer and the rotating analyser.

The intensities of light vibrations transmitted by the active and reference beam are written respectively

$$E = \frac{E_0}{2} |1 + \cos(2\Omega t + \psi)| \qquad E_r = \frac{E_{or}}{2} |1 + \cos 2\Omega t|$$

with $(X_1, A) = \Omega t$

Along the trajectory MS we can represent schematically the thick medium by a birefringent plate B' followed by a rotary power R'. Thus along the beam MSu, we have the following representation (fig. 5).

The method takes into account particular values of the amplitude A of the sinusoïdal signal and of the phase difference ψ between the two signals.

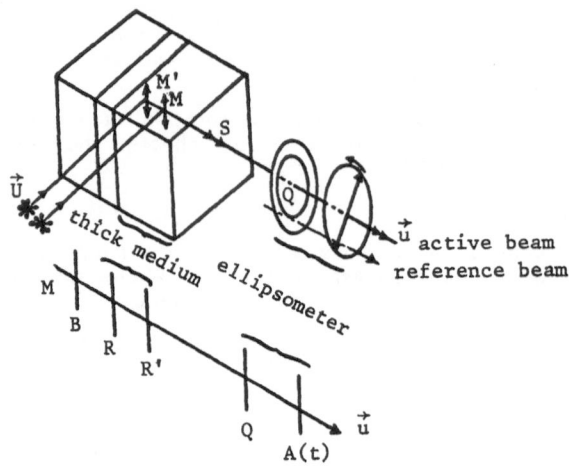

Figure 5 : Representation of elements along ray MSu

1st Step

The aim is to place a compensation at the exit of the immersion tank so as to cancel out the effects of the birefringent B' associated with the slice MS of the model.

The following results can be easily obtained

    - the polarization direction P is orientated in such a manner to get a difference of phase equal to 0 or π. Then the vibration issued from the model is rectilinearly polarized and if B' is not a half-wave plate, the polarization direction P entering B' coïncides with one of the axes of B' which are so determined. This characterization does not depend on the position of the quarter wave plate. The latter can be orientated so as to have always a value of the amplitude of the active signal sufficient for the evaluation of the phase.

    - the polarization direction P is maintained in the position defined above. The plate Q is orientated in such a manner to get a zero amplitude A. Two orthogonal positions of X_1 are thus obtained. The rectilinear vibration issuing from the thick medium MS is the bisectrix of the angle formed by these directions in a value range $\pi/2$ in which $\psi = 0$. The value of R' can then be deduced without any ambiguity :

    - the polarization direction P is now shifted of $\pi/4$. The quarter-wave plate Q being maintained in one of the positions defined above (characterized by $A = 0$), a compensator C is introduced at the exit of the model with its axes situated at $\pi/4$ of the axis of Q. The compensator is then adjusted until we obtain $\psi = 0$; then the system thick medium and compensator is equivalent to a known rotary power R'. It must be noted that this method has already been used by Gross-Petersen [14] . The advantage of our method is the use of precise criteria (value 0 or π of the phase and value 0 of sine amplitude) which do not depend on the light fluctuations encountered in scattered beam (variation of the sources and of the absorbtion in function of the path, dripting of photodetections).

In the particular case where B' is a half wave plate the phase ψ is 0 or π for any position of P. The orientation of the axis of B' with a shift equal R'/2, is determined by observing the evolution of A in function of the orientation of P and X_1 bound to remain with an angle egal $\pi/4$. A cancels out when P coïncides with the directions to be determined. It is then sufficient to place the axes of the compensator according to these directions and to adjust it to the half-wave value so that the thick medium and the compensator should be equivalent to a rotatory power R'. This value is obtained by measuring the angle formed by an incident rectilinear vibration P.

2nd Step

The $\vec{U}$ beam is shifted so as to illuminate a point M' close to M and situated along the direction of observation $\vec{u}$ (fig. 6). This optical slicing by shifting of the beam has already been used by Robert and Guillemet [5].

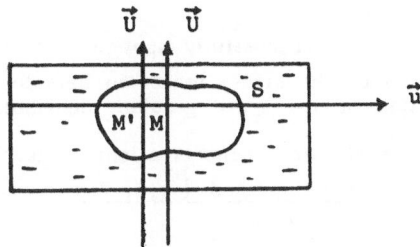

Figure 6 : Translation of principal beam $\vec{U}$ in the immersion tank

To the system "thick medium and compensator" a thin slice is thus added which we represent by a birefringent B of angular birefringence ϕ of fast axis b determined by $\alpha = (X_1, b)$ followed by a rotatory power R.

We then have on the beam scattered along the direction $\vec{u}$ the elements schematically represented in fig. 7.

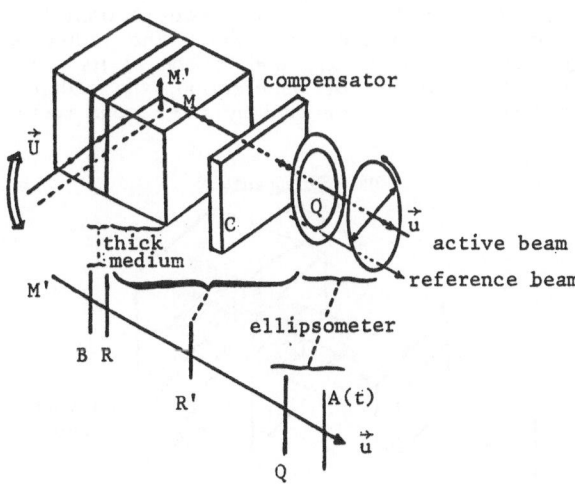

Figure 7 : Representation of elements along path MSu

As before we can orientate the direction of polarisation P until we obtain a phase difference equal to 0 and then orientate the quarter wave plate Q until we have a zero sine amplitude. We have already shown that the axes of B and the value of the rotatory power R+R' are then obtained (these values being additive) and R' being known, the value of R can also be deduced. The plate Q being maintained in a position for which $\mathscr{A}$ = 0, P is shifted of $\pi/4$, the phase difference measurement gives then $\overset{+}{-} \phi^*$ within a range of $2k\pi$. The polarisation direction P coïncides with the fast axis or with the slow axis of B according to whether the measured phase difference is positive or negative. The order of fringes k can be obtained in a simple manner by making the thickness of the thin slice tend towards zero.

The contemporary method of Robert and Royer [6] can be mentionned which used the scattering phenomenon as analyser and which gives access to the three physical parameters of the thick medium. It does not involve a compensator so the characteristics of the thin slice are deduced from those of the thick media which are bounding it.

Applications

The automatized set up has been successfully applied

- to the determination of the stress tensor in bars loaded in torsion [13]
- to the determination of the speed gradient distribution in a birefringent flow of milling yellow dye [15]
- to the measurements of stress concentrations in turbine blade [13].

5.2 Whole-field optical slicing method [7, 8, 16]

The Basis of the method

We isolate a slice of the photoelastic model by two plane sheets of light emitted from the same laser beam propagated in the direction $\vec{x}$ (fig. 8). The two-dimensional scattered light field is analysed in the direction of $\vec{z}$ orthogonal to the plane of the two illuminated sections ; at the level of the point where it is emitted (according to Rayleigh's law) the scattered light field is linearly polarized along $\vec{y}$ ($\vec{x} \wedge \vec{y} = \vec{z}$). The possibilities of interference of the radiations emitted by the two sections depend on the optical characteristics of the isolated slice. Lets imagine, for example, that the isolated slice is represented by a single birefringent plate. Then, when the birefringence is the multiple of a wavelength (in particular, zero) or when the principle direction is parallel to $\vec{y}$, the polarization of the light emitted by the first section is not modified by traversing the slice. The radiations from the two sections have the same polarization and can therefore interfere. On the other hand, the radiations can't interfere if the birefringence is an odd multiple of $\lambda/2$ and if one of its principal directions makes an angle of $\pi/4$ with the direction of polarization, the radiation emitted by the first section becoming orthogonal to that emitted by the second section.

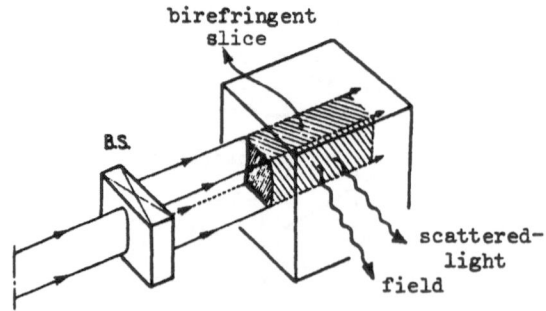

Figure 8 : Schematic of optical slicing

Correlation factor in speckle-fields

In practice the possibilities of interference of these two radiations scattered will be translated in the image plane field of the middle plane (x, y) of the slice formed with the aid of optical system of axis $\vec{z}$, by a special modulation of a speckle field.

We know, that the thin slice for the beam which traverses it along the direction of observation can be represented by a birefringent plate (ϕ^*, α) followed by a rotatory power R.

It can be shown that the correlation factor γ of the two speckle fields, in the image plane, is given by :

$$\gamma^2 = \cos^2 R - \sin^2 \alpha^* \sin (\alpha^* + R) \sin^2 \frac{\phi^*}{2}$$

It is remarquable that the correlation factor γ is a function only of the characteristics of the slice defined by the two illuminated sections independant of perturbations introduced by the medium between the slice and the model boundary. Let us add that the specific values $\gamma = 1$ and $\gamma = 0$ correspond respectively to the possibility of or lack of interference, in other words to the superpositions with respect to amplitude or to energy in the two diffused fields.

Consideration of the correlation factor

We can obtained the equality of the mean illuminations in the speckle fields for whatever birefringence encountered along the incident beam (following $\vec{x}$). In practice this is done with the aid of a separating device which is used to form the two parallel luminous sheets (fig. 9).

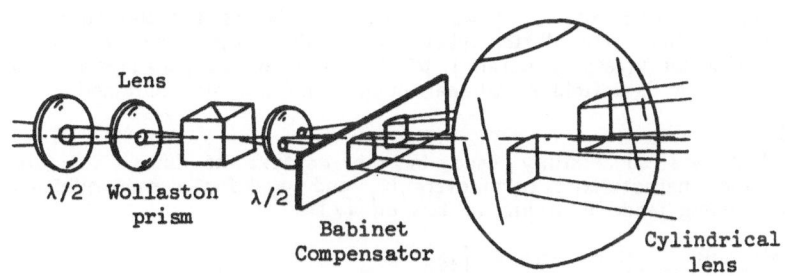

Figure 9 : Schematic of beam splitting setup for obtaining
two parallel plane beams

The apparatus consists of :

    - A half-wave plate allowing the rotation of linear polarization of the incident beam
    - A convergent lens
    - A Wollaston prism which provides two orthogonal polarized beams
    - A half-wave plate used to bring the directions of polarization of the two beams at $\pi/4$ with the vertical
    - A Babinet compensator placed horizontally and introducing a path of only one wavelength on the thickness of each beam
    - A cylindrical lens which allows the focusing of the two beams.

The intensities of the two beams are equalized by means of rotation of the half-wave plate.

It can then be shown that the contrast ρ in the speckle field takes into consideration the correlation factor

$$\rho^2 = \frac{1 + \gamma^2}{2}$$

Demonstration of contrast

A classical technique of optical filtering of the central order offers an average illuminations $<\mathcal{E}>$ proportionate to the square of contrast

$$<\mathcal{E}> = h \left[1 + \cos^2 R - \sin 2\,\alpha^* \sin 2(\alpha^* + R) \sin^2 \frac{\phi^*}{2} \right]$$

This expression is, to a constant term, identical to the illumination given in a light field rectilinear polariscope for the investigation of the slice. The fringe contrast equal to 1/3 is relatively weak.

This limitation has led us to improve the technique by using a polychromatic beam and to carry out the observation through a spectroscopic device [17]. A well known channeled spectrum appears when the interferences are possible. The spectral width of the polychromatic beam being limited, the resulting field is always a random structure.

The image of the slice thus includes in the areas where interferences are possible, speckle grains which present a sinusoïdal modulation and in the areas where interferences are impossible speckle grains stretching only in the direction of analysis of the spectroscope.

The investigation of the fringes associated to the correlation factor is carried out by a pass-band optical filtering of the negative. In theory the contrast of the filtered image is equal to the unit.

Interpretation of fringe patterns

Considering the results of § 4.2 we select a thickness for the slice so as to obtain a well-marked isocline. The latter provides us then the locus of the points where a secondary principal stress is parallel to the direction of polarization. The isochromes of parameter ϕ of a light-field rectilinear polariscope are also obtained.

Application

This method was successfully used in the context of linear fracture mechanics to determine the characteristic parameters K_1 and σ_{on} for a semi-elliptical surface crack loaded in opening made in a bar in tension [7].

6. Photoelasticity in finite strains [18]

The material used is polyurethane TM 60 A which is easily deformed under slight stress. It is obtained from a mixture of two components which makes it possible to include fine silica particles to improve the light diffusion properties.

6.1 Principle

The material is of the hyperelastic and incompressible type. The behaviour law is expressed as follows

$$\overline{\overline{T}} = - p\,\overline{\overline{I}} + G\,\overline{\overline{B}}$$

$\overline{\overline{T}}$ being the Cauchy stress tensor
$\overline{\overline{B}}$ being the Cauchy Green left strain tensor
p being a point-related hydrostatic pressure.

The incompressibility is expressed by the relation $b_1\, b_2\, b_3 = 1$.

It results that the principal directions and secondary principal directions $\overline{\overline{T}}$ and $\overline{\overline{B}}$ coïncide.

Experience shows that the loaded medium remains slightly anisotropic and satisfies a law of behaviour identical to that of classical photoelasticity

$$\overline{\overline{N}} = n_0 \overline{\overline{I}} + c_1 \overline{\overline{T}} + c_2 \left[(\text{trace } \overline{\overline{T}}) \overline{\overline{I}} - \overline{\overline{T}} \right] .$$

The relations stated in § 2 are thus valid. It results that the methods based on the use of the light diffusion phenomenon, in particular the technique of optical slicing can provide information on the orientation and the difference of secondary principal stresses. The integration process of the equilibrium equations is thus possible so as to determine the stress tensor. The law of mechanical behaviour and the condition of incompressibility allows us to obtain the hydrostatic pressure p and the strain tensor $\overline{\overline{B}}$.

6.2 Application

As an example we have studied a cylinder of circular section with a diameter of 35 mm and a length of 160 mm. The material being highly sensitive in birefringence, the condition mentioned in § 3.2 is satisfied and the required conditions are met to apply the Weller method. It is then possible to draw rapidly the comparison between the observed fringes and the theoretical ones.

The incident beam goes through the model along a diameter and the observation is carried out at right angle on the diameter plane. In the film development the refraction phenomenon is of course taken into account. Fig. 10 shows, for a torsion angle of 155°, a good agreement between the measured and calculated values.

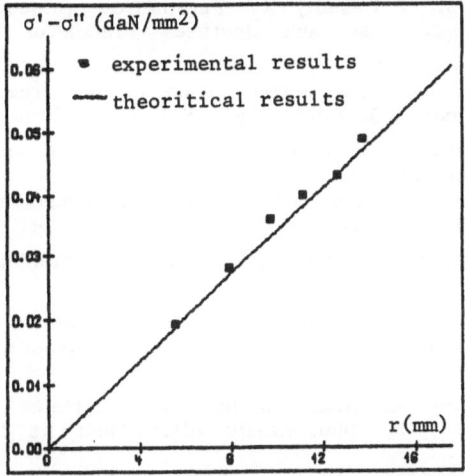

Figure 10 : Value of shear $\sigma' - \sigma''$ versus the radius

This example is sufficient to show the new possibilities offered to photoelasticity in finite strains. We conclude that the methods of optical slicing seem particularly well adapted to study complex problems.

References

[1] Drucker D. et Mindlin R. "Stress analysis by three-dimensional photoelastic methods". J. Appl. Phys., Vol. 11, 1940.

[2] Mindlin R. et Goodman L. "The optical equations of three-dimensional photoelasticity". J. Appl. Phys., 20, 1949.

[3] Lee L. "Effects of rotation of principal stresses on photoelastic retardation". Exper. Mechanics, Vol. 4, n° 10, 1964.

[4] Aben H. "Optical phenomena in photoelastic models by the rotation of principal axes". Exper. Mechanics, Vol. 6, n° 1, 1966.

[5] Robert A. et Guillemet E. "Nouvelle méthode d'utilisation de la lumière diffusée en photoélasticimétrie à trois dimensions". R.F.M., n° 5/6, p. 147-157, 1963.

[6] Robert A. et Royer J. "Principe de mesure des biréfringences à l'intérieur d'un solide transparent en vue de son application à la photoélasticimétrie tridimensionnelle". C.R. Ac. Sc. série B, t. 281, 1975, p. 373-376.

[7] Desailly R. "Méthode non-destructive de découpage optique en photoélasticimétrie tridimensionnelle - Application à la mécanique de la rupture". Thèse d'Etat n° 336, Poitiers, 1982.

[8] Desailly R. et Lagarde A. "Méthode de découpage optique de photoélasticité tridimensionnelle". Journées d'Extensométrie, Poitiers, septembre 1981, R.F.M. n° 2, 1984.

[9] Lagarde A., Brillaud J. et Desailly R. "Paramètres optiques en photoélasticimétrie tridimensionnelle". Conférences aux Journées Françaises d'Extensométrie, 1981, R.F.M. n° 4, 1983.

[10] Hickson V.M. "Errors in stress determination at the free boundaries of "Frozen Stress" photoelastic model". J. Appl. Phys., Vol. 3, n° 6, p. 176-181, 1952.

[11] Brillaud J. et Lagarde A. "Ellipsometry in scattered light and its application to the determination of optical characteristics of a thin slice in tridimensional photoelasticity". Symposium I.U.T.A.M. 'Optical Methods in Mechanics of Solids', Poitiers, septembre 1979, Ed. A. Lagarde (Sijthoff Noordoff).

[12] Brillaud J. et Lagarde A. "Méthode ponctuelle de photoélasticimétrie tridimensionnelle". R.F.M. n° 84, 1982.

[13] Brillaud J. "Mesures des paramètres caractéristiques en milieu photoélastique tridimensionnel. Réalisation d'un photoélasticimètre automatique. Applications". Thèse de Doctorat d'Etat, Poitiers, 1984.

[14] Gross-Petersen "A compensation method in scattered light photoelasticity". I.U.T.A.M. Symposium "The photoelastic effect and its applications", septembre 1975, Springer.

[15] Monnet P. "Contribution à l'étude des lois de comportement rhéo-optique de certains fluides biréfringents. Application à un écoulement non viscométrique de solutions d'alphonogele". Thèse de 3e cycle, Poitiers, 1980.

[16] Desailly R. et Lagarde A. "Sur une méthode de photoélasticimétrie tridimensionnelle à champ complet". Journal de Mécanique Appliquée". Vol. 4, n° 1, 1980.

[17] Desailly R. et Froehly C. "Whole field method in three dimensional photoelasticity : improvement in contrast fringes". Symposium I.U.T.A.M. 'Optical Methods in Mechanics of Solids", Poitiers, septembre 1979. Ed. A. Lagarde (Sijthoff Noordhoff).

[18] Brémand F. "Thèse de Doctorat de l'Université de Poitiers, Travail en cours.

An Automatic Micropolariscope Used to Study a Cracked Thread

T.P. Broadbent and H. Fessler

Department of Mechanical Engineering, University of Nottingham, University Park, Nottingham, NG7 2RD, UK

SUMMARY

A natural, three-dimensional, slightly curved crack was found in the thread of a frozen-stress model of a screwed tubular connection. The uncracked stress field in a similarly loaded cross-section was analysed using Frocht's shear difference method. From slices containing the crack, the load transmitted locally, the extent of the region where linear elastic fracture mechanics should apply and non-dimensional stress intensity factors k_{II}, k_I, and $f(r_c{}^o)$ were determined. Large variations in k_{II}, k_I and $f(r_c{}^o)$ were found in this non-uniform stress field.

INTRODUCTION

A micropolariscope has been built (1) for the examination of slices from frozen-stress models by modification of an optical transmission microscope (2) in which minimum intensity of light is measured by a photomultiplier. The positioning of the point under observation and the readings of isoclinic and fractional fringe order have been automated (1) to facilitate Stress separation using Frocht's shear difference method.

This instrument has been used to study the stress distribution near a natural crack in the internal thread of a large tube and to compare it with a corresponding uncracked part of the thread. The results show the difficulties which arise when applying to this three-dimensional, non-uniform stress field the linear fracture mechanics equations (3), (4),(5), which have been derived for uniform stress fields.

NOMENCLATURE

a, r_c, θ_c defined in Fig. 3
e, θ_t defined in Fig. 2
f material fringe value
$f(r_c{}^o)$ far field stress
$k_{I,II} = K_{I,II}/\tau_m \sqrt{\pi a}$
n fringe order
s slice thickness
τ_m local mean shear across uncracked thread $= \dfrac{1}{7.9} \displaystyle\int_0^{7.9} \tau_{rz}\, dz$

THE MICROPOLARISCOPE

In the polarising microscope shown in Fig. 1 the final intensity of the light emitted by the 100W tungsten halogen lamp is measured by a photo-multiplier which has several apertures between 3.8 and 0.11mm diameter. Because the greatest magnification of the objectives is 20 times, the smallest area of model over which the light intensity is averaged is 0.11/20 = 0.0055mm diameter; with an eyepiece magnification of 10, this is seen as a 1.1mm diameter circle. However, due to surface scratches and inhomogeneity of the model material, the smallest area of model which gives consistent results is 0.04mm diameter.

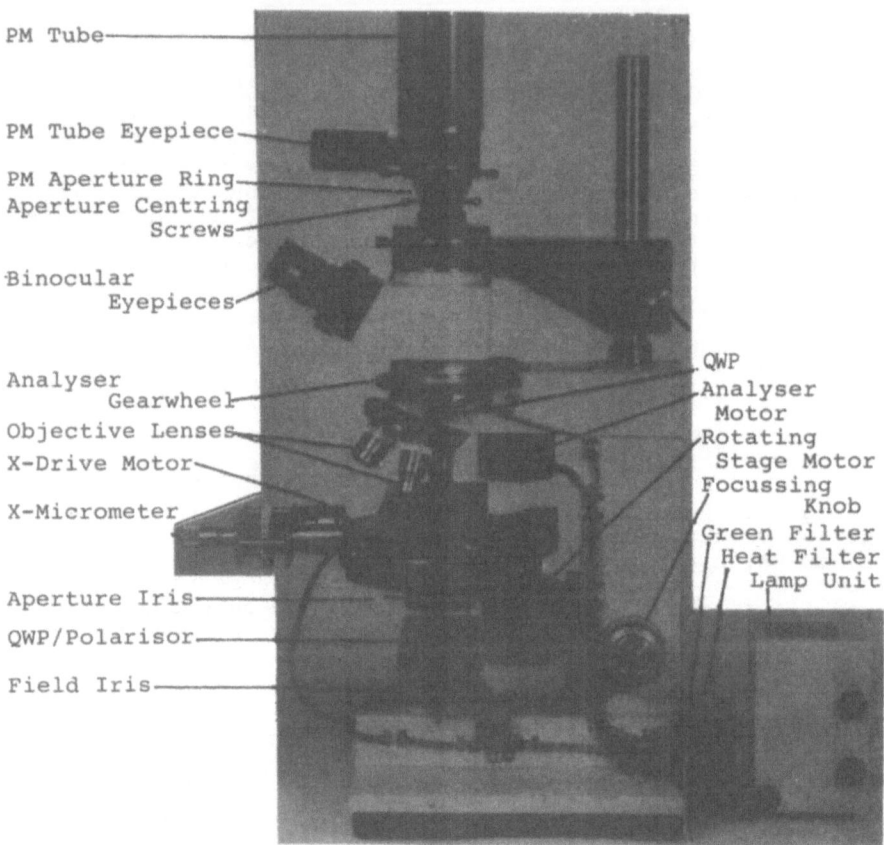

PM Tube

PM Tube Eyepiece

PM Aperture Ring
Aperture Centring
Screws

Binocular
Eyepieces

Analyser
Gearwheel
Objective Lenses
X-Drive Motor

X-Micrometer

Aperture Iris
QWP/Polarisor
Field Iris

QWP
Analyser
Motor
Rotating
Stage Motor
Focussing
Knob
Green Filter
Heat Filter
Lamp Unit

Fig. 1   Micropolariscope

Because the polariser could not be rotated, the slice had to be rotated on the stage to measure isoclinic angles. An x-y stage on the rotating stage was used to position the point under observation on the axis of the microscope. The quarter wave plates are moved out of the field for measurement of isoclinic angles. The analyser has to be rotated to use Tardy's method for fractional fringe order measurement. All these movements were motorised, using stepper motors for the two rotations and the x and y movements. The six movements are operated by a micro-processor which is controlled by a microcomputer.

The angular positions of the stage (for isoclinics) and the analyser (for fractional fringe orders), which give minimum intensity of light

transmitted through the slice, are determined by fitting 3rd order polynomials to the measurements at about 10 widely-spaced positions and
calculating the position of the minimum value of the polynomial. To
achieve accuracies of $\pm 0.05^\circ$ and ± 0.001 fringes, this process is
repeated for positions near the previous minimum. The automatic reading of one point takes about 1.2 minutes.

Integral fringe orders are measured separately and 'edited' into the
automatically obtained data.

After the positions of the corners of a grid of up to 50 x 3 points have
been entered into a Basic program, this program controls the movements
and records the co-ordinates, isoclinic and fractional fringe orders at
all grid points on a floppy disc. There are also facilities for manual
operation and direct reading the photomultiplier signal on a DVM.

TECHNIQUES

Frocht's shear difference method (6) uses the equilibrium equation in
cartesian co-ordinates

$$\frac{\partial \sigma_x}{\partial x} + \frac{\partial \tau_{xy}}{\partial y} + \frac{\partial \tau_{zx}}{\partial z} = 0 \tag{1}$$

to determine, along a straight line in the x direction, the changes in
σ_x from the known values σ_{xo} at a free boundary. In finite difference
form

$$\frac{\partial \tau_{zx}}{\partial z} = \frac{(\tau_{zx})_2 - (\tau_{zx})_4}{2 \Delta z} \tag{2}$$

where Δz is the spacing between adjacent grid lines, defined in Fig. 6a.
The optimum spacing is the minimum necessary to obtain significant
differences $(\tau_{zx})_2 - (\tau_{zx})_4$; $\Delta z = \Delta y = 0.3$mm has been used for the work
shown in Fig. 6. Slices (containing the x-y plane) have to be 1mm thick
because 0.2mm is needed between the edge of the sub-slice (containing
the z-x plane) and a grid line to account for surface irregularities,
malalignment of the grid lines and sufficient time for measurements to
be taken before time-edge effect (due to absorption of moisture from
the atmosphere) affects the readings.

Tesar's modification (7) was used to determine the shear stress gradients. With material fringe value f, fringe order n and slice thickness
s

$$\frac{\partial \tau_{zx}}{\partial z} = \frac{f}{2} \frac{\partial}{\partial z} \left(\frac{n}{s} \sin 2\alpha \right) = \frac{f}{2s} \left(2n \cos 2\alpha \frac{\partial \alpha}{\partial z} + \frac{\partial n}{\partial z} \sin 2\alpha \right) \tag{3}$$

where α is the isoclinic angle measured from the x direction to the
greater principal stress. In finite difference form, for the lines
defined in Fig. 6a,

$$\frac{\partial \tau_{zx}}{\partial z} = \frac{f}{4s \Delta z} [2 n_o(\alpha_2 - \alpha_4) \cos 2\alpha_o + (n_2 - n_4) \sin 2\alpha_o] \tag{4}$$

Having obtained σ_x by the above method, σ_y, σ_z, τ_{xy} and τ_{zx} are calculated from the readings along the 0 line, as described in Ref. 1.

MODEL AND TESTING

Fig. 2 shows the model of a screwed connection in which a crack was found after loading under eccentric tension. The screwed connection is intended for tension legs (approx. 1km long, 1.3m diameter, 65mm wall thickness) of offshore platforms.

The 'tube', coupling, dummy screwed joint and extension tubes were machined from castings of Araldite CT200 with hardener HT907 and cemented to form the two parts of the screwed connection.

Because the weight of the model does not cause any bending of the screwed connection if the axis is vertical, no measures are necessary to counteract the effect of self-weight of the model; it only reduces the tension applied to the screwed joint (see Fig. 2).

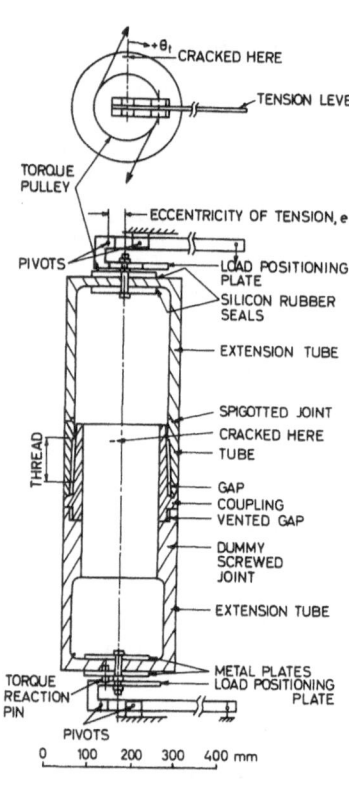

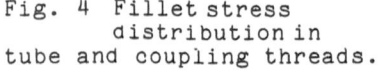

Fig. 2   Model of screwed connection with loading rig, showing angular co-ordinate θ_t.

Fig. 4   Fillet stress distribution in tube and coupling threads.

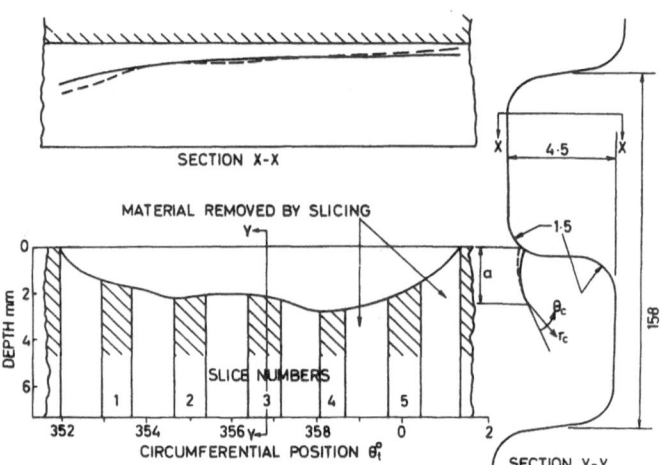

Fig. 3 Shape of crack and positions of slices. Crack front shown by full line, position of crack at surface shown dotted.

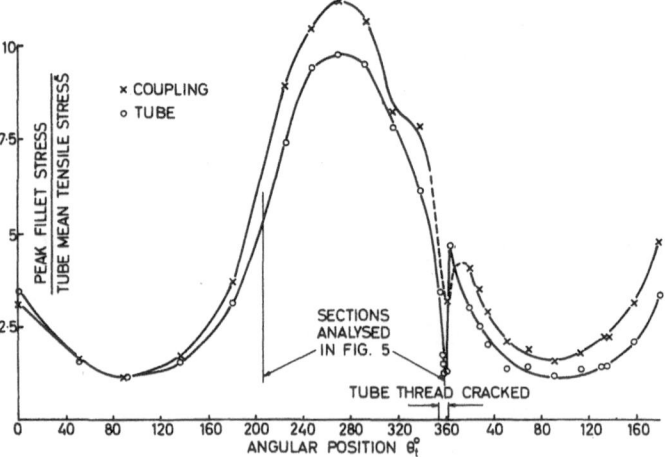

The thread form is defined in Fig. 3. It is a buttress thread, similar to that specified by the American Petroleum Institute for taper threads of tubes except that all fillet and thread tip radii have been increased and that there are radial clearances between the cylindrical surfaces of thread roots and tips as well as along the unloaded flank.

After the thread region had been thoroughly cleaned, degreased and lubricated with an inert silicon oil, the parts of the model were screwed together 'hand tight' and arranged in the rig, also shown diagramatically in Fig. 2. The assembly was aligned vertically and heated to 130°C in a hot air oven. A pure tightening torque was applied by two equal tensions for 1 hour to ensure full deformation had taken place. The torque loads were removed and an eccentric tension applied by freely hanging weights from the tension lever. The eccentricity, e was adjusted by moving the model along slots in the load-positioning plates (see Fig. 2). With the tension applied, the model was cooled at 2°C/hour to below 90°C.

The model was then cut from the extension tubes and tube and coupling were glued together at the ends to retain their relative loaded positions. Slices were cut radially along the whole length of both models at 22.5° intervals. After the crack had been discovered in the $\theta_t = 0^\circ$ slice of the tube, other radial slices were cut as close together as possible and numbered 1 to 4. The crack and the positions of the slices are shown in Fig. 3.

The mean tensile stress in the tube was determined from numerous measurements in its unthreaded part and used as the nominal stress. Stresses are presented as multiples of this nominal stress. From a calibration strip this nominal stress was obtained as 0.104N/mm^2.

THE CRACK

The axial position of the crack is shown in Fig. 2 which also defines the angular co-ordinate θ_t. The exact position, true shape and size are shown in Fig. 3 by three orthogonal views. The crack depth 'a' and the radial position of the crack at the thread surface and at its tip were measured on the faces of the slices.

The cause of cracking could not be determined. It did not originate at the point of highest fillet stress nor at the point of greatest contact stress. No flaw or inclusion could be detected; a microscopic crack could have been introduced during machining. It is assumed that crack growth stopped when the load carried by this part of the thread was reduced sufficiently by the increased flexibility due to the crack.

The crack opening at the thread surface was approximately proportional to the depth of the crack, i.e. crack opening $\simeq 0.047a$.

UNCRACKED STRESS DISTRIBUTION

Using a manual, diffused light polariscope, fillet stresses were measured along the whole tube and coupling threads at $\theta_t < 45^\circ$ intervals; the maximum values in each of these θ_t = constant planes are shown in Fig. 4 for the part near the crack. The stress distribution is sinusoidal due to the eccentricity of the tension. The reduction of fillet stress due to the crack is large, but does not extend further than the crack.

The thread surface stresses in the cracked slice No. 4 (defined in Fig. 3) are shown in Fig. 5 and compared with those in the uncracked $\theta_t = 206°$ plane. The fringe orders are defined as (stress tangential to the thread profile) - (stress perpendicular to thread profile). Hence the stress peaks on the loaded faces of the threads due to contact are positive. Due to small machining errors in the profile of the coupling threads, the contact stress differences were concentrated at the ends of the loaded face. The absence of a peak in the cracked slice near the tip and the reduction in fillet stress concentration are associated with the load reduction due to the crack. Elsewhere the surface stress distributions are very similar.

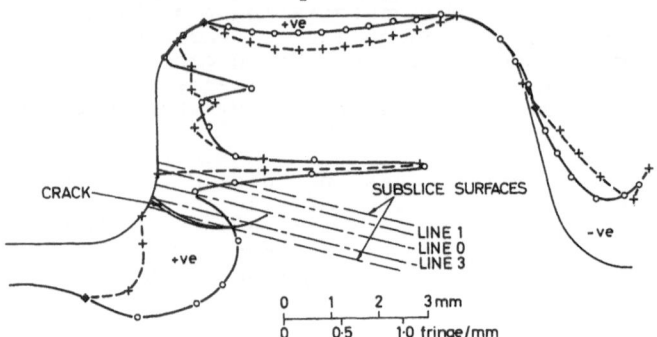

Fig. 5   Surface stress distribution at sections of thread; cracked $\theta_t = 359°$ (+ dotted line) and uncracked $\theta_t = 206°$ (o full line). Relative tension parallel to surface is positive, plotted inwards from profile.

The large variations in the 'far field stresses' represented by functions of $r_c°$ in fracture mechanics notation are apparent from these surface stress distributions.

Fig. 5 also shows the lines 1, 0, 3, along which automatic micropolariscope measurements were taken in the uncracked $\theta_t = 206°$ slice to separate the stresses along a line through the position of the crack tip and the centre of the thread fillet radius. This stress analysis is illustrated in Fig. 6. Figs. 6a and 6b show the fringe orders and isoclinic angles in the meridional plane along lines 1, 0, 3. Readings were taken at $\Delta x \simeq 0.1$mm spacing. The kinked lines were traced from the automatic plots to make them thick enough for reduction. After these readings were completed, the sub-slice was cut out of the main slice and analysed. The corresponding graphs of fringe orders and isoclinics in the x-z plane are omitted to save space. The circumferential variations were small; typical differences between lines 2 and 4 were 0.02 fringes/mm and 1° in isoclinic angle. Figs. 6a and 6b indicate the accuracy of the whole automatic process. The repeatability of slice positioning and repeat readings was 0.01mm (1). Fig. 6c shows the cartesian stresses along this line and defines their directions relative to the thread.

LINEAR FRACTURE MECHANICS

For uniform stress fields, Irwin's equations (3)

$$
\begin{bmatrix} \sigma_x \\[1em] \sigma_y \\[1em] \tau_{xy} \end{bmatrix} = \frac{K_I \cos \frac{\theta_c}{2}}{\sqrt{2\pi r_c}} \begin{bmatrix} 1 - \sin \frac{\theta_c}{2} \sin \frac{3\theta_c}{2} \\[1em] 1 + \sin \frac{\theta_c}{2} \sin \frac{3\theta_c}{2} \\[1em] \sin \frac{\theta_c}{2} \cos \frac{3\theta_c}{2} \end{bmatrix} + \frac{K_{II} \sin \frac{\theta_c}{2}}{\sqrt{2\pi r_c}} \begin{bmatrix} 2 + \cos \frac{\theta_c}{2} \cos \frac{3\theta_c}{2} \\[1em] \cos \frac{\theta_c}{2} \cos \frac{3\theta_c}{2} \\[1em] \cot \frac{\theta_c}{2} - \cos \frac{\theta_c}{2} \sin \frac{3\theta_c}{2} \end{bmatrix} + f(r_c^o)
$$

$$(5)$$

define the stresses near a crack in terms of two stress intensity
factors K_I and K_{II} and stresses which are independent of the distance
from the the crack tip. For a uniform stress field these stresses in
the uncracked component act everywhere and have been called σ_∞ and τ_∞
to indicate that they occur where $r_c \to \infty$, i.e. far from the crack tip.
For non-uniform stress fields, as described above, the appropriate value
of the r_c^o terms is not obvious.

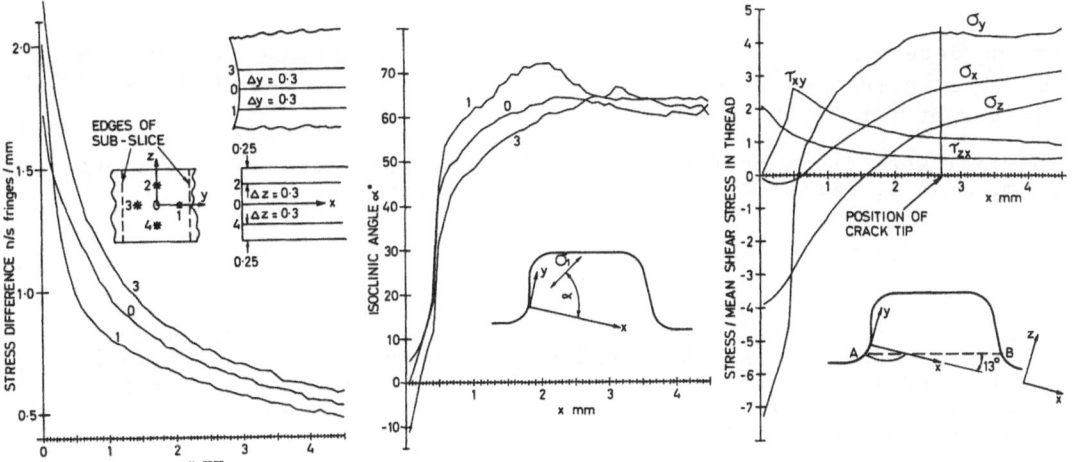

a) Fringe orders in x-y,   b) Isoclinic angles in   c) Cartesian stress
   meridional plane.   x-y, meridional plane.   distribution.

Fig. 6 Stress analysis along line 0 in uncracked slice at $\theta_t = 206^o$.

For photoelastic studies it is convenient to study the maximum shear
stresses $\hat{\tau}$ in the x-y plane. If it is assumed that the $f(r_c^o)$ terms
can be considered separately, (this is implicit in the review paper
Ref. 8)

$$
\hat{\tau} = \frac{1}{\sqrt{8\pi r_c}} \left[(K_I \sin \theta_c + 2K_{II} \cos \theta_c)^2 + (K_{II} \sin \theta_c)^2 \right]^{\frac{1}{2}} + f(r_c^o)
$$

$$(6)$$

If $\hat{\tau}$ is plotted against $1/\sqrt{r_c}$ for readings along any θ_c = constant line,
functions of K_I and K_{II} are obtained from the slopes of straight lines.
Convenient directions are:-

$$
\theta_c = 0^o \qquad \hat{\tau} = \frac{K_{II}}{\sqrt{2\pi r_c}} + f(r_c^o) \tag{7}
$$

$$\theta_c = 90^\circ \quad \hat{\tau} = \frac{1}{\sqrt{8\pi r_c}} [K_I{}^2 + K_{II}{}^2]^{\frac{1}{2}} + f(r^\circ_c) \tag{8}$$

$$\theta_c = \pm 45^\circ \quad \hat{\tau} = \frac{1}{4\sqrt{\pi r_c}} [K_I{}^2 + 4K_I K_{II} + 5K_{II}{}^2]^{\frac{1}{2}} + f(r^\circ_c) \tag{9}$$

The extent of the linear part of the $\hat{\tau} - 1/\sqrt{r_c}$ graphs indicates the range of applicability of the linear fracture mechanics approach.

An alternative to studying pre-determined θ_c directions is to use θ^*, the direction at which tangents to the fringes are perpendicular to the radius r_c, i.e. the condition $\partial\hat{\tau}/\partial\theta_c = 0$. It has been shown (9) that, if $f(r_c{}^\circ)$ is independent of θ_c,

$$(\frac{K_{II}}{K_I})^2 - \frac{4}{3} \frac{K_{II}}{K_I} \cot 2\theta^* = \frac{1}{3} \tag{10}$$

Having determined K_{II} from Equ. 7, K_I can be determined from Equ. 8 to 10. The intercepts of the $\hat{\tau} - 1/\sqrt{r_c}$ straight lines with the $\hat{\tau}$ axis give values of $f(r_c{}^\circ)$.

In model work it is preferable to produce results in non-dimensional form, i.e. to define k for the cracked region as

$$k_i = \frac{K_i}{\tau_m \sqrt{\pi a}} \tag{11}$$

where $i = I$ or II, 'a' is the crack depth and τ_m is the local mean shear stress across the UNCRACKED thread (from root to root, along the line AB in Fig. 6c) due to the thread contact force at the position θ_t under consideration. This value was chosen as a good measure of the local load on the thread.

The functions $f(r_c{}^\circ)$ are normalised by the same stress τ_m and called σ_∞ or τ_∞. Like all fracture parameters they are referred to the direction tangential to the crack tip (see Fig. 9a).

STRESS DISTRIBUTIONS NEAR CRACK

The integral fringe pattern in slice No. 4 is shown in Fig. 7. White paint on the original dark-field photograph is used to show the edge of the slice. As usual, the highest fringe order in the contact region occurs a small distance below the loaded surface. The fringe pattern suggests that k_{II} predominates (5).

The accuracy of positioning and measurement of our instrument is sufficient to obtain detailed fringe distributions. Examples are shown in Fig. 8 in the $\theta_c{}^\circ = 0$ direction for slices No. 1 to 4. No. 5 has not been used because its crack front is far from normal to the slice (see Fig. 3). The axes were chosen to suit Equ. 7. Full lines show the linear part of each stress distribution, dotted lines the non-linear part very far from and very near to the crack tip. Readings were taken at x = 0.14 or 0.16mm spacing; many 'remote' points have been omitted to avoid confusion.

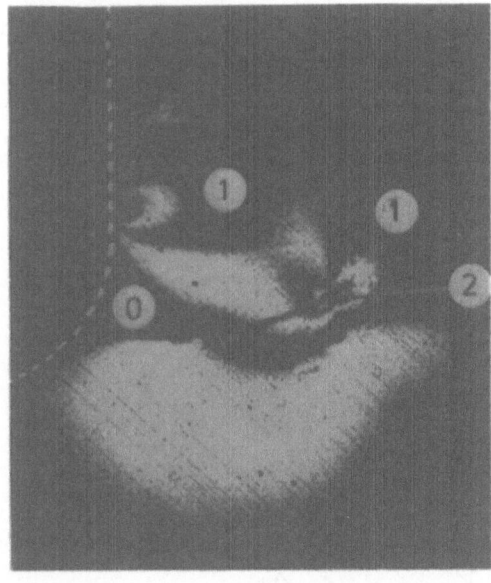

0   1   2   3   mm

Fig. 7   Fringe pattern in slice 4

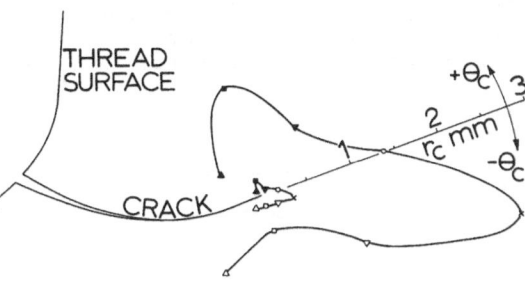

Fig. 10   Extent of linear region
in slice 4

Fig. 9 shows similar graphs for
measurements taken in 8 different
θ_c directions in slice No. 4.
Because these measurements were
not in line with the skew crack
front, it was possible to obtain
readings very near to the crack.
The horizontal lines correspond
to setting up errors of ±0.005mm
at 200 times magnification. The
unusual feature of the θ_c = +135°
curve may be associated with the
curvature of the crack. The
θ_c = +45° line coincides with a
fringe as shown in Fig. 7. The -24° readings were obtained from the
much coarser scan used to determine the shear force carried by the
uncracked section (see Fig. 11).

The extent of the linear regions estimated from Fig. 9 is shown in Fig.
10. It is not understood why it is so large in the axial direction,
where it extends almost half the distance to the unloaded flank of the
thread (see Fig. 5).

SHEAR STRESS DISTRIBUTION

The automatic instrument has also been used to determine the orthogonal
shear stresses τ_{rz} from fringe order and isoclinic measurements along
the line AB in Fig. 6c and through the uncracked parts of the cracked
slices in the z direction from the crack tips. These results are shown
in Fig. 11; the mean values τ_m are also plotted against crack length.
Because τ_m is used as a convenient measure of the load transmitted at
the section, it is the area under the τ_{rz} - z graphs, divided by the
UNCRACKED distance across the thread.

DISCUSSION

The micropolariscope is sufficiently sensitive and accurate for reliable stress analysis near a crack. Because it averages readings over
a circular area of only 0.04mm, it can take readings very near a crack
tip to ±0.001 fringe and ±0.05° isoclinic angle. The first reading

position relative to the crack is determined at an optical magnification of up to 200, i.e. easily to within ±0.005mm. Subsequent
positions are defined by micrometer screws which are rotated through
angles accurately measured by stepper motors.

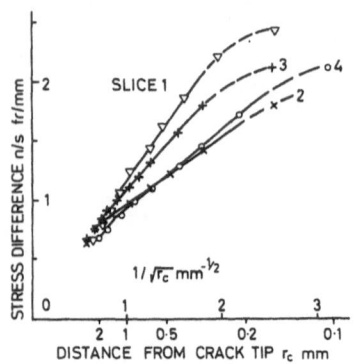

Fig. 8 Stress distributions
in $\theta_c = 0^\circ$ direction
for slices 1 to 4.

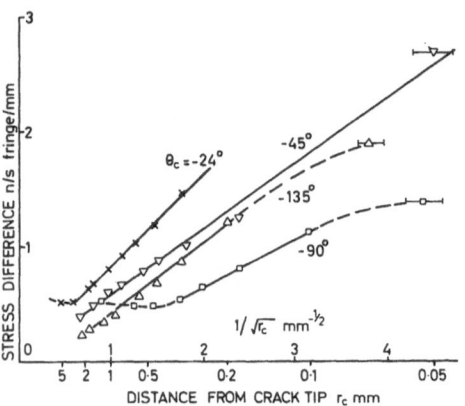

Fig. 9b Stress distributions in
slice 4

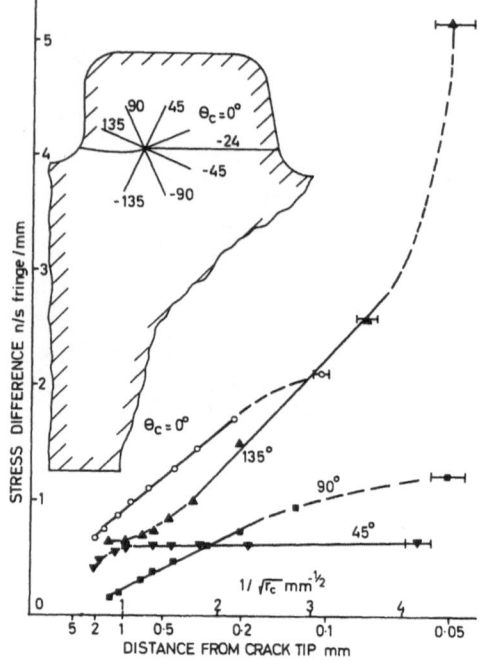

Fig. 9a Stress distributions in
slice 4.

The biggest measurement errors may
occur in the measurement of thickness of the (necessarily thin)
slices. It is difficult to
measure, say 1mm thick slices with
errors less than 0.01mm.

The irregular shape of the crack
front of this natural crack (see
Fig. 3) makes it inevitable that
the planes of the closely spaced
slices are not normal to the crack front. This makes it impossible to
obtain readings very near the crack front in the $\theta_c = 0$ direction and
introduces errors near the crack front in the other directions.

As shown in Figs. 4 and 5, the peak fillet stresses are drastically
reduced by the load reduction (see Fig. 11) associated with the crack.

Fig. 6 shows the results of the stress separation along line 0 in the
uncracked slice shown in Fig. 5. The unexpected shear stress $\tau_{zx} \neq 0$
at the free surface is attributed to errors due to the large stress
variation in the y direction near the surface within the 1mm thick sub-
slice shown in Fig. 5. These errors are unlikely to affect the stresses
at the position of the crack tip, also shown in Fig. 6c. The normalised principal stresses there are:-

$$\sigma_1 = 4.81, \qquad \sigma_2 = 2.07, \qquad \alpha = 26° \text{ (defined in Fig. 6b)}$$

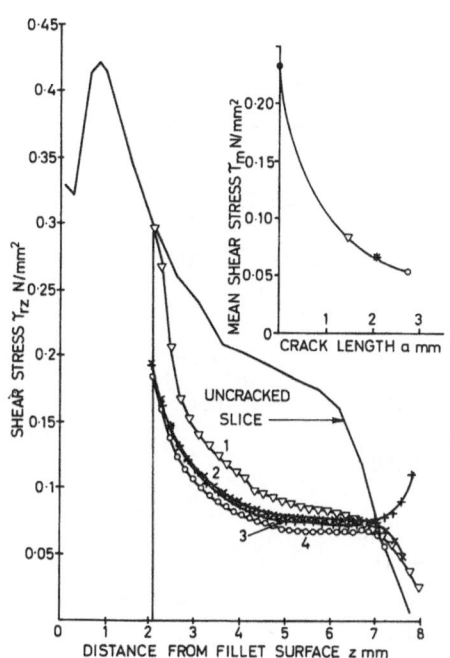

Fig. 11 Shear stresses along AB
in Fig. 6c.

$$\tau_m = \frac{1}{7.9} \int_0^{7.9} \tau_{rz} \, dz$$

A natural scale in mm is included in Figs. 8 and 9 to show how close to the crack tip measurements could be obtained. The unexpectedly high ($\theta_c = +135°$) and low ($\theta_c = +45°$) values near the crack tip were attributed to crack curvature and obliquity of the crack front, not to crack tip radius because this natural crack is very sharp. The frozen stresses are of course elastic. The uncertainties of slope of the straight lines lead to errors in the values of k_{II}, k_I and $f(r_c°)$ shown in Table 1.

Using Equ. 11 to obtain the values of k_{II} and k_I, K_{II} and K_I, given by the slopes of the straight lines in Figs. 8 and 9 and Equ. 7 to 9, have been divided by the appropriate values of τ_m and $\sqrt{a}$ shown in Fig. 11. The value of $\theta*$ for use in Equ. 10 was found to be $\theta_c = 0°$; this gave $K_I \simeq 0$. The variations in k_{II} and k_I are much larger than the likely errors in slopes of the graphs in Figs. 8 and 9.

Using Table 1, the extent of the linear regions can be compared with the distances from the crack tip to the free surfaces of the screwed tube (see Fig. 9a). The free surface fringe orders have been divided by $2\tau_m$ to give the values of τ_∞ shown.

The latter could not be related to $f(r_c°)$ obtained from the straight lines in Fig. 9 for $r_c = \infty$.

CONCLUSIONS

Although reliable, repeatable experimental techniques have been used, the results are disappointing.

From four similarly crack slices $k_{II} = 1.78 \; {}^{+0.93}_{-0.56}$

k_I values are obtained from the roots of the quadratic equations 8 and 9. The algebraically greater values give $0.98 > k_I > -0.15$.

The crack-independent stress $0.85 > f(r_c°) > -0.42$. The comparable maximum shear stress in the uncracked thread at the crack tip position is 1.37

Fracture mechanics analyses based on the above data would be of doubtful value.

Table 1   Results

| POSITION | | FREE SURFACE | | LINEAR REGION | | LEFM VALUES | | |
Slice No.	Direction θ_c°	Stress τ_{∞}	Distance mm	$\check{r}$ mm	$\hat{r}$ mm	k_I	k_{II}	$f(r_c^{\circ})$
1	0	–	–	0.3	1.25	–	2.71	-0.15
2	0	–	–	0.19	2.0	–	1.25	0.85
3	0	–	–	0.39	2.3	–	1.95	0.36
4	0	-0.08	5.9	0.19	1.4	–	1.22	0.53
"	45	-0.56	4.6	0.06	0.83	(1)	–	0.59
"	90	0	4.3	0.16	1.23	±1.86	–	0.42
"	135	5.40	2.5	0.06	0.5	+0.200 / -5.08	–	-0.13
"	-24	-1.06	6.0	0.35	2.80	–	–	0
"	-45	0	48.0	0.16	1.23	-0.149 / -4.74	–	0
"	-90	0.32	18.7	0.1	0.4	±0.98	–	-0.42
"	-135	0.32	18.3	0.15	0.94	+0.27 / -5.15	–	-0.42

(1)   No real roots of Equ.

ACKNOWLEDGEMENTS

This work is supported by a grant from the Marine Technology Directorate of SERC. The authors also thank the University's technicians for their skilled, enthusiastic assistance.

REFERENCES

1. Marston, R.E., "An automatic micropolariscope; its design, development and use for tubular joint stress analysis", Ph.D. Thesis, Dept. of Mechanical Engineering, Nottingham University, 1985
2. Vickers, M17 Polarising Microscope, Vickers Ltd., York, England.
3. Paris, P.C. and Sih, G.C., "Stress Analysis of Cracks", ASTM Spec. Tech. Pub. 381, 30-83, 1964.
4. Dally, J.W. and Sanford, R.J., "Classification of Stress-Intensity Factors from Isochromatic-Fringe Patterns", Experimental Mechanics, 1978, 441-448.
5. Gdoutos, E.E. and Theocaris, P.S., "A Photoelastic Determination of Mixedmode Stress-Intensity Factors", Experimental Mechanics, 1978, 87-96.
6. Frocht, M.M., "Photoelasticity", Vol. 1, Ch. 2 and Ch. 8, John Wiley, NY, 1960.
7. Tesar, V., "La Photoelasticimetrie et ses Applications dans la Construction Aeronautique", La Science Aerienne, 11, p.372-394, 1933.
8. Cartwright and Rooke, JSA, Vol. 10, No. 4, October 1975, pp. 217-224.
9. Smith, D.G. and Smith, C.W., "Photoelastic Determination of Mixed Mode Stress Intensity Factors", Eng. Fract. Mech. 1972, Vol. 4, pp. 357-366.